# Exit Problems for Lévy and Markov Processes with One-Sided Jumps and Related Topics

# Exit Problems for Lévy and Markov Processes with One-Sided Jumps and Related Topics

Editor

**Florin Avram**

MDPI • Basel • Beijing • Wuhan • Barcelona • Belgrade • Manchester • Tokyo • Cluj • Tianjin

*Editor*
Florin Avram
Laboratoire de Mathématiques
Appliquées, Université de Pau
France

*Editorial Office*
MDPI
St. Alban-Anlage 66
4052 Basel, Switzerland

This is a reprint of articles from the Special Issue published online in the open access journal *Risks* (ISSN 2227-9091) (available at: https://www.mdpi.com/journal/risks/special_issues/Exit_Problems_Levy_Markov).

For citation purposes, cite each article independently as indicated on the article page online and as indicated below:

LastName, A.A.; LastName, B.B.; LastName, C.C. Article Title. *Journal Name* **Year**, *Volume Number*, Page Range.

**ISBN 978-3-03928-458-0 (Hbk)**
**ISBN 978-3-03928-459-7 (PDF)**

© 2020 by the authors. Articles in this book are Open Access and distributed under the Creative Commons Attribution (CC BY) license, which allows users to download, copy and build upon published articles, as long as the author and publisher are properly credited, which ensures maximum dissemination and a wider impact of our publications.

The book as a whole is distributed by MDPI under the terms and conditions of the Creative Commons license CC BY-NC-ND.

# Contents

**About the Editor** . . . . . . . . . . . . . . . . . . . . . . . . . . . . . . . . . . . . . . . . . . . . . . . . vii

**Preface to "Exit Problems for Lévy and Markov Processes with One-Sided Jumps and Related Topics"** . . . . . . . . . . . . . . . . . . . . . . . . . . . . . . . . . . . . . . . . . . . . . . . ix

**Florin Avram, AndrasHorvath, Serge Provost and Ulyses Solon**
On the Padé and Laguerre–Tricomi–Weeks Moments Based Approximations ofthe Scale Function $W$ and of the Optimal Dividends Barrierfor Spectrally Negative LévyRisk Processes
Reprinted from: *Risks* **2019**, *7*, 121, doi:10.3390/risks7040121 . . . . . . . . . . . . . . . . 1

**Jean-François Renaud**
De Finetti's Control Problem with Parisian Ruin for Spectrally Negative Lévy Processes
Reprinted from: *Risks* **2019**, *7*, 73, doi:10.3390/risks7030073 . . . . . . . . . . . . . . . . . 25

**Mauricio Junca, Harold A. Moreno-Franco and José Luis Pérez**
Optimal Bail-Out Dividend Problem with Transaction Cost and Capital Injection Constraint
Reprinted from: *Risks* **2019**, *7*, 13, doi:10.3390/risks7010013 . . . . . . . . . . . . . . . . . 37

**Florin Avram and Dan Goreac and Jean-François Renaud**
The Løkka–Zervos Alternative for a Cramér–Lundberg Process with Exponential Jumps
Reprinted from: *Risks* **2019**, *7*, 120, doi:10.3390/risks7040120 . . . . . . . . . . . . . . . . 61

**Wenyuan Wang and Xiaowen Zhou**
Potential Densities for Taxed Spectrally Negative Lévy Risk Processes
Reprinted from: *Risks* **2019**, *7*, 85, doi:10.3390/risks7030085 . . . . . . . . . . . . . . . . . 71

**Eberhard Mayerhofer**
Three Essays on Stopping
Reprinted from: *Risks* **2019**, *7*, 105, doi:10.3390/risks7040105 . . . . . . . . . . . . . . . . 83

**Matija Vidmar**
Fluctuation Theory for Upwards Skip-Free Lévy Chains
Reprinted from: *Risks* **2018**, *6*, 102, doi:10.3390/risks6030102 . . . . . . . . . . . . . . . . 93

**Florin Avram, Danijel Grahovac and Ceren Vardar-Acar**
The $W$, $Z$ Paradigm for the First Passage of Strong Markov Processes without Positive Jumps
Reprinted from: *Risks* **2019**, *7*, 18, doi:10.3390/risks7010018 . . . . . . . . . . . . . . . . . 117

**Pavel V. Gapeev, Neofytos Rodosthenous and V. L. Raju Chinthalapati**
On the Laplace Transforms of the First Hitting Times for Drawdowns and Drawups of Diffusion-Type Processes
Reprinted from: *Risks* **2019**, *7*, 87, doi:10.3390/risks7030087 . . . . . . . . . . . . . . . . . 133

**Florin Avram and Jose-Luis Perez**
A Review of First-Passage Theory for the Segerdahl-Tichy Risk Process and Open Problems
Reprinted from: *Risks* **2019**, *7*, 117, doi:10.3390/risks7040117 . . . . . . . . . . . . . . . . 149

**Hansjörg Albrecher and Eleni Vatamidou**
Ruin Probability Approximations in Sparre Andersen Models with Completely Monotone Claims
Reprinted from: *Risks* **2019**, *7*, 104, doi:10.3390/risks7040104 . . . . . . . . . . . . . . . . 171

**Krzysztof Dębicki, Lanpeng Ji and Tomasz Rolski**
Logarithmic Asymptotics for Probability of Component-Wise Ruin in a Two-Dimensional Brownian Model
Reprinted from: *Risks* **2019**, 7, 83, doi:10.3390/risks7030083 . . . . . . . . . . . . . . . . . . . . . . **185**

# About the Editor

**Florin Avram** has worked in reverse chronological order at the Department of Mathematics, University of Pau; in the Department of Actuarial Mathematics and Statistics at Heriot-Watt University; in the Operations Research Center at MIT; in the Departmento de Estatistica, Universidade de Campinas, Brazil; the department of mathematics at Utah State in Logan, USA; Haifa University; Northeastern University; Boston (eight years); Cornell; Purdue; and Chapel Hill. His received his doctorate from Cornell University under the guidance of Murad Taqqu, and he has had four Ph.D. students: Mike Ricard (MIT), Fikri Karaesmen (Northeastern University), Martijn Pistorius (Utrecht), and Donatien Chedom Fotso (Pau).

# Preface to "Exit Problems for Lévy and Markov Processes with One-Sided Jumps and Related Topics"

Preface to Exit Problems for Lévy and Markov Processes with One-Sided Jumps and Related Topics It has long been known that exit problems for one-dimensional Lévy processes are easier when there are jumps in one direction only. In the last few years, this intuition became more precise. We know now that a great variety of identities for exit problems of spectrally-negative Lévy processes may be ergonomically expressed in terms of two "q-harmonic functions" W and Z (or scale functions, or q-martingales). See paper 1, https://www.mdpi.com/2227-9091/7/4/121 for a brief introduction to W and two numerical methods to compute it.

The reader may then get an idea of some important applications in risk theory by looking at the next six papers:

1. The paper of J. F. Renaud considers the Finetti's stochastic control problem when the controlled process is allowed to spend time under the critical level (the so-called Parisian ruin). It is shown that if the tail of the Lévy measure is log-convex, the optimal strategy is of barrier type. An interesting implied question is whether this continues to be true when this assumption is not satisfied. https://www.mdpi.com/2227-9091/7/3/73;

2. M. Junca, H.A. Moreno-Franco, and J.L. Pérez consider the optimal bail-out dividend problem with fixed transaction cost for a Lévy risk model with a constraint on the expected present value of injected capital, and establish the optimality of reflected (c1, c2)-policies. https://www.mdpi.com/2227-9091/7/1/13;

3. F. Avram, D. Goreac, and J.F. Renaud prove a so-called Løkka–Zervos alternative, for Cramér–Lundberg risk processes with exponential claims. This means that if the proportional cost of capital injections is low, then it is optimal to pay dividends and inject capital according to a double-barrier strategy, meaning that ruin never occurs; and if the cost of capital injections is high, then it is optimal to pay dividends according to a single-barrier strategy and never inject capital. Note, however, that this paper only addresses de Finetti and Shreve -Lehoczky- Gaver policies. The non-restricted stochastic control problem has been solved only recently, and, again, only with exponential claims. https://www.mdpi.com/2227-9091/7/4/120;

4. WenyuanWang and Xiaowen Zhou provide an in-depth study of spectrally negative Lévy risk process with general tax structure https://www.mdpi.com/2227-9091/7/3/85;

5. Eberhard Mayerhofer's paper https://www.mdpi.com/2227-9091/7/4/105 provides self-contained proofs concerning processes stopped at draw-down times;

6. P.V. Gapeev, N. Rodosthenous, and V.L. Chinthalapati obtain in https://www.mdpi.com/2227-9091/7/3/87 closed-form expressions for the value of the joint Laplace transform of the running maximum and minimum of a diffusion process stopped at the first time at which the associated drawdown or drawup process hits a constant level. This paper studies this problem for Lévy processes with state-dependent coefficients. The next three papers concern similar stochastic models. Note that since the essence of "W,Z" proofs is the strong Markov property applied at smooth-crossing times and variations, the results, in principle, are expected to hold for the wider class of spectrally-negative strong Markov processes.

This is established in the particular cases of certain random walks by M. Vidmar—see https://www.mdpi.com/2227-9091/6/3/102—and seems to be true for general strong Markov processes, subject to technical conditions—see the paper https://www.mdpi.com/2227-9091/7/1/18 of F. Avram, D. Grahovac, and C. Vardar-Acar.

Note, however, that computing the functions W, Z is still essentially an open problem outside the Lévy and diffusion classes. One exception is the simplest Segerdahl risk model with affine drift and exponential jumps—see https://www.mdpi.com/2227-9091/7/4/117 for this case, and also for a review of certain generalizations of the Segerdahl process.

The final two papers deal with problems concerning more general models:

1. H Albrecher, E Vatamidou https://www.mdpi.com/2227-9091/7/4/104 construct error bounds for the ruin probability of the Sparre Andersen risk process with interclaim times that belong to the class of distributions with rational Laplace transform. An exciting extension would be to Lévy perturbed Sparre Andersen risk processes.

2. Finally, K. Debicki, L. Ji and T. Rolski go multidimensional and obtain in https://www.mdpi.com/2227-9091/7/3/83 logarithmic asymptotics (large deviations) for probability of a component-wise ruin in a two-dimensional Brownian model.

**Florin Avram**
*Editor*

Article

# On the Padé and Laguerre–Tricomi–Weeks Moments Based Approximations of the Scale Function W and of the Optimal Dividends Barrier for Spectrally Negative Lévy Risk Processes

Florin Avram [1,*], Andras Horváth [2], Serge Provost [3] and Ulyses Solon [1]

[1] Laboratoire de Mathématiques Appliquées, Université de Pau, 64000 Pau, France; usolon@math.upd.edu.ph
[2] Dipartimento di Informatica, Università di Torino, Corso Svizzera 185, 10149 Torino, Italy; horvath@di.unito.it
[3] Department of Statistical and Actuarial Sciences, The University of Western Ontario, London, ON N6A5B7, Canada; provost@stats.uwo.ca
* Correspondence: florin.avram@univ-Pau.fr

Received: 28 October 2019; Accepted: 5 December 2019; Published: 11 December 2019

**Abstract:** This paper considers the Brownian perturbed Cramér–Lundberg risk model with a dividends barrier. We study various types of Padé approximations and Laguerre expansions to compute or approximate the scale function that is necessary to optimize the dividends barrier. We experiment also with a heavy-tailed claim distribution for which we apply the so-called "shifted" Padé approximation.

**Keywords:** ruin probability; Pollaczek–Khinchine formula; scale function; optimal dividends; Padé approximations; Laguerre series; Tricomi–Weeks Laplace inversion

## 1. Introduction

Let us first recall the Cramér–Lundberg risk model extended with Brownian perturbation Albrecher and Asmussen (2010); Dufresne and Gerber (1991b)

$$X_t = x + ct + \sigma B(t) - S_t, S_t = \sum_{i=1}^{N_\lambda(t)} C_i. \tag{1}$$

Here $x \geq 0$ is the initial surplus, $c \geq 0$ is the linear premium rate. The $C_i$'s, $i = 1, 2, \ldots$ are independent identically distributed (i.i.d) random variables with distribution $F(z) = F_C(z)$ representing nonnegative jumps arriving after independent exponentially distributed times with mean $1/\lambda$, and $N_\lambda(t)$ denotes the associated Poisson process counting the arrivals of claims on the interval $[0, t]$. Finally, $\sigma B(t), \sigma > 0$ is an independent Brownian perturbation. Ruin happens when, for the first time, a jump takes $X_t$ below 0.

Risk theory revolved initially around evaluating and minimizing the probability of ruin. Insurance companies are also interested in maximizing company value. This lead to the study of optimal dividend policies. As suggested by de Finetti in the 1950s de Finetti (1957)—see also Miller and Modigliani (1961)—an interesting objective is that of maximizing the expected value of the sum of discounted future dividend payments until the time of ruin.

The most important class of dividend policies is that of a constant barrier at $b$, which modifies the surplus only when $X_t > b$, by a lump payment bringing the surplus to $b$, and then keeps it there by Skorokhod reflection until the next negative jump. In financial terms, in the absence of a Brownian component, this amounts to paying out all the income while at $b$. In the case of Brownian perturbation, Skorokhod reflection means keeping the process above the barrier by minimal capital injections (whenever necessary), or below a barrier, by taking out dividends (if necessary) Skorokhod (1962).

In the presence of the barrier at $b$, the de Finetti objective (the expected value of the sum of discounted future dividend payments until ruin) has a simple expression Avram et al. (2007) in terms of the so-called "scale function" $W$:

$$V^{[b]}(x) = \mathbb{E}_x^{[b]}\left[\int_{[0,T_0^{[b]}]} e^{-qt} dU_t^b\right] = \begin{cases} \frac{W_q(x)}{W_q'(b)}, & x \le b \\ x - b + \frac{W_q(b)}{W_q'(b)}, & x > b \end{cases}, \quad (2)$$

where $T_0^{[b]}$ is the time of ruin, $q$ denotes the discount rate, $U_t^b$ the total local time at $b$ before time $t$, and $\mathbb{E}^{[b]}$ the law of the process reflected from above at $b$ and absorbed at 0 and below.

The scale function Bertoin (1998); Kyprianou (2014); Landriault and Willmot (2019); Suprun (1976) $W_q(x) : \mathbf{R} \to [0,\infty), q \ge 0$ is defined on the positive half-line by the Laplace transform

$$\widehat{W}_q(s) := \int_0^\infty e^{-sx} W_q(x) dx = \frac{1}{\kappa(s) - q}, \quad \forall s > \Phi_q, \quad (3)$$

where the "symbol" $\kappa(s)$ (also-called the cumulant generating function) is defined in Equation (8) in Section 2 where we provide the necessary background information, and $\Phi_q$ is the unique nonnegative root of the Cramér–Lundberg equation

$$\Phi_q := \sup\{s \ge 0 : \kappa(s) - q = 0\}, \quad q \ge 0. \quad (4)$$

The scale function $W_q(x)$ is continuous and increasing on $[0,\infty)$ Bingham (1976), (Bertoin 1998, Thm. VII.8), (Kyprianou 2014, Thm. 8.1). It may have, however, many inflection points (such an example is depicted in Figure 1), and these play an important role in the optimization of dividends Avram et al. (2007, 2015); Schmidli (2007). For convenience, $W_q(x)$ is extended to be 0 on $\mathbf{R}_-$. An important fact that will be exploited is that the Laplace transform of our function has a unique non-negative pole $\Phi_q$, see Equations (3) and (4).

This paper aims at computing/approximating the scale function $W_q(x)$, using its moments. The techniques being used are classic: Padé approximation and Laguerre expansions. The order $(m,n)$ Padé approximation of a function $g(x)$ is a rational function in the form

$$R(x) = \frac{a_0 + a_1 x + a_2 x^2 + \cdots + a_m x^m}{1 + b_1 x + b_2 x^2 + \cdots + b_n x^n}$$

for which $R(0) = g(0), R'(0) = g'(0), R''(0) = g''(0), \ldots, R^{(m+n)}(0) = g^{(m+n)}(0)$. In the context of probability distributions, given a density function $f(x)$ and its Laplace transform $\widehat{f}(s)$, the inverse Laplace transform of the order $(m,m)$ Padé approximant of $\widehat{f}(s)$ provides a matrix exponential approximation of $f(x)$ that matches the first $2m$ moments of $f(x)$ (including $m_0$). In Avram et al. (2011) this approach was used to approximate ruin probabilities. In this paper we develop the same approach to approximate the scale function $W_q(x)$ (Section 3). An extension of the above idea is the so-called two-point Padé approximation, which allows to match not only the moments of $W_q(x)$ but also the behavior of the function at 0, i.e., to match $W_q(0), W_q'(0), \ldots$ (Section 4). For more details on this extension see Avram et al. (2018) where ruin probabilities are approximated.

Let us draw attention now to several numeric challenges which were absent in the ruin probability problem.

1. Optimizing dividends starts by optimizing the so-called "barrier function"

$$H_D(b) := \frac{1}{W_q'(b)}, \ b \geq 0, \tag{5}$$

and the optimal dividend policy is often simply a barrier strategy at its maximum. This is the case in particular when the barrier function $H_D(b)$ is differentiable with

$$H_D'(0) > 0 \Leftrightarrow W_q''(0) < 0 \tag{6}$$

and has a unique local maximum $b^* > 0 \Longrightarrow W_q''(b^*) = 0$; then this $b^*$ yields the optimal dividend policy, and the optimal barrier function,

$$V(x) := sup_{b \geq 0} V^{b]}(x) = V^{b^*]}(x), \tag{7}$$

turns out to be the largest concave minorant of $W_q(x)$.[1]

2. **The challenge of multiple inflection points.** In the presence of several inflection points, however, the optimal policy is multiband Azcue and Muler (2005); Schmidli (2007); Avram et al. (2015); Loeffen (2008). The first numerical examples of multiband policies were produced in Azcue and Muler (2005); Loeffen (2008), with Erlang claims $Erl_{2,1}$. However, it was shown in Loeffen (2008) that multibands cannot occur when $W_q'(x)$ is increasing after its last global minimum $b^*$ (i.e., when no local minima are allowed after the global minimum).

Loeffen (2008) further made the interesting observation that for Erlang claims $ER_{2,1}$ (which are non-monotone), multiband policies may occur for volatility parameters $\sigma$ smaller than a threshold value, but barrier policies (with a non-concave value function) will occur when $\sigma$ is large enough.

Figure 1 displays the first derivative $W_q'(x)$, for $\sigma^2/2 \in \{\frac{1}{2}, 1, \frac{3}{2}, 2\}$. The last two values yield barrier policies with a non-concave value function, due to the presence of an inflection point in the interior of the interval $[0, b^*]$.

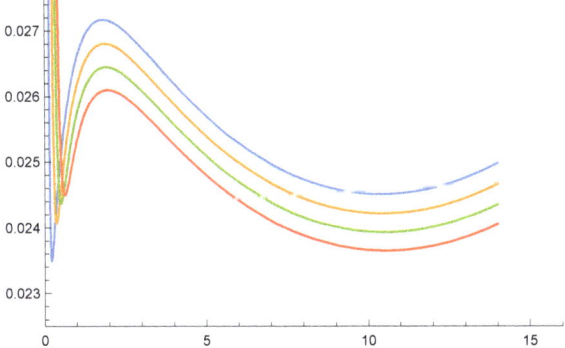

**Figure 1.** Graphs of the Loeffen example for $\kappa(s) = \frac{\sigma^2 s^2}{2} + c s + \lambda \left(\frac{1}{(s+1)^2} - 1\right), c = \frac{107}{5}, \lambda = 10, q = \frac{1}{10}, \sigma^2/2 \in \{1/2, 1, 3/2, 2\}$.

---

[1] Even when barrier strategies do not achieve the optimum, and multi-band policies must be used instead, constructing the solution must start by determining the global maximum of the barrier function Avram et al. (2015); Azcue and Muler (2005); Schmidli (2007).

Below we will investigate whether our approximations are precise enough to yield reasonable approximations for $W_q'''(0)$ and the root(s) of $W_q'''(\cdot)$.

Special features. While our methods consist essentially of Padé approximation and Laguerre–Tricomi–Weeks Laplace transform inversion, we found that exploiting the special features of our problem is useful. These are:

1.  including known values of $W_q(0)$, $W_q'(0)$ (using thus two-point Padé approximations).
2.  shifting the approximations around $\Phi_q$ specified in Equation (4), which transforms $W_q(x)$ into a survival probability. As a consequence, we end up using a certain judicious choice of the Laguerre exponential decay parameter of Equation (40), which is usually left to be tuned by the user in the Laguerre–Tricomi–Weeks method Weideman (1999).

Contents. We briefly review classical ruin theory in Section 2. Padé approximations are provided in Section 3, where we also spell out the simplest algorithm for the computation of the scale function. In Section 4, we derive low-order Padé and two-point Padé approximations of $W_q(x)$, reminiscent of the de Vylder approximation of the ruin probability. Some of these approximations appeared already in Gerber et al. (2008), where however the Padé method and the fact that they can be easily extended to higher orders is not mentioned. Section 5 offers our personal strategy for inverting Laplace transforms of interest in probability, in the presence of uncertainty. Section 5.1 implements the Laguerre–Tricomi–Weeks Laplace transform inversion with a certain judicious choice of the exponential decay parameter (40), which is believed to be new. Section 6 presents numeric experiments with mixed exponential claims. Section 7 presents experiments with Pareto claims; since these have a heavy tail and, consequently, finitely many moments, we apply a "shifted" Padé approximation of the claim distribution. Section 8 includes a computer program required to obtain test cases with exact rational answers, using the Wiener–Hopf factorization; of course, this is quite convenient for the initial testing of the precision of our algorithms. Finally, Section 9 reviews a more general version of the Laguerre–Tricomi–Weeks Laplace inversion method, which may be of interest for further experiments.

## 2. A Short Review of Classical Ruin Theory

The process defined in Equation (1) is a particular example of spectrally negative Lévy processes, with finite mean, which are defined by assuming instead of Equation (1) that $S_t$ is a subordinator with $\sigma$-finite Lévy measure $\nu(dx)$ that integrates $x$, but having possibly infinite activity near the origin $\nu(0, \infty) = \infty$ Bertoin (1998) (for Equation (1), the Lévy measure is given by $\nu(dx) = \lambda F(dx) \Longrightarrow \nu(0, \infty) = \lambda$). A spectrally negative Lévy process is characterized by its Lévy–Khintchine/Laplace exponent/cumulant generating function/symbol $\kappa(s)$ defined by $E_0[e^{sX(t)}] = e^{t\kappa(s)}$, with $\kappa(s)$ of the particular form

$$\kappa(s) = cs + \int_0^\infty (e^{-sx} - 1)\nu(dx) + \frac{\sigma^2 s^2}{2} := s\left(c - \widehat{\overline{\nu}}(s) + \frac{\sigma^2}{2}s\right). \tag{8}$$

Some concepts of interest in classical risk theory are:

- First passage times below and above a level $a$

$$T_{a,-(+)} := \inf\{t > 0 : X(t) < (>)a\}.$$

- The first first passage quantity to be studied historically was the eventual ruin probability:

$$\Psi(x) := P[T_{0,-} < \infty | X(0) = x]. \tag{9}$$

In order that the eventual ruin probability not be identically 1, the parameter

$$p := c - \lambda m_1 = \kappa'(0), \text{ where } m_1 = \int_0^\infty zF(dz),$$

which is called drift or profit rate, must be assumed positive.

The Laplace transform of the ruin probability is explicit, given by the so-called Pollaczek–Khinchine formula, which states that the Laplace transform of $\overline{\Psi}(x) = 1 - \Psi(x)$ is:

$$\widehat{\overline{\Psi}}(s) = \int_0^\infty e^{-sx}\overline{\Psi}(x)dx = \frac{c - \lambda m_1}{\kappa(s)} = \frac{\kappa'(0)}{\kappa(s)}. \tag{10}$$

The roots with negative real part of the Cramér–Lundberg equation

$$\kappa(s) = q, q \geq 0 \tag{11}$$

are important, when such roots exist. They will be denoted by $-\gamma_1, -\gamma_2, \cdots, -\gamma_N, N \geq 0$, and ordered by their absolute values $|\gamma_1| \leq |\gamma_2| \leq \ldots \leq |\gamma_N|$. $\gamma_1 > 0$ is called the adjustment coefficient, and furnishes the Cramér–Lundberg asymptotic approximation

$$\Psi(u) \sim \frac{\kappa'(0)}{-\kappa'(-\gamma_1)}e^{-\gamma_1 u}.$$

**Laplace transform inversion.** As explained here, the first passage theory for Lévy processes with one-sided jumps reduces essentially to inverting the Laplace transform Equation (3). This applies not only to the well-known ruin probabilities, but also to intricate optimization problems involving dividends, capital gains, liquidation of subsidiary companies, etc.

**Remark 1.** *If the claims have a phase-type or matrix exponential distribution, then the Cramér–Lundberg equation has a finite number of roots, and Laplace transform inversion reduces to finding roots of denominators and to partial fractions, operations which are essentially exact with the current computing resources. For example, the ruin probability is provided by an exact formula (which extends the Cramér–Lundberg asymptotic approximation). With distinct roots, this is*

$$\Psi(u) = \sum_{i=1}^N \frac{\kappa'(0)}{-\kappa'(-\gamma_i)}e^{-\gamma_i u}. \tag{12}$$

Similar formulas hold for the scale function and related quantities (derivatives and integrals of the scale function, etc).

This refocuses the question of ruin probabilities and similar quantities to the harder cases of

1. non-matrix exponential claims,
2. when $\gamma_1$ does not exist (the non-Cramér case), and
3. when not all moments exist, which we will call "heavy tails".

For non-matrix exponential jumps, one can start by a Padé approximation of the Laplace transform, which is perhaps the oldest method of Laplace transform inversion. For heavy-tailed claims, whose moment generating function is not analytic at the origin, one may apply, as it will be shown in Section 7, a "shifted" Padé approximation before Laplace transform inversion—see, for example, Avram et al. (2018). In this paper, we also compare the precision of the classic Padé and two-point Padé Laplace transform inversion methods with the Laguerre–Tricomi–Weeks inversion, as applied to the optimal dividends barrier problem. As test cases, we consider mixed exponentials, for which exact answers are available for the case of rational Wiener–Hopf factorization roots described in Section 8.

## 3. Padé/matrix exponential Approximations of the Scale Function

The essence of classic ruin theory is the availability of explicit "output Laplace transforms" expressed in terms of "input Laplace transforms", Equations see (3), (10) and (16). Equivalently, "output moments" (i.e., Taylor coefficients of functions of interest) are expressed in terms of "input data moments". In the case of insufficient data, Padé approximations of the output function seem to deserve special attention, because they involve only a few input moments, and seem to extract at low orders the essence of the data. Note, as pointed out in Avram et al. (2011, 2018); Avram and Pistorius (2014), that the classical ruin theory approximations of Cramér–Lundberg, De Vylder and Renyi are all first order Padé approximations. Slightly increasing the order yields more sophisticated moments based approximations Avram et al. (2018); Avram and Pistorius (2014); Ramsay (1992). Similar approximations, which could be useful in dividends optimization, are presented in Section 4—see also Hu et al. (2017) for a related application to reinsurance.

With more reliable data, higher order Padé approximations are just as easy to obtain, using computer systems such as Maple, Mathematica, Sage, Matlab, etc. Let $v_k = \lambda m_k, k \geq 1$ denote the moments of the Lévy measure[2], and let

$$v_{2,\sigma} = v_2 + \sigma^2. \tag{13}$$

The simplest Padé approximation of the scale function and optimal dividend barrier requires implementing the following algorithm, which is valid for claim distributions having $2n$ moments.

1. Obtain the power series expansion of the Laplace exponent in terms of the moments of the Lévy measure

$$\kappa(s) = s\left(c - \widehat{\overline{v}}(s) + \frac{\sigma^2}{2}s\right) = (c - v_1)s + v_{2,\sigma}\frac{s^2}{2} + \sum_{k=3}^{\infty} v_k \frac{(-s)^k}{k!}. \tag{14}$$

2. Construct a Padé approximation

$$\widehat{W}_q(s) = \frac{1}{(c - v_1)s + v_{2,\sigma}s^2/2 - v_3 s^3/6 + \cdots - q} \sim \frac{P_{n-1}(s)}{Q_n(s)} = \frac{\sum_{i=0}^{n-1} a_i s^i}{cs^n + \sum_{i=0}^{n-1} b_i s^i},$$

i.e., find $a_i, b_i, i = 0, \ldots, n-1$, by solving the linear system $P_{n-1}(s)(\kappa(s) - q) - Q_n(s) = O(s^{n+1})$. As emphasized in Figure 2, the series expansion of $\kappa(s)$ is a good approximation only for small $s$; since the most important part of the approximation is the root $\Phi_q$ in Equation (25), simple Padé approximation will only work for small $q$.

---

[2] Note that only moments starting from 1 are required, so this may be applied to processes whose subordinator part has infinite activity as well.

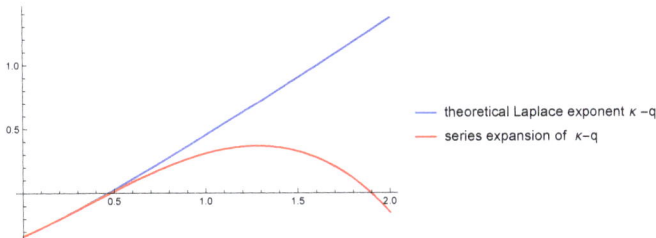

**Figure 2.** The series expansion of $\kappa(s)$ has multiple contact with $\kappa(s)$ at $s = 0$, but increases asymptotically at a smaller rate. Despite that, it yields quite reasonable approximations of $\Phi_q$, for $q$ small enough.

3. Factor the denominator as

$$cs^n + \sum_{i=0}^{n-1} b_i s^i = c(s - \Phi_q) \prod_{i=1}^{n}(s + \gamma_i). \qquad (15)$$

This operation is the essential numerical difficulty, but may be achieved nowadays with arbitrarily high precision.

4. Then, partial fractions plus inversion quickly yield an approximate Laplace transform inverse

$$W_q(x) \sim C_0 e^{\Phi_q x} + \sum_{i=1}^{n} C_i e^{-\gamma_i x}.$$

5. The dividend barrier is obtained by finding the largest nonnegative root of $W''(x) = 0$.

**Remark 2.** *Comparing Equation* (15) *with the fundamental relation of Equation* (44), *we see that the Padé approximation of the output Laplace transform may be viewed as an approximate Wiener–Hopf factorization.*

**Remark 3.** *Note that, as already mentioned in Remark* 1, *Padé inversion cannot be improved upon when the density of the claims is matrix exponential of some order n, (or, equivalently, when its Laplace transform is rational), since it results into the exact decomposition of the scale function into exponentials*

$$W_q(x) = \Phi'_q e^{\Phi_q x} + \sum_{i=1}^{n} C_i e^{-\gamma_i x},$$

*where the $C_i$ are partial fraction decomposition coefficients (and where the first coefficient $\Phi'_q$ follows from the normalization constant chosen in Equation* (3)).

The coefficients $C_i$ are less important than the roots, and in fact one may easily improve on Padé, for example, by recomputing them so that the approximation is optimal in $L_2$ sense. Positivity of the $C_i$'s, or, better, of their "Coxian linear combinations" can be further ensured by linear programming, if desired.

The preceding observations make Laplace inversion by rational approximation of the Laplace transform a quite attractive tool in ruin theory and related fields, from both a pedagogical and numerical point of view Avram et al. (2011, 2018); Avram and Pistorius (2014). One drawback is the appearance of non-admissible roots (for example, with $\Re(\gamma_i) < 0$), which will imply non-admissible ruin probabilities for very large initial reserves $x$. It may be argued, however, that if the admissibility range is large enough, in practice this may not cause problems. Moreover, this problem may be fixed by convenient mathematical tools, implemented in packages such as BUTools and SOPE Bobbio et al. (2005); Dumitrescu et al. (2016); Horváth and Telek (2016).

For the sake of simplicity, we will mostly take $\sigma = 0$ from now on. In this case, the Pollaczek–Khinchine formula, Equation (10), may be simply expressed in terms of the so-called "excess distribution $f_e$" of the jumps, and of the "intensity" $\rho$:

$$\widehat{\overline{\Psi}}(s) = \frac{c(1-\rho)}{\kappa(s)} = \frac{1-\rho}{s(1-\rho\widehat{f_e}(s))} \Leftrightarrow$$

$$\widehat{\Psi}(s) = \int_0^\infty e^{-sx}\Psi(x)dx = \frac{1}{s}\left(1 - \frac{1-\rho}{1-\rho\widehat{f_e}(s)}\right) = \rho\frac{\widehat{\overline{F_e}}(s)}{1-\rho\widehat{f_e}(s)} = \rho\frac{\widehat{\overline{F_e}}(s)}{1-\rho\frac{\widehat{\overline{F}}(s)}{m_1}}, \quad (16)$$

where $\widehat{f}(s) = \int_0^\infty e^{-sx}dF(x), \overline{F}(x) = 1 - F(x), m_1 = \int_0^\infty xdF(x), f_e(x) = \overline{F}(x)/m_1, \widehat{f_e}(s) = \int_0^\infty e^{-sx} f_e(x)dx = (1 - \widehat{f}(s))/(m_1 s), \rho = \lambda m_1/c, \widehat{\overline{F_e}}(s) = (1 - \widehat{f_e}(s))/s$. The last formula is especially convenient for cases where the excess distribution is simply related to the original one, as in the case of the Pareto distribution.

We would like to note however that the Padé approach works with no problems in the presence of Brownian motion, the only modifications being the form of $\kappa(s)$ and the second moment $\nu_2$ of the associated Lévy measure—see Equation (13).

## 4. Two-Point Padé Approximations, with Low Order Examples

One may obtain better results by incorporating into the Padé approximation the following initial values, which can be derived easily from the Laplace transform:

$$W_q(0) = \lim_{s\to\infty} s\widehat{W_q}(s) = \begin{cases} \frac{1}{c}, & \text{if } X \text{ is of bounded variation/Cramér–Lundberg} \\ 0, & \text{if } X \text{ is of unbounded variation} \end{cases}, \quad (17)$$

$$W_q'(0) = \lim_{s\to\infty} s\left(\frac{s}{\kappa(s)-q} - W_q(0)\right) = \begin{cases} \frac{q+\nu(0,\infty)}{c^2}, & \text{if } X \text{ is of bounded variation} \\ \frac{2}{\sigma^2}, & \text{if } \sigma > 0, \\ \infty, & \text{if } \sigma = 0 \text{ and } \nu(0,\infty) = \infty \end{cases}. \quad (18)$$

Furthermore, when the jump distribution has a density $f$, it holds that :[3]

$$W_q''(0_+) = \lim_{s\to\infty} s\left(s\left(\frac{s}{\kappa(s)-q} - W_q(0)\right) - W_q'(0_+)\right)$$
$$= \begin{cases} \frac{1}{c}\left(\left(\frac{\lambda+q}{c}\right)^2 - \frac{\lambda}{c}f(0)\right), & \text{if } X \text{ is of bounded variation} \\ -c\left(\frac{2}{\sigma^2}\right)^2, & \text{if } \sigma > 0 \end{cases} \quad (19)$$

Further derivatives at 0 could be computed, but we stop at order 2, since $W_q''(0)$ already requires estimating $f_C(0)$, which is a rather delicate task starting from real data.

We will provide in Proposition 1 below a couple of two-point Padé approximations, when $n = 2$. Before that, it is worth recalling the case of exponential claims.

**Example 1.** *The Cramér–Lundberg model with exponential jumps.* Consider the Cramér–Lundberg model with exponential jump sizes with mean $1/\mu$, jump rate $\lambda$, premium rate $c > 0$, and Laplace exponent $\kappa(s) = s\left(c - \frac{\lambda}{\mu+s}\right)$, assuming $\kappa'(0) = c - \frac{\lambda}{\mu} > 0$. Let $\gamma = \mu - \lambda/c$ denote the adjustment coefficient, and let $\rho = \frac{\lambda}{c\mu}$.

---

[3] This equation is important in establishing the nonnegativity of the optimal dividends barrier.

Solving $\kappa(s) - q = 0 \Leftrightarrow cs^2 + s(c\mu - \lambda - q) - q\mu = 0$ for $s$ yields two distinct solutions $\gamma_2 \leq 0 \leq \gamma_1 = \Phi_q$ given by

$$\gamma_1 = \frac{1}{2c}\left(-(\mu c - \lambda - q) + \sqrt{(\mu c - \lambda - q)^2 + 4\mu q c}\right),$$

$$\gamma_2 = \frac{1}{2c}\left(-(\mu c - \lambda - q) - \sqrt{(\mu c - \lambda - q)^2 + 4\mu q c}\right).$$

The W scale function is:

$$W_q(x) = \frac{A_1 e^{\gamma_1 x} - A_2 e^{\gamma_2 x}}{c(\gamma_1 - \gamma_2)} \Leftrightarrow \widehat{W}_q(s) = \frac{s + \mu}{cs^2 + s(c\mu - \lambda - q) - q\mu} \quad (20)$$

where $A_1 = \mu + \gamma_1, A_2 = \mu + \gamma_2$.

Furthermore, it is well-known and easy to check that the function $W'_q(x) = H_D(x)^{-1}$ is in this case unimodal with global minimum at

$$b^* = \frac{1}{\gamma_1 - \gamma_2}\begin{cases} \log \frac{(\gamma_2)^2 A_2}{(\gamma_1)^2 A_1} = \log \frac{(\gamma_2)^2(\mu+\gamma_2)}{(\gamma_1)^2(\mu+\gamma_1)} & \text{if } W''_q(0) < 0 \Leftrightarrow (q+\lambda)^2 - c\lambda\mu < 0 \\ 0 & \text{if } W''_q(0) \geq 0 \Leftrightarrow (q+\lambda)^2 - c\lambda\mu \geq 0 \end{cases}, \quad (21)$$

since $W''_q(0) = \frac{(\gamma_1)^2(\mu+\gamma_1) - (\gamma_2)^2(\mu+\gamma_2)}{c(\gamma_1 - \gamma_2)} \sim (q+\lambda)^2 - c\lambda\mu$ and that the optimal strategy for the de Finetti problem is the barrier strategy at level $b^*$ Avram et al. (2007).

**Proposition 1.** 1. To secure both the values of $W_q(0)$ and $W'_q(0)$, take into account Equations (17) and (18), i.e., use the Padé approximation

$$\widehat{W}_q(s) \sim \frac{\sum_{i=0}^{n-1} a_i s^i}{cs^n + \sum_{i=0}^{n-1} b_i s^i}, a_{n-1} = 1, b_{n-1} = ca_{n-2} - \lambda - q.$$

This yields

$$\widehat{W}_q(s) \sim \frac{\frac{1}{m_1} + s}{cs^2 + s\left(\frac{c}{m_1} - \lambda - q\right) - \frac{q}{m_1}}. \quad (22)$$

In view of Equation (20), this yields the same result as approximating the claims by exponential claims, with $\mu = \frac{1}{m_1}$.

2. To ensure $W_q(0) = \frac{1}{c}$, we must only impose the behavior specified in Equation (17), i.e., use the Padé approximation

$$\widehat{W}_q(s) \sim \frac{\sum_{i=0}^{n-1} a_i s^i}{cs^n + \sum_{i=0}^{n-1} b_i s^i}, a_{n-1} = 1.$$

For $n = 2$, this yields

$$\widehat{W}_q(s) \sim \frac{\frac{2m_1}{m_2} + s}{cs^2 + \frac{s(2cm_1 - 2\lambda m_1^2 - m_2 q)}{m_2} - \frac{2m_1 q}{m_2}} = \frac{\frac{1}{\tilde{m}_1} + s}{cs^2 + s\left(\frac{c}{\tilde{m}_1} - \lambda \frac{m_1}{\tilde{m}_1} - q\right) - \frac{q}{\tilde{m}_1}}, \quad (23)$$

where we denoted by $\tilde{m}_1 = \frac{m_2}{2m_1}$ the first moment of the excess density $f_e(x)$. For exponential claims this coincides with Equation (22) (since $f_e(x) = f(x)$). This is the De Vylder B) method (Gerber et al. 2008, (5.6–5.7)), derived therein by assuming exponential claims, with $\mu = \frac{2m_1}{m_2}$, and simultaneously modifying $\lambda$ to fit the first two moments of the risk process.

3. When the pure Padé approximation is applied, the first step yields

$$\widehat{W}_q(s) \sim \frac{s + \frac{3m_2}{m_3}}{s^2\left(c + \lambda(\frac{3m_2^2}{2m_3} - m_1)\right) + s\left(c\frac{3m_2}{m_3} - \frac{3m_1 m_2}{m_3}\lambda - q\right) - \frac{3m_2}{m_3}q}$$

$$= \frac{s + \frac{1}{\tilde{m}_3}}{s^2\left(c + \lambda(\frac{\tilde{m}_2}{\tilde{m}_3} - 1)\right) + s\left(c\frac{1}{\tilde{m}_3} - \frac{m_1}{\tilde{m}_3}\lambda - q\right) - \frac{1}{\tilde{m}_3}q}, \quad (24)$$

where $\tilde{m}_i = \frac{m_i}{i\, m_{i-1}}$ is a so-called "normalized moment" *Bobbio et al. (2005)*.

This is the De Vylder A) method (*Gerber et al. 2008*, (5.2–5.4)), derived therein by assuming exponential claims, with $\mu = \frac{3m_2}{m_3}$, and simultaneously modifying both $\lambda$ and $c$ to fit the first three moments of the risk process.

**Lemma 1.** *In the case of exponential claims, the three approximations given above are exact.*

**Proof.** It suffices to check that for exponential claims all the normalized moments are equal to $m_1 = \mu^{-1}$. □

In particular, the optimal barrier for exponential claims obtained by the explicit Equation (21) is the same. For example, $\mu = 2/5, \lambda = 9/10, c = 1, q = 1/10$ yields

$$W_q(x) \sim 0.652989\; 2.71828^{0.0659646x} - 0.152989\; 2.71828^{-1.51596x}$$

and $b^* = 3.04576$.

## 5. Two Numerical Methods for Computing $W_q(x)$: The Tijms–Padé and Laguerre–Tricomi–Weeks Approximations

Computing $W_q(x)$ reduces in principle to the case $q = 0$, if $\Phi_q$ may be accurately computed, via the well-known relation

$$W_q(x) = e^{x\Phi_q}\, W_0^{(\Phi_q)}(x), \quad (25)$$

where $W_0^{(\Phi_q)}(x)$ denotes the 0-scale function with respect to the "Esscher transformed" measure $P^{(\Phi_q)}$ (in general, the transform $P^{(r)}$ of the measure $P$ of a Lévy process with Laplace exponent $\kappa(s)$ is the measure of the Lévy process with Laplace exponent $\kappa(s+r) - \kappa(r)$, with $r$ in the domain of $\kappa(\cdot)$) (*Albrecher and Asmussen 2010*, Proposition 4.2), (*Kyprianou 2014*, 3.3 pg.83)).[4]

The Esscher transform removes the exponential growth, or, equivalently, the unique positive pole of $\widehat{W}_q(s)$. When working with the Esscher transform, let

$$G(x) := W_0^{(\Phi_q)}(\infty) - W_0^{(\Phi_q)}(x)$$

denote the function after the behavior at $\Phi_q$ and $\infty$ has been exploited[5] We construct then a mixed exponential/rational approximation

$$RG(x) := \sum_i A_i e^{-\gamma_i x} \Leftrightarrow \widehat{RG}(s) := \frac{a(s)}{\prod(s + \gamma_i)} \sim \widehat{G}(s). \quad (26)$$

---

[4] Equation (25) is easy to check by taking the Laplace transform, but quite important, numerically, since $W_0^{(\Phi_q)}(x)$ is a monotone bounded function (with values in the interval $(\lim_{s\to\infty} \frac{s}{\kappa(s)}, \frac{1}{\kappa'(\Phi_q)})$).

[5] Since $W^{(\Phi_q)}(x)$ still converges to a constant when $x \to \infty$, this non-zero limit at $\infty$ must be removed first.

For general rational approximations, we will use the name of "Esscher–Tijms type algorithms", and for the Padé case we will use the name of "Esscher–Tijms–Padé approximation ", in reference to the fact that the special positive pole, supposed to be known somehow exactly, has been removed.

**Remark 4.** *Clearly, a very precise approximation of the unique positive pole $\Phi_q$ and the "moments" $\kappa^{(n)}(\Phi_q)$ are an essential for determining $W_q, q > 0$, which was missing in the ruin probability problem. $\Phi_q$ is quite easy to compute under parametric models where the Laplace transform of the jumps is explicit, as the unique positive root of a convex function. It may also be estimated empirically by the plug-in method, but this seems not yet studied in this context. We took therefore the shortcut of assuming that $\Phi_q$ is available with infinite precision. Conceptually, this amount to studying the Esscher transform.*

The Padé approximation of the Esscher transform is the first method we have investigated. An interesting alternative to the Esscher transform is to consider the simplified Laplace transform:

$$\tilde{W}(s) := (s - \Phi_q)\widehat{W}_q(s) = \frac{\Phi_q}{q}\phi^-(s), \tag{27}$$

whose positive pole has been removed and thus coincides with the negative Wiener–Hopf factor $\phi^-(s)$ (see Section 8, (44)), up to the constant $\frac{\Phi_q}{q}$.

Subsequently, we may apply a Padé (or more sophisticated rational approximation Nakatsukasa et al. (2018)), with the goal of approximating the dominant poles of the presumably imprecisely known transform.

Note that a Padé of $\tilde{W}(s)$ will then yield immediately a rational approximation of $\widehat{W}_q(s)$ and a mixed exponential approximation of $W_q(x)$. The Padé approximation requires solving

$$\tilde{W}_q(s) := (s - \Phi_q)\widehat{W}_q(s) = \frac{s - \Phi_q}{(c - v_1)s + v_{2,\sigma}s^2/2 - v_3 s^3/6 + \cdots - q} \sim \frac{P_n(s)}{Q_n(s)} + o(s^n).$$

For the Cramér–Lundberg case, also fixing

$$w_0 = \tilde{W}_q(0) = \frac{1}{c}, w_0' = \tilde{W}_q'(0) = \frac{1}{c}\left(\frac{q+\lambda}{c} - \Phi_q\right) \tag{28}$$

leads to the linear system

$$\left(\sum_{i=0}^{n-1} a_i s^i + s^n\right)\left(\kappa(s) - q\right) = \left(s - \Phi_q\right)c\left(s^n + (a_{n-1} + \Phi_q - \frac{\lambda+q}{c})s^{n-1} + \sum_{i=0}^{n-2} b_i s^i\right) + o(s^n).$$

**Remark 5.** *The "Tijms-two-point Padé approximation " with $n = 1$ and $w_0$ satisfied by Equation (28) is obtained by solving:*

$$\frac{s - \Phi_q}{(c - v_1)s + \cdots - q} \sim \frac{s + a}{c(s + \gamma)} \Leftrightarrow (s - \Phi_q)c(s + \gamma) \sim (ps - q)(s + a_0) \Longrightarrow \gamma = a\frac{q}{c\Phi_q}, a = \frac{q - c\Phi_q}{p - q/\Phi_q}. \tag{29}$$

*This is exact for the Cramér–Lundberg process with exponential jumps of rate $\mu$, since then $a = \mu$ and $-\gamma$ reduces to the second root of the Cramér–Lundberg equation.*

*Applying Equation (21) to Equation (29) with $A_1 = \frac{a}{\Phi_q + \gamma}, A_2 = 1 - A_1$ yields then an approximation for the optimal barrier.*

Starting with order two, we may also satisfy $w'_0$. We must now solve

$$\frac{s - \Phi_q}{(c - v_1)s + \cdots - q} \sim \frac{s^2 + a_1 s + a_0}{c(s^2 + (a_0 + \Phi_q - \frac{\lambda+q}{c})s + b_0)}$$

$$\Leftrightarrow (s - \Phi_q)c(s^2 + (a_0 + \Phi_q - \frac{\lambda+q}{c})s + b_0) \sim (ps + v_2 s^2/2 - q)(s^2 + a_1 s + a_0)$$

$$\Rightarrow b_0 = a_0 \frac{q}{c\Phi_q}, a_1 q = a_0(p - q/\Phi_q + c\Phi_q) + c\Phi_q(\Phi_q - \frac{\lambda+q}{c}), a_0 = \ldots$$

Tedious computations or the help of Mathematica reveal that this is exact for the Cramér–Lundberg process with mixed exponential jumps of order 2.

In the next subsection we propose a method similar in spirit to the above: Obtaining first the simplest exponential approximation, and refining this subsequently (analogously to Step 3) above by means of applying the Laguerre–Tricomi–Weeks expansion.

### 5.1. Laguerre–Tricomi–Weeks Laplace Transform Inversion with Prescribed Exponent

In this section we obtain first an exponential approximation of the transformed scale function of the form $W^{(\Phi_q)}(x) \sim W^{(\Phi_q)}(\infty) - C \frac{\alpha}{2} e^{-\frac{\alpha}{2} x}$, by a first order Padé approximation[6] of its Laplace transform

$$\widehat{W^{(\Phi_q)}}(s) = \frac{1}{\kappa(s + \Phi_q) - q} = \frac{1}{\kappa(s + \Phi_q) - \kappa(\Phi_q)} := \frac{1}{\kappa^{(\Phi_q)}(s)}. \tag{30}$$

The exponent $\alpha/2$ thus obtained is the "second step output" form the previous section.

Next, apply the Tricomi–Weeks method for Laplace transform inversion (see, for example, Abate et al. (1998)), i.e., search for a Laguerre series expansion

$$G(x) := W^{(\Phi_q)}(\infty) - W^{(\Phi_q)}(x) \sim C \sum_{n=0}^{\infty} B_n\, e^{-\alpha x/2} L(n, \alpha x) \tag{31}$$

where, following tradition, C normalizes the sum following it to be a pdf[7], and where

$$L_n(x) = \frac{e^x}{n!} \frac{d^n}{dx^n}(e^{-x} x^n) = \sum_{k=0}^{n} \binom{n}{k} \frac{(-x)^k}{k!}, \quad x \geq 0, n = 0, 1, \ldots, \tag{32}$$

denote the Laguerre polynomials, which are orthogonal with respect to the weight $e^{-x/2}$. The Laplace transform of the Laguerre polynomial is

$$\hat{L}(n, s) = \frac{(s-1)^n}{s^{n+1}}, n = 0, 1, \ldots, \tag{33}$$

The Laguerre–Tricomi–Weeks method is based on the fact that Equation (31) is equivalent to

$$\hat{G}(s) := \frac{W^{(\Phi_q)}(\infty)}{s} - \widehat{W}^{(\Phi_q)}(s) \sim C \sum_{n=0}^{\infty} B_n \frac{(s - \alpha/2)^n}{(s + \alpha/2)^{n+1}} \Leftrightarrow$$

$$\hat{H}(s) := (s + \alpha/2) \hat{G}(s) \sim C \sum_{n=0}^{\infty} B_n \frac{(s - \alpha/2)^n}{(s + \alpha/2)^n}. \tag{34}$$

---

[6] Higher-order rational approximations may be considered as well.
[7] C can also be included in the coefficients $B_n$, but introducing it does render the computation of Equation (40) more convenient.

Now the last RHS may be obtained using the "collocation transformation" $z = \frac{s-\alpha/2}{s+\alpha/2} \Leftrightarrow s = \frac{\alpha}{2}\frac{1+z}{1-z}$ and a power series expansion of the LHS, yielding

$$H\left(\frac{\alpha}{2}\frac{1+z}{1-z}\right) = \frac{\alpha}{1-z}G\left(\frac{\alpha}{2}\frac{1+z}{1-z}\right) \sim C\sum_{n=0}^{\infty} B_n z^n. \quad (35)$$

In conclusion, extracting the Taylor coefficients $CB_n$ of $H(\frac{\alpha}{2}\frac{1+z}{1-z})$ yields approximations to $W^{(\Phi_q)}$ by substituting it into Equation (31). After multiplication by $e^{\Phi_q x}$, we obtain approximations of $W_q(x)$.

**Determining $\alpha$.** Previous work on choosing $\alpha$ involved the radius of convergence of a Laguerre–Tricomi–Weeks expansion which is slightly more general than Equation (35) Weideman (1999)—see also Section 9. We now introduce a different method.

Recall that our Laplace transform $\widehat{W^{(\Phi_q)}}(s)$ has been shifted by $\Phi_q$, so that 0 is the largest pole, and that the growth at $\infty$ has been removed by subtracting

$$W^{(\Phi_q)}(\infty) = \lim_{s \to 0} s\widehat{W^{(\Phi_q)}}(s) = \frac{s}{\kappa(s+\Phi_q)-q} = \frac{1}{\kappa'(\Phi_q)} = \Phi'_q. \quad (36)$$

Now, let

$$C := \lim_{s \to 0} \widehat{G}(s) = \lim_{s \to 0} \frac{\kappa(s+\Phi_q)-q-s\kappa'(\Phi_q)}{s\kappa'(\Phi_q)(\kappa(s+\Phi_q)-q)} = \lim_{s \to 0} \frac{s^2\kappa''(\Phi_q)/2}{s\kappa'(\Phi_q)(s\kappa'(\Phi_q))} = \frac{\kappa''(\Phi_q)}{2(\kappa'(\Phi_q))^2} \quad (37)$$

denote the mass of our measure.

A reasonable $\alpha$ may be obtained by approximating $G(x)$ via a first order exponential approximation of $C$

$$G(x) \sim C\frac{\alpha}{2}e^{-\frac{\alpha}{2}x} \Leftrightarrow \widehat{G}(s) \sim C\frac{\alpha/2}{s+\alpha/2} \quad (38)$$

(or, equivalently, by fitting the first two moments of $\widehat{G}(s)$), so that the Laguerre exponent in Equation (31) is fitted at order $n=0$.

Fitting moments may be achieved by Padé's method. At order 1, this yields

$$\widehat{G}(s) = \frac{\kappa(s+\Phi_q)-q-s\kappa'(\Phi_q)}{s\kappa'(\Phi_q)(\kappa(s+\Phi_q)-q)} \sim \frac{C\alpha/2}{\alpha/2+s} \Leftrightarrow$$

$$\alpha/2 = \lim_{s\to 0} \frac{s(\kappa(s+\Phi_q)-q-s\kappa'(\Phi_q))}{(\kappa(s+\Phi_q)-q)(Cs\kappa'(\Phi_q)-1)+s\kappa'(\Phi_q)} =$$

$$\lim_{s\to 0} \frac{s(\kappa(s+\Phi_q)-q-s\kappa'(\Phi_q))}{(s\kappa'(\Phi_q)+s^2\kappa''(\Phi_q)/2+s^3\kappa'''(\Phi_q)/6+\ldots)(Cs\kappa'(\Phi_q)-1)+s\kappa'(\Phi_q)} = \quad (39)$$

$$\lim_{s\to 0} \frac{s(\kappa(s+\Phi_q)-q-s\kappa'(\Phi_q))}{s^3(\frac{\kappa''(\Phi_q)}{2}C\kappa'(\Phi_q)-\kappa'''(\Phi_q)/6)} = \frac{\frac{\kappa''(\Phi_q)}{2}}{C\kappa'(\Phi_q)\frac{\kappa''(\Phi_q)}{2}-\frac{\kappa'''(\Phi_q)}{6}} = \frac{\frac{\kappa''(\Phi_q)}{2}}{\frac{\kappa''(\Phi_q)\kappa''(\Phi_q)}{2\kappa'(\Phi_q)\cdot 2}-\frac{\kappa'''(\Phi_q)}{6}}$$

Finally, a bit of algebra yields

$$\alpha/2 = \frac{\frac{\kappa''(\Phi_q)}{2}}{\frac{(\kappa''(\Phi_q)/2)^2}{\kappa'(\Phi_q)}-\frac{\kappa'''(\Phi_q)}{6}} = \frac{6\kappa'(\Phi_q)\kappa''(\Phi_q)}{3(\kappa''(\Phi_q))^2-2\kappa'(\Phi_q)\kappa'''(\Phi_q)}. \quad (40)$$

**Remark 6.** *It may also be checked that*

$$G(0) = \lim_{s\to\infty} \frac{1-s\frac{\kappa'(\Phi_q)}{\kappa(s+\Phi_q)-q}}{\kappa'(\Phi_q)} = \begin{cases} \frac{1}{\kappa'(\Phi_q)} & \text{unbounded variation} \\ \frac{1}{\kappa'(\Phi_q)}-\frac{1}{c} & \text{bounded variation} \end{cases},$$

and

$$G'(0) = \lim_{s \to \infty} s\widehat{G}'(s) = \begin{cases} \frac{2}{3\kappa'(\Phi_q)} & \text{unbounded variation} \\ \frac{1}{2\kappa'(\Phi_q)} - \frac{1}{2c} & \text{bounded variation} \end{cases}.$$

This will be useful for developing higher order two-point Padé approximations of $G(s)$.

**Remark 7.** *It is possible that the choice in Equation (40) for the Laguerre–Tricomi–Weeks exponential parameter (often left to the user) has previously been proposed, but we have been unable to find this result in the literature. Note the natural generalization to the case of Laplace transforms with a finite number of positive poles (instead of a single one, as in our case).*

We show in the next sections that the approximations obtained by the Tijms–Padé and Laguerre–Tricomi–Weeks methods are accurate enough so that the resulting inverse $W_q$, $W'_q$ retain the features of the intricate Azcue and Loeffen examples Azcue and Muler (2005); Loeffen (2008).

### 5.2. Combining the Tijms–Padé and Laguerre–Tricomi–Weeks Approximations

We end this section by proposing a more general strategy for inverting $\widehat{W}_q(s)$ and other Laplace transforms of interest in probability, given the uncertainty inherent in real world data. This consists in refining the rational approximation of Equation (26) by constructing the quotient of the original Laplace transform and its approximation

$$\tilde{G}(s) := \frac{\widehat{G}(s)}{\widehat{R}G(s)} = \widehat{G}(s) \frac{\prod(s+\gamma_i)}{a(s)} \tag{41}$$

and by applying to it an inversion method with good convergence properties like Laguerre–Tricomi–Weeks. The final approximation of $W_q(x)$ will be then a convolution of the inverse Laplace transforms $RG(x)$ and $\tilde{G}(x)$. This may be viewed as a to a two stage filter, aimed at removing uncertainties of different orders of magnitude.

## 6. Mixed Exponential Claims

We will now consider examples with rational roots. The idea for obtaining "rational first passage probabilities", which appeared essentially in Dufresne and Gerber (1991a), is to replace the additive parametrization of Equation (8) with that provided by the so-called Wiener–Hopf factorization of Equation (44). For Cramér–Lundberg processes, it is enough to specify the negative roots $\gamma_i$, $i \geq 1$ and poles $\beta_i$ (for example, by specifying the Wiener–Hopf factor, Equation (44), and also the positive root $\Phi_q$ of the symbol. The additive decomposition of the model's symbol, Equation (8), may be recovered using its behavior when $s \to \infty$, and partial fractions. This is automated in the Mathematica program RatC which is available upon request from the authors.

This package is useful for providing testing cases for our approximations, since in such examples the case of mixed exponential claims is reduced to exact rational root-solving and partial fractions (included automatically in Mathematica's command InverseLaplaceTransform).

**Example 2.** *A Cramér–Lundberg model with exponential mixture jumps of order two. This example illustrates the computational steps of the previous section, and the fact that the second order Tijms approximation for mixed exponential claims of order 2 is exact (and so is in particular the optimal barrier).*

*We chose $\Phi_q = \frac{1}{3}$, $c = 1/2$ and a negative Wiener–Hopf factor*

$$\phi_-(s) = \frac{\left(\frac{s}{2}+1\right)(s+1)}{\left(\frac{2s}{3}+1\right)(2s+1)},$$

which is input into our Rat program—see Section 8—by specifying the interspersed roots and poles exr = (1/2, 3/2); exc = (1, 2).

This corresponds—see Section 8—to a Cramér–Lundberg process with cumulant generating function

$$\kappa(s) = cs + \frac{8}{48}(\frac{1}{s+1} - 1) + \frac{21}{48}(\frac{2}{s+2} - 1), c = \frac{1}{2},$$

where we used $\lambda = \frac{29}{48}$, and claim density

$$b(x) = \frac{8}{29}e^{-x} + \frac{21}{29}2e^{-2x},$$

with mean $m_1 = \frac{37}{58}$ and $\rho = \frac{\lambda m_1}{c} = \frac{37}{48}$. The resulting scale function with $q = 1/16$ is

$$W_q(x) = -\frac{3}{11}e^{-3x/2} - \frac{9e^{-x/2}}{5} + \frac{224e^{x/3}}{55},$$

see Figure 3, and the optimal barrier is $b^* = 0.642265$. Recall from Remark 5 that the Tijms–Padé approximation is exact at order 2.

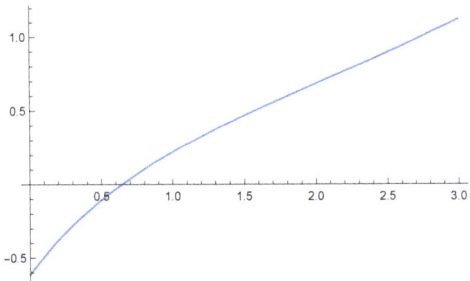

**Figure 3.** $W''(x)$.

After an Esscher shift of $\frac{1}{3}$, the scale function transform becomes

$$\widehat{W^{(\Phi_q)}}(s) = \frac{8(4+3s)(7+3s)}{s(5+6s)(11+6s)},$$

with dominant non-zero pole $-\frac{5}{6}$. After the removal of the pole 0, this becomes

$$\hat{G}(s) = \frac{72(57s+97)}{55(6s+5)(6s+11)}.$$

The Padé approximation of order $(0,1)$ is $\frac{677448}{55(6177s+5335)}$, the Laguerre exponent is $\alpha/2 = 5335/6177 = 0.863688$, and the largest error with $n = 30$ is $6 \times 10^{-16}$—see Figure 4.

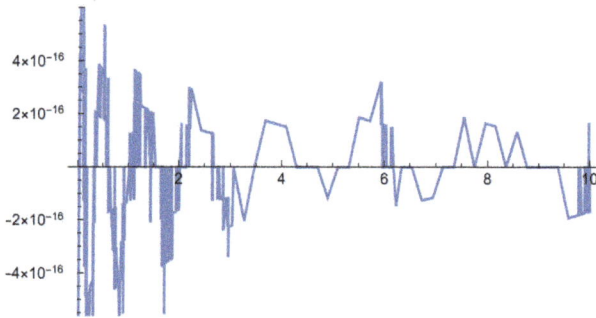

**Figure 4.** Relative errors of the Laguerre–Tricomi–Weeks inversion with mixed exponential claims of order 2.

**Example 3.** *A perturbed Cramér–Lundberg model with exponential mixtures jumps of order three. Our next example is produced by taking $\Phi_q = \frac{1}{3}$ and a negative Wiener–Hopf factor*

$$\phi_-(s) = \frac{\left(\frac{s}{2}+1\right)(s+1)}{\left(\frac{2s}{5}+1\right)\left(\frac{2s}{3}+1\right)(2s+1)}, \tag{42}$$

*which corresponds to the cumulant generating function:*

$$\kappa(s) = s^2 + \frac{7s}{6} + \frac{1}{2(s+1)} + \frac{7}{8(s+2)} - \lambda, \lambda = \frac{15}{16}.$$

*Note that the only impact of the presence of Brownian motion with $\sigma > 0$ is that the degree of the numerator in Equation (42) equals the degree of the denominator $-1$.*
*The resulting scale function with $q = \frac{5}{16}$ is*

$$W_q(x) = -\frac{9}{68}e^{-5x/2} - \frac{3}{22}e^{-3x/2} - \frac{9e^{-x/2}}{20} + \frac{672e^{x/3}}{935}.$$

*The input to the Laguerre–Tricomi–Weeks inversion is*

$$\widehat{G}(s) = \frac{9}{11(6s+11)} + \frac{27}{34(6s+17)} + \frac{27}{10(6s+5)},$$

*its Padé approximation of order $(0,1)$ is $\frac{4639156488}{935(8004369s+7505245)}$, the Laguerre exponent is $\alpha/2 = 0.937644$, and the largest error with $n = 40$ is $4 \times 10^{-14}$.[8] Other exponents do better however. For example, the larger exponent $\alpha/2 = \frac{6\kappa'(\Phi_q)\kappa''(\Phi_q)}{3(\kappa''(\Phi_q))^2 - \kappa'(\Phi_q)\kappa'''(\Phi_q)} = 1.00688$, where we have erased the 2 in the denominator, has a smaller larger error when $n = 40$ of $4 \times 10^{-15}$.*

**Example 4.** *A Cramér–Lundberg model with exponential mixtures jumps of order three. Our next example is produced by taking $\Phi_q = \frac{1}{3}, c = 1$ and a negative Wiener–Hopf factor*

$$\phi_-(s) = \frac{\left(\frac{s}{3}+1\right)\left(\frac{s}{2}+1\right)(s+1)}{\left(\frac{2s}{5}+1\right)\left(\frac{2s}{3}+1\right)(2s+1)}$$

---

[8] Beyond $n = 40$, the precision needs to be changed to obtain better results.

with poles $-\frac{1}{2}, -\frac{3}{2}, -\frac{5}{2}$. This corresponds to a Cramér–Lundberg process with cumulant generating function:

$$\kappa(s) = cs + \frac{1}{4(s+1)} + \frac{7}{8(s+2)} + \frac{25}{8(s+3)} - \lambda, c = 1, \lambda = \frac{83}{48}.$$

The scale function with $q = \frac{5}{48}$ is

$$W_q(x) = -\frac{9}{136}e^{-5x/2} - \frac{9}{44}e^{-3x/2} - \frac{9e^{-x/2}}{8} + \frac{448e^{x/3}}{187}$$

and the optimal barrier is $b^* = 0.866289$. The Padé Tijms approximation is $b^* = 0.876898$. The input to the Laguerre–Tricomi–Weeks inversion is

$$\widehat{G}(s) = \frac{216\left(261s^2 + 1155s + 1202\right)}{187(6s+5)(6s+11)(6s+17)},$$

the Padé approximation of order $(0,1)$ is $\frac{312077664}{187(1278399s+1123870)}$, the Laguerre exponent is $\alpha/2 = 0.879123$, and the largest error with $n = 40$ is $4 \times 10^{-14}$—see Figure 5.

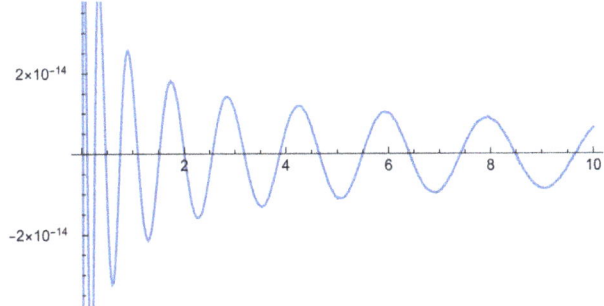

**Figure 5.** Relative errors of the Laguerre–Tricomi–Weeks inversion with mixed exponential claims of order 3.

Again, the exponent $\alpha/2 = \frac{6\kappa'(\Phi_q)\kappa''(\Phi_q)}{3(\kappa''(\Phi_q))^2 - \kappa'(\Phi_q)\kappa'''(\Phi_q)} = 1.138$ does better, with the largest error with $n = 40$ of $6 \times 10^{-16}$. This suggests the importance of optimizing $\alpha$, which is a difficult problem Giunta et al. (1989); Weideman (1999). Recall however our proposal to circumvent it by starting with a higher order Padé approximation of $G(s)$—see (41).

## 7. Padé Approximation of Heavy-Tailed Claims

In this section we experiment with heavy-tailed claims in the context of dividends barrier. To this end we consider hereby the risk process with Pareto II/Lomax claim size distribution (also known as American Pareto) with survival function given by $\overline{F}(y) = \left(1 + \frac{y}{\beta}\right)^{-\alpha}$, $y > 0$, $\alpha, \beta > 0$, mean $m_1 = \frac{\beta}{\alpha-1}$ and $F_e(y) = \left(1 + \frac{y}{\beta}\right)^{-(\alpha-1)}$, $y > 0$. The Laplace transform of the density is Nadarajah and Kotz (2006)

$$\widehat{f}(s) = \int_0^\infty e^{-sx} dF(x) = \alpha(\beta s)^\alpha e^{\beta s} \Gamma(-\alpha, \beta s) = \alpha U(1, 1-\alpha, \beta s) = 1 - (\beta s)^\alpha e^{\beta s} \Gamma(1-\alpha, \beta s), s \geq 0$$

where $\Gamma(\alpha, s) = \int_s^\infty t^{\alpha-1} e^{-t} dt$ is the incomplete gamma function and

$$U[a, a+c, z] = \frac{1}{\Gamma[a]} \int_0^\infty e^{-zt} t^{a-1} (t+1)^{c-1} dt, \operatorname{Re}[z] > 0, \operatorname{Re}[a] > 0$$

is Tricomi's Hypergeometric U function (Abramowitz and Stegun 1965, 13.2.5), where we used $U(1, 1 + \tilde{\lambda}, v) = e^v v^{-\lambda} \Gamma(\lambda, v)$ (see, (Abramowitz and Stegun 1965, 6.5.22)). It follows that the Laplace transform of the survival function is

$$\widehat{\overline{F}}(s) = \beta U(1, 2 - \alpha, \beta s). \tag{43}$$

As discussed in Albrecher et al. (2010), the model is suitable for obtaining precise quantities by numerical inverse Laplace transform using the fixed Talbot (FT) algorithm with parameter $M = 200$. We will refer to results obtained by the FT algorithm as "exact".

We apply $\alpha = 3/2, \beta = 1$ with which the distribution is of infinite variance. One can still obtain moment based Padé approximations by "shifting" the claim distribution as described in Avram et al. (2018). The resulting approximate claim density function approximations of order $K = 4, 6,$ and 8 are the following (the shift factor we applied is 1.6):

$$f_4(x) = 0.3674 e^{-4.8697x} + 0.7439 e^{-2.1699x} + 0.3412 e^{-0.8795x} + 0.0465 e^{-0.2583x}$$

$$f_6(x) = 0.1053 e^{-6.6998x} + 0.4679 e^{-3.5877x} + 0.5442 e^{-1.9191x} + 0.2898 e^{-0.9714x} +$$
$$0.08279 e^{-0.4298x} + 0.0099 e^{-0.1339x}$$

$$f_8(x) = 0.02944 e^{-8.3839x} + 0.2211 e^{-4.9512x} + 0.4433 e^{-2.9899x} + 0.424 e^{-1.7882x} +$$
$$0.2512 e^{-1.0331x} + 0.1008 e^{-0.555x} + 0.02605 e^{-0.2566x} + 0.003042 e^{-0.08227x}$$

which match 2K moments (including $m_0$) of the shifted density.

Based on the Laplace transform of the above approximations it is straightforward to obtain the Laplace transform of $W_q(x)$ and then by symbolic inverse Laplace transform $W_q(x)$ itself. We studied the model with $\lambda = 1, c = 9/4, q = 1/10$. We considered both the perturbed model with $\sigma = 1$ and the non-perturbed one. In the following we present figures where dots represent values obtained by numerical Laplace inversion while solid lines are given by the order 4 approximation. Figures 6 and 7 gives the result for the perturbed case and Figures 8 and 9 for the non-perturbed model. Numerical inverse Laplace transform requires about 60 seconds on a modern laptop to obtain a single point of $W_q(x)$ while the Padé approximation is immediate. Order 4 Padé approximation provides results that are not distinguishable from the exact values. The right panels of Figures 7 and 9 show that increasing the order of the approximation improves accuracy.

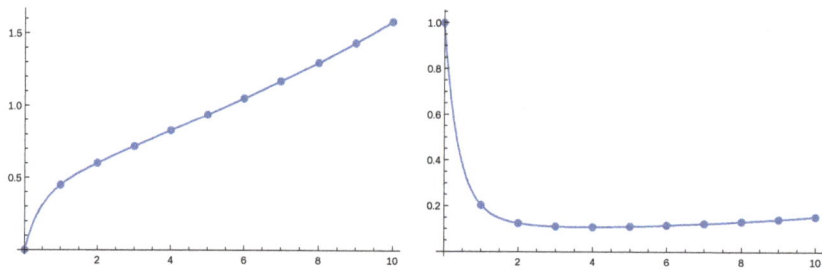

**Figure 6.** The scale function $W_q(x)$ (on the left) and its derivative (on the right) in the case of the perturbed model.

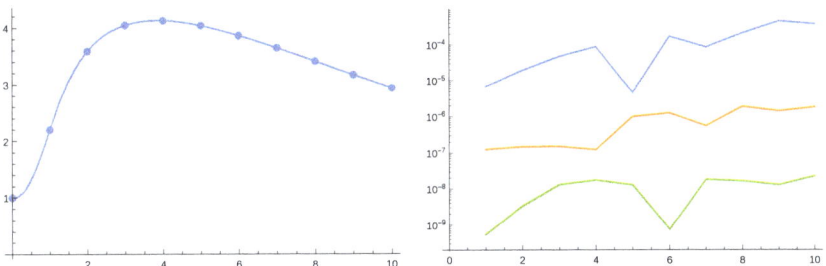

**Figure 7.** On the left the expected value of total discounted dividends, $V^{b]}(x)$, as function of $b$ with $x = 1$ in the case of the perturbed model. On the right the absolute error of approximations of $V^{b]}(x)$ of order 4 (blue), 6 (orange) and 8 (green).

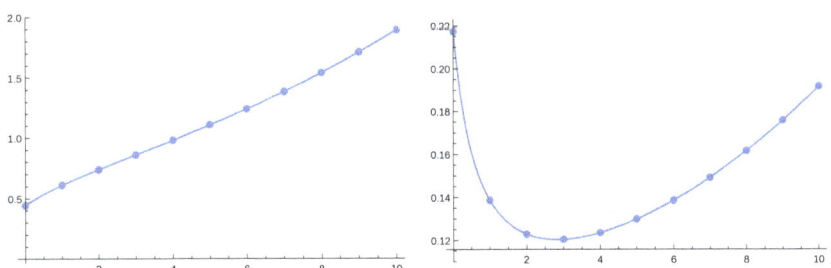

**Figure 8.** The scale function $W_q(x)$ (on the left) and its derivative (on the right) in the case of the unperturbed model.

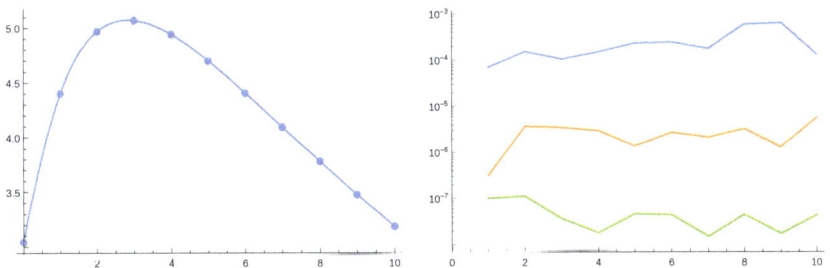

**Figure 9.** On the left the expected value of total discounted dividends, $V^{b]}(x)$, as function of $b$ with $x = 1$ in the case of the unperturbed model. On the right the absolute error of approximations of $V^{b]}(x)$ of order 4 (blue), 6 (orange) and 8 (green).

## 8. The Wiener–Hopf Factorization and Risk Theory with Rational Roots

Lévy processes are naturally parameterized via the additive decomposition of the symbol Equation (14). The roots and poles of the symbol play an important role and this has lead eventually to the discovery of the well-known Wiener–Hopf factorization Kyprianou (2014)

$$q\widehat{W}_q(s) = \frac{q}{\kappa(s) - q} = -\phi_q^+(s)\phi_q^-(s), q \geq 0. \tag{44}$$

With mixed exponential claims of rates $\beta_i$ for example, this holds with $\phi_q^+(s) = \frac{1}{1-s/\Phi_q}$ and

$$\phi^-(s) = \frac{\prod_{i=1}^n (1 + s/\beta_i)}{\prod_{i=1}^n (1 + s/\gamma_i)}, \tag{45}$$

where $\Phi_q$ and $-\gamma_i$ are, respectively, the nonnegative and negative roots of the Cramér–Lundberg equation.

One may bypass the numerical difficulties of finding roots by parameterizing a risk process by the poles and roots of its Wiener–Hopf factor $\phi^-(s)$. To further simplify matters, one may choose them as integers; then, the parameters of the process will be rational.

The simplest illustration is provided in the tables of (Dufresne and Gerber 1991a, p. 24): The input parameters therein are the interspersed non-zero integer roots $\gamma_i$ (denoted there by $r_i$) and the poles $\beta_i$ of the symbol $\kappa(s)$; essentially, this is equivalent to specifying Equation (45). The output parameter is the limit of Equation (45) when $s \to \infty$, yielding

$$\prod \frac{\gamma_i}{\beta_i} = 1 - \rho,$$

where $\rho := \Psi(0) = \lambda m_1 / c$ (Dufresne and Gerber (1991a) gives instead the safety load $\theta := \rho^{-1} - 1$). The other output parameters are partial fraction decomposition coefficients, given by the well-known partial fractions decomposition formulas, see Equation (12) for an example.[9] Once these coefficients are known, it is easy to show that

$$\begin{cases} m_1^{-1} = f_e(0) = \lim_{s \to \infty} s f_e^*(s) = \rho^{-1}(\sum \mu_i - \sum \gamma_i) \Leftrightarrow \\ \tilde{\lambda} := \frac{\lambda}{c} = \frac{\rho}{m_1} = \sum \mu_i - \sum \gamma_i \end{cases}$$

Note that the Levy model of Equation (1) is overdetermined, since by scaling the time, the symbol will be multiplied by an arbitrary factor; thus, one of the parameters, for example, $c$, may be specified at will.

*A Computer Program "Rat" That Outputs a Spectrally Negative Lévy Process with Given Roots and Poles of Its Symbol*

The first program inputs consists of

1. a vector of length $N$ containing the negatives $\gamma_1, \ldots, \gamma_N$ of the Cramér–Lundberg roots with negative real part, and
2. a vector $\beta_1, \ldots, \beta_n$, of length $n = \begin{cases} N & \sigma = 0 \\ N - 1 & \sigma > 0 \end{cases}$, containing the diagonal of the triangular matrix $B$ representing the Coxian distribution (the rest of the parameters of this law are only provided indirectly, via the Cramér–Lundberg roots).

The algorithm must distinguish between four cases, depending on whether or not $q = 0$ and $\sigma = 0$, and this is achieved by different inputs. The case $q > 0$ is specified via the presence of a third input parameter $\Phi_q$ (which is 0 if $q = 0$), and the cases $\sigma > 0 / \sigma = 0$ are distinguished by having $N = n + 1$ and $N = n$, respectively.

We will now introduce an artificial normalization constant $q_{c,\sigma}$, defined by

$$q_{c,\sigma} = \begin{cases} \frac{q}{\Phi_q(1-\rho_q)} := \frac{q}{\Phi_q} \prod \frac{\beta_i}{\gamma_i}, & q > 0 \\ \frac{1}{\Phi_q'(0)(1-\rho_q)} = \frac{1}{\Phi_q'(0)} \prod \frac{\beta_i}{\gamma_i}, & q = 0 \end{cases} \qquad (46)$$

(the factor $1 - \rho_q$ being incorporated just for convenience, see Equation (48)). Note that this definition allows one to deal with the case $q = 0 = \Phi_q$ as a limiting case of $q > 0$. $q_{c,\sigma}$ may be thought of at

---

[9] Nowadays, the simplest way to obtain them is by solving linear systems with Mathematica, Maple, Sage, etc.

first as the leading coefficient of $\kappa(s)$ (which is $c$ when $\sigma = 0$ and $\frac{\sigma^2}{2}$ when $\sigma > 0$); for a more precise statement, see Equation (48).

When $q > 0$, we will write the symbol as

$$\kappa(s) - q = q_{c,\sigma}\left(\frac{\tilde{\sigma}^2}{2}s^2 + \tilde{c}s - \Phi_q(1-\rho_q) + \sum_{j=1}^{n}\tilde{A}_j\left(\frac{\beta_j}{\beta_j + s} - 1\right)\right), \quad (47)$$

where $\tilde{c} = \frac{c}{q_{c,\sigma}}, \tilde{\sigma}^2 = \frac{\sigma^2}{q_{c,\sigma}}, \tilde{A}_j = \frac{A_j}{q_{c,\sigma}}$.

With this parameterization, results when $q \to 0$ will have a limit, provided that $q_{c,\sigma}$ is kept constant. If we assume (w.l.o.g.) that the polynomial part inside the parentheses in Equation (47) is monic (i.e., $\tilde{c} = 1$, if $\sigma = 0$, and $\tilde{\sigma}^2/2 = 1$, else), then the parameter $q_{c,\sigma}$ equals either the original parameter $c$ or $\frac{\sigma^2}{2}$.

**Remark 8.** *In terms of $q_{c,\sigma}$, we may rewrite the Wiener–Hopf factorization as*

$$\widehat{W}_q(s) = \frac{1}{\kappa(s) - q} = \frac{\Phi_q}{q(s - \Phi_q)}\phi_q^-(s) := \frac{\Phi_q}{q}(1-\rho_q)\frac{\tilde{\phi}_q^-(s)}{s - \Phi_q} = \frac{1}{q_{c,\sigma}}\frac{\tilde{\phi}_q^-(s)}{s - \Phi_q}$$

$$\Leftrightarrow \kappa(s) - q = q_{c,\sigma}\frac{s - \Phi_q}{\tilde{\phi}_q^-(s)} \quad (48)$$

*where the last factor in Equation (48) may be represented as a quotient of monic polynomials. This equation motivated the introduction of $q_{c,\sigma}$.*

**Remark 9.** *The various components of the cumulant generating function are readily obtained: $\sigma$, $c$ and $q$ by extracting the polynomial part; the remaining jump part $j(s)$ then yields $\lambda = -\lim_{s\to\infty} j(s)$.*

**Remark 10.** *Note that if the condition*

$$\gamma_{n+1} > \beta_n > \gamma_n > \ldots > \gamma_2 > \beta_1 > \gamma_1 \geq 0 \quad (49)$$

*is satisfied, then the nonnegativity of the density of the jumps is ensured—see, for example, Kuznetsov et al. (2012) (this condition is also necessary with hyperexponential jumps).*

*If this condition is not satisfied, then the user must deal somehow with the nonnegativity by restricting the position of the Cramér–Lundberg roots (not an easy task).*

The output of our program consists of several first passage characteristics:

1. Eventual ruin probabilities
2. Cumulants generating function (Laplace exponent) of the Levy process $\kappa(s)$
3. Homogeneous scale function $W^{(q)}(y)$, for one fixed $q$

They are most easily described as combinations of exponentials.

## 9. Further Background on the Laguerre–Tricomi–Weeks Method

We now proceed to review an extension of the Laguerre expansion, the "Laguerre–Tricomi–Weeks" method of inverting the Laplace transform, initially proposed by Tricomi in 1935 and McCully (see McCully (1960)), which is considered as one of the most efficient methods of inverting the Laplace transform Weideman (1999). We mention also that an exact explicit closed-form Laguerre series expansion formula was proposed recently in Zhang and Cui (2019).

Consider a given function

$$f(x) = e^{\beta x}g(\alpha x), \quad g \in L^2 \quad (50)$$

(note that the transformation Equation (50) corresponds to a linear transformation for the Laplace transform $\hat{f}(s) = \hat{g}(\frac{s-\beta}{\alpha})$ and its singularities). First, one attempts to construct a Laguerre expansion

$$f(x) = e^{\beta x} \sum_{n=0}^{\infty} B_n e^{-\alpha x/2} L_n(\alpha x) \Leftrightarrow f(x/\alpha) e^{-\beta x/\alpha} = \sum_{n=0}^{\infty} B_n e^{-x/2} L_n(x)$$

$$\Leftrightarrow \hat{f}(\alpha s - \alpha/2 + \beta) = s^{-1} \sum_{n=0}^{\infty} B_n (1 - \alpha/s)^n \qquad (51)$$

with two judiciously chosen parameters $\alpha > 0$ and $\beta > s_f$, where $s_f$ is the maximum real part of the singular points of $f^*(s)$.

After the change of variables $1 - \alpha/s = z$, the coefficients $B_n$ may be found from the Taylor expansion

$$\frac{\alpha}{1-z} f^* \left( \frac{\alpha}{1-z} - \alpha/2 + \beta \right) = \sum_{n=0}^{\infty} B_n z^n, \ |z| < R_\Phi \qquad (52)$$

where $R_\Phi$ is the radius of convergence of $\Phi(z) = \frac{\alpha}{1-z} f^* \left( \frac{\alpha}{1-z} - \alpha/2 + \beta \right)$, or, equivalently,

$$\alpha f^* \left( \frac{\alpha}{1-z} - \alpha/2 + \beta \right) = \sum_{n=0}^{\infty} b_n z^n, \ \text{with } b_0 = B_0, b_n = B_n - B_{n-1}, n \geq 1.$$

Judicious choices of $\beta$ (related to maximizing $R_\Phi$) and $\alpha$ (related to finding a minimal circle including the singular points of $\hat{f}(s)$) may turn this into one of the most effective inversion methods Weideman (1999), Abate et al. (1998).

Some particular choices are $\beta = \alpha/2$, $\beta = 0$ and $\beta = -\alpha/2$, in which case the Taylor expansion for the coefficients $B_n$ becomes, respectively,

$$\frac{\alpha}{1-z} f^* \left( \frac{\alpha}{1-z} \right) = \sum_{n=0}^{\infty} B_n z^n \Leftrightarrow f(x/\alpha) = \sum_{n=0}^{\infty} B_n L_n(x) \qquad (53)$$

$$\frac{\alpha}{1-z} f^* \left( \frac{\alpha}{2} \cdot \frac{1+z}{1-z} \right) = \sum_{n=0}^{\infty} B_n z^n \Leftrightarrow f(x/\alpha) = \sum_{n=0}^{\infty} B_n e^{-x/2} L_n(x) \qquad (54)$$

$$\frac{\alpha}{1-z} f^* \left( \alpha \frac{z}{1-z} \right) = \sum_{n=0}^{\infty} B_n z^n \Leftrightarrow f(x/\alpha) = \sum_{n=0}^{\infty} B_n e^{-x} L_n(x). \qquad (55)$$

As noted by Keilson and Nunn Keilson and Nunn (1979) (see also Keilson et al. (1980, 1981)), and Abate et al. (1998), respectively, in the second and third cases, the expansion in the of powers Equation (32) implies simple relations between the Laguerre transform and another interesting technique, the Erlang transform, which is a useful tool for the solving integral equations of the convolution type. We refer to these papers for a further discussion of the relative advantages of the Erlang and Laguerre transforms, and the spaces of the functions within which they converge.

**Author Contributions:** F.A., A.H. and S.P. contributed equally to Conceptualization, Methodology, Formal Analysis, Investigation, Writing-Original Draft Preparation, Writing-Review, Project Administration, Supervision and Editing, and U.S. contributed to Software, Validation and Data Curation.

**Funding:** This research received no external funding.

**Conflicts of Interest:** The authors declare no conflict of interest.

## References

Abate, Joseph, Gagan L. Choudhury, and Ward Whitt. 1998. Numerical inversion of multidimensional laplace transforms by the Laguerre method. *Performance Evaluation* 31: 229–43. [CrossRef]

Abramowitz, Milton, and Irene A. Stegun. 1965. *Handbook of Mathematical Functions: With Formulas, Graphs, and Mathematical Tables*. Mineola: Dover Publications, vol. 55.

Albrecher, Hansjörg, and Sören Asmussen. 2010. *Ruin Probabilities*. Singapore: World Scientific, vol. 14.

Albrecher, Hansjörg, Florin Avram, and Dominik Kortschak. 2010. On the efficient evaluation of ruin probabilities for completely monotone claim distributions. *Journal of Computational and Applied Mathematics* 233: 2724–36. [CrossRef]

Avram, Florin, Abhijit Datta Banik, and Andras Horváth. 2018. Ruin probabilities by padé's method: Simple moments based mixed exponential approximations (Renyi, De Vylder, Cramér–Lundberg), and high precision approximations with both light and heavy tails. *European Actuarial Journal* 9: 273–99. [CrossRef]

Avram, Florin, D. Fotso Chedom, and Andras Horváth. 2011. On moments based Padé approximations of ruin probabilities. *Journal of Computational and Applied Mathematics* 235: 3215–28. [CrossRef]

Avram, Florin, Zbigniew Palmowski, and Martijn R. Pistorius. 2007. On the optimal dividend problem for a spectrally negative Lévy process. *The Annals of Applied Probability* 17: 156–80. [CrossRef]

Avram, Florin, Zbigniew Palmowski, and Martijn R. Pistorius. 2015. On Gerber–Shiu functions and optimal dividend distribution for a Lévy risk process in the presence of a penalty function. *The Annals of Applied Probability* 25: 1868–1935. [CrossRef]

Avram, Florin, and Martijn R. Pistorius. 2014. On matrix exponential approximations of ruin probabilities for the classic and Brownian perturbed Cramér-Lundberg processes. *Insurance: Mathematics and Economics* 59: 57–64. [CrossRef]

Azcue, Pablo, and Nora Muler. 2005. Optimal reinsurance and dividend distribution policies in the Cramér-Lundberg model. *Mathematical Finance* 15: 261–308. [CrossRef]

Bertoin, Jean. 1998. *Lévy Processes*. Cambridge: Cambridge University Press, vol. 121.

Bingham, Nicholas H. 1976. Continuous branching processes and spectral positivity. *Stochastic Processes and Their Applications* 4: 217–42. [CrossRef]

Bobbio, Andrea, Andras Horváth, and Miklos Telek. 2005. Matching three moments with minimal acyclic phase type distributions. *Stochastic Models* 21: 303–26. [CrossRef]

De Finetti, Bruno 1957. Su un'impostazione alternativa della teoria collettiva del rischio. In *Transactions of the XVth International Congress of Actuaries*. New York: Mallon, vol. 2, pp. 433–43.

Dufresne, Francois, and Hans U. Gerber. 1991a. Rational ruin problems—A note for the teacher. *Insurance: Mathematics and Economics* 10: 21–29. [CrossRef]

Dufresne, Francois, and Hans U. Gerber. 1991b. Risk theory for the compound Poisson process that is perturbed by diffusion. *Insurance: Mathematics and Economics* 10: 51–59. [CrossRef]

Dumitrescu, Bogdan, Bogdan C. Şicleru, and Florin Avram. 2016. Modeling probability densities with sums of exponentials via polynomial approximation. *Journal of Computational and Applied Mathematics* 292: 513–25. [CrossRef]

Gerber, Hans U., Elias S. W. Shiu, and Nathaniel Smith. 2008. Methods for estimating the optimal dividend barrier and the probability of ruin. *Insurance: Mathematics and Economics* 42: 243–54. [CrossRef]

Giunta, Giulio Giuliano Laccetti, and Mariarosaria Rizzardi. 1989. More on the Weeks method for the numerical inversion of the Laplace transform. *Numerische Mathematik* 54: 193–200. [CrossRef]

Horváth, Gabor, and Miklos Telek. 2016. Butools 2: A rich toolbox for Markovian performance evaluation. Paper presented at 10th EAI International Conference on Performance Evaluation Methodologies and Tools, Taormina, Italy, October 25–28.

Hu, Xiang, Baige Duan, and Lianzeng Zhang. 2017. De Vylder approximation to the optimal retention for a combination of quota-share and excess of loss reinsurance with partial information. *Insurance: Mathematics and Economics* 76: 48–55. [CrossRef]

Keilson, Julian, and W. R. Nunn. 1979. Laguerre transformation as a tool for the numerical solution of integral equations of convolution type. *Applied Mathematics and Computation* 5: 313–59. [CrossRef]

Keilson, Julian, W. R. Nunn, and Ushio Sumita. 1980. *The Laguerre Transform*. Technical Report. Alexandria: Center for Naval Analyses Alexandria va Operations Evaluation Group.

Keilson, Julian, W. R. Nunn, and Ushio Sumita. 1981. The bilateral laguerre transform. *Applied Mathematics and Computation* 8: 137–74. [CrossRef]

Kuznetsov, Alexey, Andreas E. Kyprianou, and Victor Rivero. 2012. The theory of scale functions for spectrally negative Lévy processes. In *Lévy Matters II*. Berlin/Heidelberg: Springer, pp. 97–186.

Kyprianou, Andreas, 2014. *Fluctuations of Lévy Processes with Applications: Introductory Lectures*. Berlin/Heidelberg: Springer Science & Business Media.

Landriault, David, and Gordon E. Willmot. 2019. On series expansions for scale functions and other ruin-related quantities. *Scandinavian Actuarial Journal* 1–15. doi:10.1080/03461238.2019.1663444. [CrossRef]

Loeffen, Ronnie Lambertus, 2008. Stochastic Control for Spectrally Negative Lévy Processes. Ph.D. Thesis, University of Bath, Bath, UK.

McCully, Joseph 1960. The Laguerre transform. *Siam Review* 2: 185–91. [CrossRef]

Miller, Merton H., and Franco Modigliani. 1961. Dividend policy, growth, and the valuation of shares. *The Journal of Business* 34: 411–33. [CrossRef]

Nadarajah, Saralees, and Samuel Kotz. 2006. On the Laplace transform of the pareto distribution. *Queueing Systems* 54: 243–44. [CrossRef]

Nakatsukasa, Yuji, Olivier Sète, and Lloyd N Trefethen. 2018. The AAA algorithm for rational approximation. *SIAM Journal on Scientific Computing* 40: A1494–522. [CrossRef]

Ramsay, Colin M. 1992. A practical algorithm for approximating the probability of ruin. *Transactions of the Society of Actuaries* 44: 443–61.

Schmidli, Hanspeter. 2007. *Stochastic Control in Insurance*. Berlin/Heidelberg: Springer Science & Business Media.

Skorokhod, Anatolii Vladimirovich 1962. Stochastic equations for diffusion processes in a bounded region. II. *Theory of Probability & Its Applications* 7: 3–23. doi:10.1137/1107002. [CrossRef]

Suprun, V. N. 1976. Problem of destruction and resolvent of a terminating process with independent increments. *Ukrainian Mathematical Journal* 28: 39–51. [CrossRef]

Weideman, Jacob Andre C. 1999. Algorithms for parameter selection in the weeks method for inverting the Laplace transform. *SIAM Journal on Scientific Computing* 21: 111–28. [CrossRef]

Zhang, Zhimin, and Zhenyu Cui. 2019. Laguerre Series Expansion for Scale Functions and Its Applications in Risk Theory. Working Paper. Available online: https://www.researchgate.net/publication/336613949_Laguerre_series_expansion_for_scale_functions_and_its_applications_in_risk_theory (accessed on 28 October 2019).

© 2019 by the authors. Licensee MDPI, Basel, Switzerland. This article is an open access article distributed under the terms and conditions of the Creative Commons Attribution (CC BY) license (http://creativecommons.org/licenses/by/4.0/).

Article

# De Finetti's Control Problem with Parisian Ruin for Spectrally Negative Lévy Processes

Jean-François Renaud

Département de mathématiques, Université du Québec à Montréal (UQAM), Montréal, QC H2X 3Y7, Canada; renaud.jf@uqam.ca

Received: 12 June 2019; Accepted: 26 June 2019; Published: 3 July 2019

**Abstract:** We consider de Finetti's stochastic control problem when the (controlled) process is allowed to spend time under the critical level. More precisely, we consider a generalized version of this control problem in a spectrally negative Lévy model with exponential Parisian ruin. We show that, under mild assumptions on the Lévy measure, an optimal strategy is formed by a barrier strategy and that this optimal barrier level is always less than the optimal barrier level when classical ruin is implemented. In addition, we give necessary and sufficient conditions for the barrier strategy at level zero to be optimal.

**Keywords:** stochastic control; spectrally negative Lévy processes; optimal dividends; Parisian ruin; log-convexity; barrier strategies

## 1. Introduction and Main Result

In the 1950s, Bruno de Finetti (1957) formulated the following stochastic control problem: find the dividend strategy maximizing the expected present value of the dividend payments associated with an insurance surplus process. Presently, this control problem is known as de Finetti's optimal dividends problem. Another active field of research in insurance mathematics is the analysis of Parisian implementation delays in the recognition of default (see e.g., (Landriault et al. 2011; Loeffen et al. 2013)) and/or in the design of dividend strategies (see e.g., (Albrecher et al. 2011; Dassios and Wu 2009)). In what follows, we formulate and solve an extension of de Finetti's control problem with Parisian ruin.

### 1.1. Problem Formulation

On a filtered probability space $(\Omega, \mathcal{F}, \{\mathcal{F}_t, t \geq 0\}, \mathbb{P})$, let $X = \{X_t, t \geq 0\}$ be a spectrally negative Lévy process with Laplace exponent $\theta \mapsto \psi(\theta)$ and with $q$-scale functions $\{W^{(q)}, q \geq 0\}$ given by

$$\int_0^\infty e^{-\theta x} W^{(q)}(x) dx = (\psi(\theta) - q)^{-1},$$

for all $\theta > \Phi(q) = \sup\{\lambda \geq 0 \colon \psi(\lambda) = q\}$. Recall that

$$\psi(\theta) = \gamma \theta + \frac{1}{2}\sigma^2 \theta^2 + \int_0^\infty \left( e^{-\theta z} - 1 + \theta z \mathbf{1}_{(0,1]}(z) \right) \nu(dz),$$

where $\gamma \in \mathbb{R}$ and $\sigma \geq 0$, and where $\nu$ is a $\sigma$-finite measure on $(0, \infty)$, called the Lévy measure of $X$, satisfying

$$\int_0^\infty (1 \wedge x^2) \nu(dx) < \infty.$$

For more details on spectrally negative Lévy processes and scale functions, see e.g., (Kuznetsov et al. 2012; Kyprianou 2014).

In what follows, we will use the following notation: the law of $X$ when starting from $X_0 = x$ is denoted by $\mathbb{P}_x$ and the corresponding expectation by $\mathbb{E}_x$. We write $\mathbb{P}$ and $\mathbb{E}$ when $x = 0$.

Let the spectrally negative Lévy process $X$ be the underlying surplus process. A dividend strategy $\pi$ is represented by a non-decreasing, left-continuous and adapted stochastic process $L^\pi = \{L_t^\pi, t \geq 0\}$, where $L_t^\pi$ represents the cumulative amount of dividends paid up to time $t$ under this strategy, and such that $L_0^\pi = 0$. For a given strategy $\pi$, the corresponding controlled surplus process $U^\pi = \{U_t^\pi, t \geq 0\}$ is defined by $U_t^\pi = X_t - L_t^\pi$. The stochastic control problem considered in this paper involves the time of Parisian ruin (with rate $p > 0$) for $U^\pi$ defined by

$$\sigma_p^\pi = \inf\left\{t > 0 : t - g_t^\pi > \mathbf{e}_p^{g_t^\pi} \text{ and } U_t^\pi < 0\right\},$$

where $g_t^\pi = \sup\{0 \leq s \leq t : U_s^\pi \geq 0\}$, with $\mathbf{e}_p^{g_t^\pi}$ an independent random variable, following the exponential distribution with mean $1/p$, associated with the corresponding excursion below 0 (see (Baurdoux et al. 2016) for more details). Please note that, without loss of generality, we have chosen 0 to be the critical level.

**Remark 1.** *Recall that $X$ and $L^\pi$ are adapted to the filtration. Set $\mathcal{F}_\infty = \bigvee_{t \geq 0} \mathcal{F}_t$, i.e., the smallest $\sigma$-algebra containing $\mathcal{F}_t$, for all $t \geq 0$. It is implicitly assumed that $\mathcal{F}_\infty$ is strictly less than $\mathcal{F}$ and that all exponential clocks are independent of $\mathcal{F}_\infty$.*

A strategy $\pi$ is said to be admissible if a dividend payment is not larger than the current surplus level, i.e., $L_{t+}^\pi - L_t^\pi \leq U_t^\pi$, for all $t < \sigma_p^\pi$, and if no dividends are paid when the controlled surplus is negative, i.e., $t \mapsto L_t^\pi \mathbf{1}_{(-\infty,0)}(U_t^\pi) \equiv 0$. The set of admissible dividend strategies will be denoted by $\Pi_p$. These two conditions are motivated by the following interpretation: if $U^\pi$ enters the interval $(-\infty, 0)$, then a period of *financial distress* begins. Consequently, dividend payments should not cause an excursion under the critical level nor should they be made during those critical periods.

Fix a discounting rate $q \geq 0$. The value function associated with an admissible dividend strategy $\pi \in \Pi_p$ is defined by

$$v_\pi(x) = \mathbb{E}_x\left[\int_0^{\sigma_p^\pi} e^{-qt} dL_t^\pi\right], \quad x \in \mathbb{R}.$$

The goal is to find the optimal value function $v_*$ defined by

$$v_*(x) = \sup_{\pi \in \Pi_p} v_\pi(x)$$

and an optimal strategy $\pi_* \in \Pi_p$ such that

$$v_{\pi_*}(x) = v_*(x),$$

for all $x \in \mathbb{R}$. Because of the Parisian nature of the time of ruin considered in this control problem, we have to deal with possibly negative starting capital.

### 1.2. Main Result and Organization of the Paper

Let us introduce the family of horizontal barrier strategies, also called reflection strategies. For $b \in \mathbb{R}$, the (horizontal) barrier strategy at level $b$ is the strategy denoted by $\pi^b$ and with cumulative amount of dividends paid until time $t$ given by $L_t^b = \left(\sup_{0 < s \leq t} X_s - b\right)_+$, for $t > 0$. If $X_0 = x > b$, then $L_{0+}^b = x - b$. Please note that, if $b \geq 0$, then $\pi_b \in \Pi_p$. The corresponding value function is thus given by

$$v_b(x) = \mathbb{E}_x\left[\int_0^{\sigma_p^b} e^{-qt} dL_t^b\right],$$

for all $x \in \mathbb{R}$, where $\sigma_p^b$ is the time of Parisian ruin (with rate $p > 0$) for the controlled process $U_t^b = X_t - L_t^b$.

Before stating the main result of this paper, recall that the tail of the Lévy measure is the function $x \mapsto \nu(x, \infty)$, where $x \in (0, \infty)$, and that a function $f \colon (0, \infty) \to (0, \infty)$ is log-convex if the function $\log(f)$ is convex on $(0, \infty)$.

**Theorem 1.** *Fix $q \geq 0$ and $p > 0$. If the tail of the Lévy measure is log-convex, then an optimal strategy for the control problem is formed by a barrier strategy.*

The original version of de Finetti's optimal dividends problem, i.e., when the time of ruin is the first passage time below the critical level (intuitively, when $p \to \infty$), has been extensively studied. In a spectrally negative Lévy model, following the work of Avram et al. (2007), an important breakthrough was made by Loeffen (2008); in the latter paper, a sufficient condition, on the Lévy measure $\nu$, is given for a barrier strategy to be optimal. This condition was further relaxed by Loeffen and Renaud (2010); in this other paper, it is shown that if the tail of the Lévy measure is log-convex then a barrier strategy is optimal for de Finetti's optimal dividends problem with an affine penalty function at ruin (if we set $S = K = 0$ in that paper, we recover the classical problem). To the best of our knowledge, this still stands as the mildest condition for the optimality of a barrier strategy in a spectrally negative Lévy model. Finally, note that Czarna and Palmowski (2014) have considered de Finetti's control problem with deterministic Parisian delays.

The rest of the paper is organized as follows. First, we provide an alternative interpretation of the value function and we fill the gap between the models with classical ruin and no ruin. Then, we compute the value function of an arbitrary horizontal barrier strategy and find the optimal barrier level $b_p^*$ (see the definition in (9)). Finally, we derive the appropriate verification lemma for this control problem and prove that, under our assumption on the Lévy measure, the barrier strategy at level $b_p^*$ is optimal.

## 2. More on the Value Function

Please note that for $\pi \in \Pi_p$ and $x < 0$, using the strong Markov property and the spectral negativity of $X$, we can easily verify that

$$v_\pi(x) = \mathbb{E}_x \left[ e^{-q\tau_0^+} \mathbf{1}_{\{\tau_0^+ < \mathbf{e}_p\}} \right] v_\pi(0) = e^{\Phi(p+q)x} v_\pi(0), \tag{1}$$

where $\tau_0^+ = \inf \{t > 0 \colon X_t > 0\}$ and where $\mathbf{e}_p$ is an independent exponentially distributed random variable with mean $1/p$, thanks to the well-known fluctuation identity (see e.g., (Kyprianou 2014))

$$\mathbb{E}_x \left[ e^{-r\tau_b^+} \mathbf{1}_{\{\tau_b^+ < \infty\}} \right] = e^{-\Phi(r)(b-x)}, \quad x \leq b, \tag{2}$$

where $\tau_b^+ = \inf \{t > 0 \colon X_t > b\}$.

Interestingly, we can show that (see the proof of Lemma 1 below), for any $\pi \in \Pi_p$, we have

$$v_\pi(x) = \mathbb{E}_x \left[ \int_0^\infty e^{-qs - p \int_0^s \mathbf{1}_{(-\infty,0)}(U_r^\pi) dr} dL_s^\pi \right]. \tag{3}$$

Using this last identity, we can argue that using Parisian ruin with rate $p$ fills the gap between the model with classical ruin (no delay, $p \to \infty$) and the model with no ruin (infinite delays, $p \to 0$). Indeed, using (3), we see directly that

$$v_\pi(x) = \mathbb{E}_x \left[ \int_0^\infty e^{-qs - p \int_0^s \mathbf{1}_{(-\infty,0)}(U_r^\pi) dr} dL_s^\pi \right] \xrightarrow[p \to 0]{} \mathbb{E}_x \left[ \int_0^\infty e^{-qs} dL_s^\pi \right].$$

On the other hand, as

$$\int_0^\infty e^{-qs-p\int_0^s \mathbf{1}_{(-\infty,0)}(U_r^\pi)dr} dL_s^\pi = \int_0^\infty e^{-qs} \mathbf{1}_{\{s \leq \sigma_\infty^\pi\}} dL_s^\pi + \int_0^\infty e^{-qs-p\int_0^s \mathbf{1}_{(-\infty,0)}(U_r^\pi)dr} \mathbf{1}_{\{s > \sigma_\infty^\pi\}} dL_s^\pi,$$

where $\sigma_\infty^\pi := \inf\{t > 0 : U_t^\pi < 0\}$, we obtain

$$v_\pi(x) = \mathbb{E}_x\left[\int_0^\infty e^{-qs-p\int_0^s \mathbf{1}_{(-\infty,0)}(U_r^\pi)dr} dL_s^\pi\right] \xrightarrow[p \to \infty]{} \mathbb{E}_x\left[\int_0^{\sigma_\infty^\pi} e^{-qs} dL_s^\pi\right].$$

**Remark 2.** *Note also that the expression of the value function given in (3) tells us that the current control problem amounts to a control problem with no ruin and in which the dividend payments are penalized by the occupation time of the surplus process. Indeed, from this point of view, the discount factor increases with the time spent below zero by the surplus process.*

## 3. Horizontal Barrier Strategies

Before computing the value function of an arbitrary barrier strategy at level $b$, we have to define another family of scale functions, also called second $q$-scale functions of $X$.

### 3.1. Second Family of Scale Functions

The so-called second scale functions are defined by: for each $q, \theta \geq 0$ and for $x \in \mathbb{R}$, let

$$Z_q(x, \theta) = e^{\theta x} \left(1 - (\psi(\theta) - q) \int_0^x e^{-\theta y} W^{(q)}(y) dy\right). \tag{4}$$

Please note that for $x \leq 0$ or for $\theta = \Phi(q)$, we have $Z_q(x, \theta) = e^{\theta x}$. The second scale functions have appeared in the literature in various forms; see e.g., (Albrecher et al. 2016; Avram et al. 2015; Ivanovs and Palmowski 2012).

In what follows, $Z_q'(x, \theta)$ will represent the derivative with respect to the first argument. Consequently, for $x > 0$, we have $Z_q'(x, \theta) = \theta Z_q(x, \theta) - (\psi(\theta) - q)W^{(q)}(x)$ and, for $x < 0$, we have $Z_q'(x, \theta) = \theta e^{\theta x}$.

In this paper, we will encounter the function $Z_q$ when $\theta = \Phi(p+q)$, that is the function

$$Z_q(x, \Phi(p+q)) = e^{\Phi(p+q)x} \left(1 - p \int_0^x e^{-\Phi(p+q)y} W^{(q)}(y) dy\right),$$

from which we deduce that, for $x > 0$,

$$Z_q'(x, \Phi(p+q)) = \Phi(p+q) Z_q(x, \Phi(p+q)) - p W^{(q)}(x). \tag{5}$$

Consequently, set $Z_q'(0, \Phi(p+q)) = \Phi(p+q) - pW^{(q)}(0)$. Since we assume that $p > 0$, we have that $\Phi(p+q) > \Phi(q)$ and we can write

$$Z_q(x, \Phi(p+q)) = p \int_0^\infty e^{-\Phi(p+q)y} W^{(q)}(x+y) dy, \quad x \in \mathbb{R}. \tag{6}$$

Then, for $x > 0$, we have

$$Z_q'(x, \Phi(p+q)) = p \int_0^\infty e^{-\Phi(p+q)y} W^{(q)'}(x+y) dy, \tag{7}$$

which is well defined since $W^{(q)}$ is differentiable almost everywhere (see e.g., Lemma 2.3 in (Kuznetsov et al. 2012)). Clearly, $x \mapsto Z_q(x, \Phi(p+q))$ is a non-decreasing continuous function. In fact, it will be proved in Appendix B that if the tail of the Lévy measure is log-convex, then $Z_q'(\cdot, \Phi(p+q))$ is a log-convex function on $(0, \infty)$.

## 3.2. Value Function of a Barrier Strategy

Here is the value of an arbitrary admissible barrier strategy:

**Proposition 1.** *For $q, b \geq 0$, the value function associated with $\pi_b$ is given by*

$$v_b(x) = \begin{cases} \dfrac{Z_q(x, \Phi(p+q))}{Z_q'(b, \Phi(p+q))} & \text{for } x \in (-\infty, b], \\ x - b + v_b(b) & \text{for } x \in (b, \infty). \end{cases} \tag{8}$$

**Proof.** This result has appeared before in the literature. See for example Equation (15) in (Albrecher and Ivanovs 2014) or Equation (46) in (Avram and Zhou 2016). Nevertheless, we provide an alternative proof in Appendix A. □

## 3.3. Optimal Barrier Level

As defined in (Loeffen 2008; Loeffen and Renaud 2010), the optimal barrier level in de Finetti's classical control problem is given by

$$b_\infty^* = \sup\left\{b \geq 0 : W^{(q)\prime}(b) \leq W^{(q)\prime}(x), \text{ for all } x \geq 0\right\}.$$

Similarly, let us define the candidate for the optimal barrier level for the current version of this control problem by

$$b_p^* = \sup\left\{b \geq 0 : Z_q'(b, \Phi(p+q)) \leq Z_q'(x, \Phi(p+q)), \text{ for all } x \geq 0\right\}. \tag{9}$$

**Proposition 2.** *Fix $q \geq 0$ and $p > 0$. Suppose the tail of the Lévy measure is log-convex. Then, $0 \leq b_p^* \leq b_\infty^*$ and $b_p^* > 0$ if and only if*

$$\Phi(p+q) - pW^{(q)}(0) < \frac{p}{\Phi(p+q)} W^{(q)\prime}(0+). \tag{10}$$

*Equivalently, $b_p^* > 0$ if and only if one of the following three cases hold:*

(a) $\sigma > 0$ and $(\Phi(p+q))^2 / p < 2/\sigma^2$;
(b) $\sigma = 0$ and $\nu(0, \infty) = \infty$;
(c) $\sigma = 0$, $\nu(0, \infty) < \infty$ and

$$\frac{c\Phi(p+q)}{p}\left(\Phi(p+q) - \frac{p}{c}\right) < \frac{q + \nu(0, \infty)}{c},$$

*where $c = \gamma + \int_0^1 x\nu(dx)$.*

**Proof.** See the proof in Appendix B. □

First of all, note from Proposition 2 that the optimal barrier level $b_p^*$, when Parisian ruin with rate $p$ is implemented, is always lower than the optimal barrier level $b_\infty^*$ when classical ruin is used.

In cases (a) and (c), the value of $b_p^*$ can be either positive or zero, depending on the parameters of the model. It is clear from the condition in (10) that, when $q > 0$, if the Parisian rate $p$ is small enough (large delays), then $b_p^* = 0$; in words, if Parisian delays are infinite (no ruin), then it is better to start paying out dividends right away. However, when $q = 0$ (no discounting), if Parisian delays are infinite (no ruin), then $b_p^* > 0$ if and only if $\mathbb{E}[X_1] > 0$.

Also, in case (a), if $X_t = ct + \sigma B_t$ is a Brownian motion with drift, then

$$\Phi(p+q) = \frac{1}{\sigma^2}\left(\sqrt{c^2 + 2\sigma^2(p+q)} - c\right)$$

and we can verify that $b_p^* = 0$ as soon as the Brownian coefficient $\sigma$ is large enough.

**Remark 3.** *In Section 4 of (Avram and Minca 2017), economic principles for evaluating the efficiency of a surplus process are discussed. One of them is that the optimal barrier level be equal to zero.*

Interestingly, the condition in (c) can be re-written as follows:

$$\frac{c\Phi(p+q)}{p}\mathbb{E}\left[\int_0^{\sigma_\infty^0} e^{-qt}dL_t^0\right] < \mathbb{E}\left[\int_0^{\sigma_p^0} e^{-qt}dL_t^0\right] = v_0(0).$$

Indeed, when $\sigma = 0$ and $\nu(0,\infty) < \infty$, it is known (see Equation (3.14) in (Avram et al. 2007)) that

$$\mathbb{E}\left[\int_0^{\sigma_\infty^0} e^{-qt}dL_t^0\right] = \frac{c}{q + \nu(0,\infty)}$$

and, from Proposition 1, we have

$$\mathbb{E}\left[\int_0^{\sigma_p^0} e^{-qt}dL_t^0\right] = \frac{1}{\Phi(p+q) - pW^{(q)}(0+)}.$$

## 4. Verification Lemma and Proof of the Main Result

Define the operator $\Gamma$ associated with $X$ by

$$\Gamma v(x) = \gamma v'(x) + \frac{\sigma^2}{2}v''(x) + \int_0^\infty \left(v(x-z) - v(x) + v'(x)z\mathbf{1}_{(0,1]}(z)\right)\nu(dz), \quad (11)$$

where $v$ is a function defined on $\mathbb{R}$ such that $\Gamma v(x)$ is well defined. We say that a function $v$ is sufficiently smooth if it is continuously differentiable on $(0,\infty)$ when $X$ is of bounded variation and twice continuously differentiable on $(0,\infty)$ when $X$ is of unbounded variation.

Next is the verification lemma of our stochastic control problem. As the controlled process is now allowed to spend time below the critical level, it is different from the classical verification lemma (see (Loeffen 2008)).

**Lemma 1.** *Let $\Gamma$ be the operator defined in (11). Suppose that $\hat{\pi} \in \Pi_p$ is such that $v_{\hat{\pi}}$ is sufficiently smooth and that, for all $x \in \mathbb{R}$,*

$$\left(\Gamma - q - p\mathbf{1}_{(-\infty,0)}\right)v_{\hat{\pi}}(x) \leq 0$$

*and, for all $x > 0$, $v'_{\hat{\pi}}(x) \geq 1$. In this case, $\hat{\pi}$ is an optimal strategy for the control problem.*

**Proof.** Set $w := v_{\tilde{\pi}}$ and let $\pi \in \Pi_p$ be an arbitrary admissible strategy. As $w$ is sufficiently smooth, applying an appropriate change-of-variable/version of Ito's formula to the joint process $\left(t, \int_0^t \mathbf{1}_{(-\infty,0)}(U_r^\pi) dr, U_t^\pi\right)$ yields

$$\begin{aligned}
&e^{-qt-p\int_0^t \mathbf{1}_{(-\infty,0)}(U_r^\pi)dr} w(U_t^\pi) - w(U_0^\pi) \\
&= \int_0^t e^{-qs-p\int_0^s \mathbf{1}_{(-\infty,0)}(U_r^\pi)dr} \left[(\Gamma-q)w(U_s^\pi) - p\mathbf{1}_{(-\infty,0)}(U_s^\pi) w(U_s^\pi)\right] ds \\
&\quad - \int_0^t e^{-qs-p\int_0^s \mathbf{1}_{(-\infty,0)}(U_r^\pi)dr} w'(U_{s-}^\pi) dL_s^\pi + M_t^\pi \\
&\quad + \sum_{0 < s \leq t} e^{-qs-p\int_0^s \mathbf{1}_{(-\infty,0)}(U_r^\pi)dr} \left[w(U_{s-}^\pi - \Delta L_s^\pi) - w(U_{s-}^\pi) + w'(U_{s-}^\pi) \Delta L_s^\pi\right],
\end{aligned} \tag{12}$$

where $M^\pi = \{M_t^\pi, t \geq 0\}$ is a (local) martingale.

Consider an independent (of $\mathcal{F}_\infty$) Poisson process with intensity measure $p\,dt$ and jump times $\{T_i^p, i \geq 1\}$. Therefore, we can write

$$e^{-p\int_0^s \mathbf{1}_{(-\infty,0)}(U_r^\pi)dr} = \mathbb{P}_x\left(T_i^p \notin \{r \in (0,s] : U_r^\pi < 0\}, \text{ for all } i \geq 1 | \mathcal{F}_\infty\right) = \mathbb{E}_x\left[\mathbf{1}_{\{\sigma_p^\pi > s\}} | \mathcal{F}_\infty\right]$$

and consequently

$$\mathbb{E}_x\left[\int_0^t e^{-qs-p\int_0^s \mathbf{1}_{(-\infty,0)}(U_r^\pi)dr} dL_s^\pi\right] = \mathbb{E}_x\left[\int_0^t e^{-qs} \mathbb{E}_x\left[\mathbf{1}_{\{\sigma_p^\pi > s\}} | \mathcal{F}_\infty\right] dL_s^\pi\right] = \mathbb{E}_x\left[\int_0^{\sigma_p^\pi \wedge t} e^{-qs} dL_s^\pi\right],$$

where we used the definition of a Riemann-Stieltjes integral and the monotone convergence theorem for conditional expectations.

Now, as for all $x \in \mathbb{R}$,

$$\left(\Gamma - q - p\mathbf{1}_{(-\infty,0)}\right) w(x) \leq 0$$

and, for all $x > 0$, $w'(x) \geq 1$, using standard arguments (see e.g., (Loeffen 2008)) and our definition of an admissible strategy, e.g., that $L^\pi$ is identically zero when $U^\pi$ is below zero, we get

$$w(x) \geq \mathbb{E}_x\left[\int_0^\infty e^{-qs-p\int_0^s \mathbf{1}_{(-\infty,0)}(U_r^\pi)dr} dL_s^\pi\right] = \mathbb{E}_x\left[\int_0^{\sigma_p^\pi} e^{-qs} dL_s^\pi\right] = v_\pi(x).$$

This concludes the proof. □

The rest of this section is devoted to proving Theorem 1, i.e., proving that an optimal strategy for the control problem is formed by the barrier strategy at level $b^* := b_p^*$.

By the definition of $b^*$ given in (9), for $0 \leq x \leq b^*$, we have

$$v'_{b^*}(x) = \frac{Z'_q(x, \Phi(p+q))}{Z'_q(b^*, \Phi(p+q))} \geq 1.$$

By the definition of $v_{b^*}$, for $x > b^*$, we have $v'_{b^*}(x) = 1$. This means $v'_{b^*}(x) \geq 1$, for all $x \geq 0$. Please note that for any $x \in \mathbb{R}$, we have

$$\begin{aligned}
(\Gamma - q - p) e^{\Phi(p+q)x} &= e^{\Phi(p+q)x}\left(\gamma \Phi(p+q) + \frac{\sigma^2}{2}\Phi^2(p+q)\right) \\
&\quad + e^{\Phi(p+q)x}\left[\int_0^\infty \left(e^{-\Phi(p+q)z} - 1 + \Phi(p+q)z\mathbf{1}_{(0,1]}(z)\right) \nu(dz) - (q+p)\right] \\
&= e^{\Phi(p+q)x}\left[\psi(\Phi(p+q)) - (q+p)\right] = 0.
\end{aligned} \tag{13}$$

Consequently, for $x < 0$, we have

$$(\Gamma - q - p) Z_q(x, \Phi(p+q)) = 0$$

and, for $x \geq 0$, using (6), we have

$$(\Gamma - q) Z_q(x, \Phi(p+q)) = p \int_0^\infty e^{\Phi(p+q)y} (\Gamma - q) W^{(q)}(x+y) dy = 0,$$

since $(\Gamma - q) W^{(q)}(x) = 0$ for all $x > 0$ (see e.g., (Biffis and Kyprianou 2010)). Please note that under our assumption, $W^{(q)}$ is sufficiently smooth. Indeed, by Theorem 1.2 in (Loeffen and Renaud 2010), if the tail of the Lévy measure is log-convex, then $W^{(q)\prime}$ is log-convex. Therefore, $W^{(q)\prime\prime}(x)$ exists and is continuous for almost all $x \in (0, \infty)$; see e.g., (Roberts and Varberg 1973).

As a consequence, and since $v_{b^*}$ is smooth in $x = b^*$, we have

$$\left(\Gamma - q - p\mathbf{1}_{(-\infty,0)}\right) v_{b^*}(x) = 0, \quad \text{for } x \leq b^*.$$

All that is now left to verify is that $(\Gamma - q) v_{b^*}(x) \leq 0$, for all $x > b^*$. It can be done following the same steps as in the proof of Theorem 2 in (Loeffen 2008), thanks to the fact that, under our assumption on the Lévy measure, the function $Z_q'(\cdot, \Phi(p+q))$ is sufficiently smooth (see the details in Appendix B). The details are left to the reader.

**Funding:** This research was funded by the Natural Sciences and Engineering Research Council of Canada (NSERC).

**Acknowledgments:** Let me thank two anonymous referees for their diligence and comments which improved this final version of the paper. Special thanks to Ronnie Loeffen, for providing the interpretation in Remark 2 and other comments which improved the final presentation, and to Florin Avram, for pointing out important literature which had been overlooked in a previous version. With their comments and suggestions, they both contributed to Section 2.

**Conflicts of Interest:** The author declares no conflict of interest.

## Appendix A. Proof of Proposition 1

To prove this result, we can adapt the methodology used in the proof of Proposition 1 of (Renaud and Zhou 2007) (for the case $k = 1$ in that paper). Let us define $\kappa^p$ as the time of Parisian ruin with rate $p$ for $X$ or, said differently, the time of Parisian ruin when the pay-no-dividend strategy, i.e., the strategy $\pi$ with $L_t^\pi \equiv 0$, is implemented. More precisely, define

$$\kappa^p = \inf\left\{t > 0 \colon t - g_t > e_p^{g_t} \text{ and } X_t < 0\right\},$$

where $g_t = \sup\{0 \leq s \leq t \colon X_s \geq 0\}$. Let us also define, for $a \in \mathbb{R}$, the stopping time

$$\tau_a^+ = \inf\{t > 0 \colon X_t > a\}.$$

It is known that (see e.g., Equation (16) in (Lkabous and Renaud 2019)), for $x \leq a$,

$$\mathbb{E}_x\left[e^{-q\tau_a^+} \mathbf{1}_{\{\tau_a^+ < \kappa^p\}}\right] = \frac{Z_q(x, \Phi(p+q))}{Z_q(a, \Phi(p+q))}. \tag{A1}$$

As in Renaud and Zhou (2007), we can show that

$$\left(v_b(b) + \frac{1}{n}\right) \mathbb{E}_{b-1/n}\left[e^{-q\tau_b^+} \mathbf{1}_{\{\tau_b^+ < \kappa^p\}}\right] \\ \leq v_b(b) \leq \left(v_b(b) + \frac{1}{n}\right) \mathbb{E}_b\left[e^{-q\tau_{b+1/n}^+} \mathbf{1}_{\{\tau_{b+1/n}^+ < \kappa^p\}}\right] + o(1/n). \tag{A2}$$

The result for $x = b$ follows by taking a limit and then the result for $0 \leq x \leq b$ follows by using again the identity in (A1). Finally, if $x < 0$, then using (1) we have

$$v_b(x) = e^{\Phi(p+q)x}\frac{Z_q(0,\Phi(p+q))}{Z_q'(b,\Phi(p+q))} = \frac{Z_q(x,\Phi(p+q))}{Z_q'(b,\Phi(p+q))}.$$

**Appendix B. Proof of Proposition 2**

Recall from (7) that, for $x \in (0,\infty)$, we have

$$Z_q'(x,\Phi(p+q)) = p\int_0^\infty e^{-\Phi(p+q)y}W^{(q)\prime}(x+y)dy. \tag{A3}$$

By Theorem 1.2 in (Loeffen and Renaud 2010), if the tail of the Lévy measure is log-convex, then $W^{(q)\prime}$ is log-convex. Using the properties of log-convex functions, as presented in (Roberts and Varberg 1973), we can deduce that $x \mapsto pe^{-\Phi(p+q)y}W^{(q)\prime}(x+y)$ is log-convex on $(0,\infty)$, for any fixed $y \in (0,\infty)$. Then, as Riemann integrals are limits of partial sums, we have that $x \mapsto Z_q'(x,\Phi(p+q))$ is also a log-convex function on $(0,\infty)$. In particular, $Z_q'(\cdot,\Phi(p+q))$ is convex on $(0,\infty)$, so we can write, for some fixed $c > 0$,

$$Z_q'(x,\Phi(p+q)) = Z_q'(c,\Phi(p+q)) + \int_c^x Z_q''^-(y,\Phi(p+q))dy,$$

where $Z_q''^-(\cdot,\Phi(p+q))$ is the left-hand derivative of $Z_q'(\cdot,\Phi(p+q))$. Since $Z_q''^-(\cdot,\Phi(p+q))$ is increasing and $\lim_{x\to\infty} Z_q'(x,\Phi(p+q)) = \infty$, we have that the function $Z_q'(\cdot,\Phi(p+q))$ is ultimately strictly increasing. This proves that $b_p^*$ is well-defined.

It is known that $W^{(q)\prime}$ is strictly increasing on $(b_\infty^*,\infty)$; see (Loeffen and Renaud 2010). Then, using together the representations of $Z_q'(x,\Phi(p+q))$ given in (5) and (7), we obtain

$$\begin{aligned}Z_q''(x,\Phi(p+q)) &= \Phi(p+q)p\int_0^\infty e^{-\Phi(p+q)y}W^{(q)\prime}(x+y)dy - pW^{(q)\prime}(x)\\ &> pW^{(q)\prime}(x)\int_0^\infty \Phi(p+q)e^{-\Phi(p+q)y}dy - pW^{(q)\prime}(x) = 0,\end{aligned} \tag{A4}$$

for all $x > b_\infty^*$. In other words, $x \mapsto Z_q'(x,\Phi(p+q))$ is strictly increasing on $(b_\infty^*,\infty)$. Consequently, $b_p^* \leq b_\infty^*$.

The rest of the proof is similar to Lemma 3 in (Kyprianou et al. 2012), where a function closely related to one of the representations of $Z_q'(x,\Phi(p+q))$ appears. For simplicity, set $g(x) = Z_q'(x,\Phi(p+q))$. Using (5), we can write, for $x > 0$,

$$g'(x) = \Phi(p+q)\left(g(x) - \frac{p}{\Phi(p+q)}W^{(q)\prime}(x)\right).$$

It follows that $g'(x) > 0$ (resp. $g'(x) < 0$) if and only if $g(x) > \frac{p}{\Phi(p+q)}W^{(q)\prime}(x)$ (resp. $g(x) < \frac{p}{\Phi(p+q)}W^{(q)\prime}(x)$). This means $g(b) > \frac{p}{\Phi(p+q)}W^{(q)\prime}(b)$ for $b < b_p^*$ and $g(b) < \frac{p}{\Phi(p+q)}W^{(q)\prime}(b)$ for $b > b_p^*$. If $b_p^* > 0$ then $g(b_p^*) = (p/\Phi(p+q))W^{(q)\prime}(b_p^*)$.

We deduce that $b_p^* > 0$ if and only if $g(0+) < (p/\Phi(p+q))W^{(q)\prime}(0+)$, where $g(0+) = \Phi(p+q) - pW^{(q)}(0)$. Written differently, we have $b_p^* > 0$ if and only if

$$\Phi(p+q) - pW^{(q)}(0) < \frac{p}{\Phi(p+q)}W^{(q)\prime}(0+).$$

If $\sigma > 0$, then $W^{(q)}(0) = 0$ and $W^{(q)\prime}(0+) = 2/\sigma^2$, which implies that $b_p^* > 0$ if and only if

$$\frac{(\Phi(p+q))^2}{p} < \frac{2}{\sigma^2}.$$

If $\sigma = 0$ and $\nu(0,\infty) = \infty$, then $W^{(q)\prime}(0+) = \infty$, which implies that $b_p^* > 0$. Finally, if $\sigma = 0$ and $\nu(0,\infty) < \infty$, then $W^{(q)}(0) = 1/c$, where $c > 0$ is the drift, and $W^{(q)\prime}(0+) = (q + \nu(0,\infty))/c^2$, which implies that $b_p^* > 0$ if and only if

$$\Phi(p+q) - \frac{p}{c} < \frac{p}{\Phi(p+q)} \frac{q + \nu(0,\infty)}{c^2}.$$

## References

Albrecher, Hansjörg, and Jevgenijs Ivanovs. 2014. Power identities for Lévy risk models under taxation and capital injections. *Stochastic Systems* 4: 157–72. [CrossRef]

Albrecher, Hansjörg, Eric C. K. Cheung, and Stefan Thonhauser. 2011. Randomized observation periods for the compound Poisson risk model dividends. *ASTIN Bulletin* 41: 645–72.

Albrecher, Hansjörg, Jevgenijs Ivanovs, and Xiaowen Zhou. 2016. Exit identities for Lévy processes observed at Poisson arrival times. *Bernoulli* 22: 1364–82. [CrossRef]

Avram, Florin, and Andreea Minca. 2017. On the central management of risk networks. *Advances in Applied Probability* 49: 221–37. [CrossRef]

Avram, Florin, and Xiaowen Zhou. 2016. On fluctuation theory for spectrally negative Lévy processes with Parisian reflection below, and applications. *Theory of Probability and Mathematical Statistics* 95: 17–40. [CrossRef]

Avram, Florin, Zbigniew Palmowski, and Martijn R. Pistorius. 2007. On the optimal dividend problem for a spectrally negative Lévy process. *The Annals of Applied Probability* 17: 156–80. [CrossRef]

Avram, Florin, Zbigniew Palmowski, and Martijn R. Pistorius. 2015. On Gerber-Shiu functions and optimal dividend distribution for a Lévy risk process in the presence of a penalty function. *The Annals of Applied Probability* 25: 1868–935. [CrossRef]

Baurdoux, Erik J., Juan Carlos Pardo, José Luis Pérez, and Jean-François Renaud. 2016. Gerber-Shiu distribution at Parisian ruin for Lévy insurance risk processes. *Journal of Applied Probability* 53: 572–84. [CrossRef]

Biffis, Enrico, and Andreas E. Kyprianou. 2010. A note on scale functions and the time value of ruin for Lévy insurance risk processes. *Insurance: Mathematics and Economics* 46: 85–91. [CrossRef]

Czarna, Irmina, and Zbigniew Palmowski. 2014. Dividend problem with Parisian delay for a spectrally negative Lévy risk process. *Journal of Optimization Theory and Applications* 161: 239–56. [CrossRef]

Dassios, Angelos, and Shanle Wu. 2009. On barrier strategy dividends with Parisian implementation delay for classical surplus processes. *Insurance: Mathematics and Economics* 45: 195–202. [CrossRef]

De Finetti, Bruno. 1957. Su un' impostazione alternativa dell teoria collettiva del rischio. *Transactions of the XVth International Congress of Actuaries* 2: 433–43.

Ivanovs, Jevgenijs, and Zbigniew Palmowski. 2012. Occupation densities in solving exit problems for Markov additive processes and their reflections. *Stochastic Processes and Their Applications* 122: 3342–60. [CrossRef]

Kuznetsov, Alexey, Andreas E. Kyprianou, and Victor Rivero. 2012. *The Theory of Scale Functions for Spectrally Negative Lévy Processes*. Lévy Matters. Berlin/Heidelberg: Springer.

Kyprianou, Andreas E. 2014. *Fluctuations of Lévy Processes with Applications—Introductory Lectures*, 2nd ed. Heidelberg: Springer.

Kyprianou, Andreas E., Ronnie Loeffen, and José-Luis Pérez. 2012. Optimal control with absolutely continuous strategies for spectrally negative Lévy processes. *Journal of Applied Probability* 49: 150–66. [CrossRef]

Landriault, David, Jean-François Renaud, and Xiaowen Zhou. 2011. Occupation times of spectrally negative Lévy processes with applications. *Stochastic Processes and Their Applications* 121: 2629–41. [CrossRef]

Lkabous, Mohamed Amine, and Jean-François Renaud. 2019. A unified approach to ruin probabilities with delays for spectrally negative Lévy processes. *Scandinavian Actuarial Journal*. [CrossRef]

Loeffen, Ronnie L. 2008. On optimality of the barrier strategy in de Finetti's dividend problem for spectrally negative Lévy processes. *The Annals of Applied Probability* 18: 1669–80. [CrossRef]

Loeffen, Ronnie L., and Jean-François Renaud. 2010. De Finetti's optimal dividends problem with an affine penalty function at ruin. *Insurance: Mathematics and Economics* 46: 98–108. [CrossRef]

Loeffen, Ronnie, Irmina Czarna, and Zbigniew Palmowski. 2013. Parisian ruin probability for spectrally negative Lévy processes. *Bernoulli* 19: 599–609. [CrossRef]

Renaud, Jean-François, and Xiaowen Zhou. 2007. Distribution of the present value of dividend payments in a Lévy risk model. *Journal of Applied Probability* 44: 420–27. [CrossRef]

Roberts, A. W., and D. E. Varberg. 1973. Convex functions. In *Pure and Applied Mathematics*. New York and London: Academic Press, vol. 57.

© 2019 by the author. Licensee MDPI, Basel, Switzerland. This article is an open access article distributed under the terms and conditions of the Creative Commons Attribution (CC BY) license (http://creativecommons.org/licenses/by/4.0/).

Article

# Optimal Bail-Out Dividend Problem with Transaction Cost and Capital Injection Constraint

Mauricio Junca [1], Harold A. Moreno-Franco [2] and José Luis Pérez [3],*

[1] Department of Mathematics, Universidad de los Andes, Bogotá 11711, Colombia; mj.junca20@uniandes.edu.co
[2] Department of Mathematics and Statistics, Universidad del Norte, Barranquilla 080003, Colombia; hamoreno@uninorte.edu.co
[3] Department of Probability and Statistics, Centro de Investigación en Matemáticas A.C., Guanajuato 36000, Mexico
* Correspondence: jluis.garmendia@cimat.mx

Received: 18 December 2018; Accepted: 29 January 2019; Published: 31 January 2019

**Abstract:** We consider the optimal bail-out dividend problem with fixed transaction cost for a Lévy risk model with a constraint on the expected present value of injected capital. To solve this problem, we first consider the optimal bail-out dividend problem with transaction cost and capital injection and show the optimality of reflected $(c_1, c_2)$-policies. We then find the optimal Lagrange multiplier, by showing that in the dual Lagrangian problem the complementary slackness conditions are met. Finally, we present some numerical examples to support our results.

**Keywords:** dividend payment; optimal control; capital injection constraint; spectrally negative Lévy processes; reflected Lévy processes; scale functions

**MSC:** 60G51; 91B30

## 1. Introduction

De Finetti introduced in 1957 the expected net present value (NPV) of dividends paid by an insurance company as a criterion for assessing its stability. According to this model, the maximum of the expected NPVs, if it exists, can be a proxy for the insurance company's value. In some cases (e.g., due to regulatory issues), the insurance company has to ensure the negative balance protection and therefore must be rescued by injecting capital. Hence, the company aims to maximize the total amount of expected dividend payments minus the total expected cost of capital injection while permanently keeping the surplus process non-negative.

Usually, spectrally negative Lévy processes (Lévy processes with only downward jumps) are used to model the underlying surplus process of an insurance company, which increases with premium payments and decreases with insurance payouts. The optimization problem for this model was studied by Avram et al. (2007), who proved that it is optimal to inject capital when the process is below zero and pay dividends when the process is above a suitably chosen threshold.

In this paper, we focus on the case when the insurance company pays a fixed transaction cost each time a dividend payment is made. The fixed transaction cost makes the continuous payment of dividends no longer feasible, which implies that only lump sum dividend payments are possible. In this case, a strategy is assumed to have the form of impulse control; whenever dividends are accrued, a constant transaction cost $\delta > 0$ is incurred. Unlike the barrier strategies described above, which are typically optimal for the case without transaction cost, we pursue the optimality of the reflected $(c_1, c_2)$-policies. In these strategies, the surplus process is brought down to $c_1$ whenever it exceeds the level $c_2$ for some $0 \le c_1 < c_2 < \infty$, and pushes the surplus to 0 whenever it goes below 0.

Previously, the version of the de Finetti's optimal dividend problem with fixed transaction cost and without bail-outs was solved for the spectrally negative case by Loeffen (2009) and for the dual model (i.e., spectrally positive Lévy processes) by Bayraktar et al. (2014b).

In this paper, we also propose a model to maximize the value of the insurance company by means of the dividend payments while keeping the expected present value of the capital injection bounded. The idea of introducing this constraint is to bound the budget needed for the company to survive and therefore to reduce the risk faced (e.g., operational risk).

Specifically, we solve the following two problems:

1. We find the solution to the optimal bail-out dividend problem with fixed transaction cost for the case of spectrally negative Lévy processes. We show that a reflected $(c_1, c_2)$-policy is optimal (see Theorem 1). We use scale functions to characterize the optimal thresholds as well as the value function. We prove the optimality of the proposed policy by means of a verification theorem.
2. We solve the constrained dividend maximization problem with capital injection on the set of strategies such that the expected net present value of injected capital must be bounded by a given constant. This is an offshoot of Hernández et al. (2018) for the bail-out case. Using the previous results, in Theorems 2 and 3, we present the solution when the surplus of the company is modeled by a spectrally negative Lévy process.

This paper is organized as follows. In Section 2, we introduce the problem. In Section 3, we provide a review of scale functions and some fluctuation identities of spectrally negative Lévy processes as well as their reflected versions. In Section 4, we solve the optimal bail-out dividend problem with fixed transaction cost for the case of a spectrally negative Lévy process. In Section 5, we present the solution for the constrained bail-out dividend problem. In Section 6, we illustrate our main results by giving some numerical examples.

## 2. Formulation of the Problem

Let $X = \{X_t : t \geq 0\}$ be a Lévy process defined on a probability space $(\Omega, \mathcal{F}, \mathbb{P})$, and let $\mathbb{F} := \{\mathcal{F}_t : t \geq 0\}$ be the completed and right-continuous filtration generated by $X$. Recall that a Lévy process is a process that has càdlàg paths and stationary and independent increments. For $x \in \mathbb{R}$, we denote by $\mathbb{P}_x$ the law of $X$, where $X_0 = x$. For convenience, we take $\mathbb{P}_0 \equiv \mathbb{P}$, when $x = 0$. The expectation operator associated with $\mathbb{P}_x$ is denoted by $\mathbb{E}_x$. We take $\mathbb{E}_0 \equiv \mathbb{E}$, where $\mathbb{E}$ is the expectation operator associated with $\mathbb{P}$.

We henceforth assume that the insurance company's surplus $X$ is modeled by a spectrally negative process, i.e., a Lévy process that only has negative jumps. We omit the case when $X$ has monotone trajectories to avoid trivial cases.

The Laplace exponent of $X$ is given by

$$\psi(\theta) := \log \mathbb{E}[\theta X_1] = \gamma\theta + \frac{\sigma^2}{2}\theta^2 - \int_{(0,\infty)} \left(1 - e^{-\theta z} - \theta z \mathbf{1}_{\{0 < z \leq 1\}}\right)\Pi(dz), \quad \theta \geq 0,$$

where $\gamma \in \mathbb{R}$, $\sigma \geq 0$, and the Lévy measure of $X$, $\Pi$, is a measure defined on $(0, \infty)$ satisfying

$$\int_{(0,\infty)} (1 \wedge z^2)\Pi(dz) < \infty.$$

As is well-known, the process $X$ has bounded variation paths if and only if $\sigma = 0$ and $\int_{(0,1]} z\Pi(dz) < \infty$. In this case, $X$ can be written as

$$X_t = ct - \widetilde{S}_t, \quad t \geq 0, \tag{1}$$

where $c := \gamma + \int_{(0,1]} z\Pi(\mathrm{d}z)$ and $\tilde{S} = \{\tilde{S}_t : t \geq 0\}$ is a drift-less subordinator. Since we omit the case when $X$ has monotone paths, it is necessary that the constant $c$ is greater than zero. Note that the Laplace exponent of $X$, with $X$ as in Equation (1), is given as follows,

$$\psi(\theta) = c\theta - \int_{(0,\infty)} (1 - e^{-\theta z}) \Pi(\mathrm{d}z), \quad \theta \geq 0.$$

*De Finetti's Problem with Fixed Transaction Cost and Capital Injection*

Let $\pi = \{L^\pi, R^\pi\}$ be a strategy, where $L^\pi$ is left-continuous $\mathbb{P}_x$-a.s., and $R^\pi$ is right-continuous $\mathbb{P}_x$-a.s. Additionally, we assume that $L^\pi$ and $R^\pi$ are non-negative, and non-decreasing $\mathbb{P}_x$-a.s., start at zero and are adapted to the filtration $\mathbb{F}$. Then, the controlled process, $X^\pi$, associated with the strategy $\pi$, is the following

$$X_t^\pi = X_t - L_t^\pi + R_t^\pi, \quad t \geq 0.$$

For each $t \geq 0$, the quantities $L_t^\pi$ and $R_t^\pi$ represent the cumulative amounts that the insurance company has paid to its shareholders and has injected, respectively.

The set of admissible policies $\Theta$ consists of those policies $\pi$ for which $X^\pi$ is non-negative and for $x \geq 0$,

$$\mathbb{E}_x \left[ \int_0^\infty e^{-qt} \, \mathrm{d}R_t^\pi \right] < \infty.$$

When there is a fixed transaction cost $\delta > 0$, we only consider the class of admissible strategies $\pi = \{L^\pi, R^\pi\} \in \Theta$ such that

$$L_t^\pi = \sum_{0 \leq s \leq t} \Delta L_s^\pi, \quad t \geq 0,$$

where $\Delta L_t^\pi := L_{t+}^\pi - L_t^\pi$. We denote this class by $\Theta_\delta$ and in the case $\delta = 0$, we take $\Theta_0 \equiv \Theta$.

Given an initial capital $x \geq 0$ and a policy $\pi = \{L^\pi, R^\pi\} \in \Theta_\delta$, with $\delta \geq 0$, we define the expected NPV as follows,

$$v_{\delta,\Lambda}^\pi(x) := \mathbb{E}_x \left[ \int_0^\infty e^{-qt} \, \mathrm{d} \left( L_t^\pi - \delta \sum_{0 \leq s \leq t} \mathbf{1}_{\{\Delta L_s^\pi > 0\}} \right) - \Lambda \int_0^\infty e^{-qt} \, \mathrm{d}R_t^\pi \right], \quad (2)$$

where $q > 0$, $\delta \geq 0$, and $\Lambda > 0$ is the unit cost per capital injected.

**Remark 1.** *Note that in the case of proportional transaction cost the expected NPV changes to*

$$\mathbb{E}_x \left[ \int_0^\infty e^{-qt} \, \mathrm{d} \left( \beta L_t^\pi - \delta \sum_{0 \leq s \leq t} \mathbf{1}_{\{\Delta L_s^\pi > 0\}} \right) - \Lambda \int_0^\infty e^{-qt} \, \mathrm{d}R_t^\pi \right],$$

*where $0 < \beta < 1$, so by changing $\delta$ and $\Lambda$ appropriately we can recover Equation (2).*

Hence, the value function we aim to find is

$$V_{\delta,\Lambda}(x) := \sup_{\pi \in \Theta_\delta} v_{\delta,\Lambda}^\pi(x). \quad (3)$$

**Remark 2.** *Since we want to avoid this function taking the value $-\infty$, we assume that $\psi'(0+) = \mathbb{E}[X_1] > -\infty$. We also assume that $\Lambda \geq 1$, otherwise the value function will go to infinity since large amounts of dividends will be paid, given that the company will inject capital at a cheaper cost to bail out.*

Note that the problem in Equation (3) was studied by Avram et al. (2007) under the assumption $\delta = 0$ (see Section 3.2). Therefore, we focus on the optimal control problem when $\delta > 0$ (see Section 4).

## 3. Preliminaries

In this section, we revise the scale functions of spectrally negative Lévy processes and their properties (see, e.g., Kuznetsov et al. (2013); Kyprianou (2014)). We also recall well known results regarding optimal dividend strategies with capital injection for spectrally one-sided Lévy processes when the transaction cost is equal to 0 (i.e., $\delta = 0$).

For each $q \geq 0$, there exists a map $W^{(q)} : \mathbb{R} \longrightarrow [0, \infty)$, called $q$-scale function, satisfying $W^{(q)}(x) = 0$ for $x \in (-\infty, 0)$, and strictly increasing on $[0, \infty)$, which is defined by its Laplace transform:

$$\int_0^\infty e^{-\theta x} W^{(q)}(x) dx = \frac{1}{\psi(\theta) - q}, \quad \theta > \Phi(q), \tag{4}$$

where

$$\Phi(q) := \sup\{\lambda \geq 0 : \psi(\lambda) = q\}.$$

We also define, for $x \in \mathbb{R}$,

$$\overline{W}^{(q)}(x) := \int_0^x W^{(q)}(y) dy, \quad Z^{(q)}(x) := 1 + q\overline{W}^{(q)}(x),$$

$$\overline{Z}^{(q)}(x) := \int_0^x Z^{(q)}(z) dz = x + q \int_0^x \int_0^z W^{(q)}(w) dw dz.$$

Since $W^{(q)}$ is equal to zero on $(-\infty, 0)$, we have

$$\overline{W}^{(q)}(x) = 0, \quad Z^{(q)}(x) = 1 \quad \text{and} \quad \overline{Z}^{(q)}(x) = x, \quad x \leq 0.$$

**Remark 3.**

1. By Equation (8.26) of Kyprianou (2014), the left- and right-hand derivatives of $W^{(q)}$ always exist on $\mathbb{R}\backslash\{0\}$. In addition, as in, e.g., (Chan et al. 2011, Theorem 3), if $X$ is of unbounded variation or the Lévy measure is atomless, we have $W^{(q)} \in C^1(\mathbb{R}\backslash\{0\})$.
2. From Lemmas 3.1–3.2 of Kuznetsov et al. (2013), we know

$$W^{(q)}(0) = \begin{cases} 0, & \text{if } X \text{ is of unbounded variation,} \\ \dfrac{1}{c}, & \text{if } X \text{ is of bounded variation,} \end{cases}$$

$$W^{(q)\prime}(0+) := \lim_{x \downarrow 0} W^{(q)\prime}(x) = \begin{cases} \dfrac{2}{\sigma^2}, & \text{if } \sigma > 0, \\ \infty, & \text{if } \sigma = 0 \text{ and } \Pi(0, \infty) = \infty, \\ \dfrac{q + \Pi(0, \infty)}{c^2}, & \text{if } \sigma = 0 \text{ and } \Pi(0, \infty) < \infty. \end{cases}$$

3. From Lemma 3.3 of Kuznetsov et al. (2013), $W_{\Phi(q)}(x) := e^{-\Phi(q)x} W^{(q)}(x) \nearrow \psi'(\Phi(q))^{-1}$, as $x \uparrow \infty$.

Due to Remark 3, we make the following assumption throughout the paper.

**Assumption 1.** We assume that either $X$ has unbounded variation or $\Pi$ is absolutely continuous with respect to the Lebesgue measure. Under this assumption, it holds that $W^{(q)}$ is $C^1$ in $(0, \infty)$.

We give the following properties related to $Z^{(q)}$ and $\overline{W}^{(q)}$ for later use.

**Remark 4.**

(i) By Proposition 5.5 in Hernández et al. (2018), we have that $Z^{(q)}$ is a strictly log-convex function on $(0, \infty)$, for $q > 0$.
(ii) From Lemma 1 in Avram et al. (2007), it is known that $\overline{W}^{(q)}$ is a log-concave function on $(0, \infty)$.

We define the stopping times $\tau_{a^-}$ and $\tau_{a^+}$, respectively, as follows,

$$\tau_a^- := \inf\{t > 0 : X_t < a\} \quad \text{and} \quad \tau_a^+ := \inf\{t > 0 : X_t > a\}, \quad a \in \mathbb{R};$$

here and further on, let $\inf \varnothing = \infty$. By Theorem 8.1 in Kyprianou (2014), we have that

$$\mathbb{E}_x\left[e^{-q\tau_a^+} \mathbf{1}_{\{\tau_a^+ < \tau_b^-\}}\right] = \frac{W^{(q)}(x-b)}{W^{(q)}(a-b)},$$

$$\mathbb{E}_x\left[e^{-q\tau_b^-} \mathbf{1}_{\{\tau_a^+ > \tau_b^-\}}\right] = Z^{(q)}(x-b) - Z^{(q)}(a-b)\frac{W^{(q)}(x-b)}{W^{(q)}(a-b)}, \quad \text{for } a > b \text{ and } x \leq a. \quad (5)$$

### 3.1. Reflected Lévy Processes

Let $S = \{S_t : t \geq 0\}$ and $R^0 = \{R_t^0 : t \geq 0\}$ be defined, respectively, as

$$S_t := \sup_{0 \leq s \leq t}(X_s \vee 0) \quad \text{and} \quad R_t^0 := \sup_{0 \leq s \leq t}(-X_s \vee 0). \quad (6)$$

We denote $\hat{Y} := S - X$ and $Y := X + R^0$, which are strong Markov processes. Observe that the process $R^0$ pushes $X$ upwards whenever it attempts to down-cross the level 0; as a result the process $Y$ only takes values on $[0,\infty)$. An introduction to the theory of Lévy processes and their reflected processes can be encountered in Bertoin (1998); Kyprianou (2014).

Let $\hat{\tau}_a$ be defined as $\hat{\tau}_a = \inf\{t > 0 : \hat{Y}_t \in (a,\infty)\}$, with $a > 0$. Then, by Proposition 2 in Pistorius (2004),

$$\mathbb{E}_{-x}\left[e^{-q\hat{\tau}_a}\right] = Z^{(q)}(a-x) - qW^{(q)}(a-x)\frac{W^{(q)}(a)}{W^{(q)\prime}(a)}, \quad x \in [0,a].$$

We define for $a > 0$,

$$H(a) := \mathbb{E}_0\left[e^{-q\hat{\tau}_a}\right] = Z^{(q)}(a) - q\frac{[W^{(q)}(a)]^2}{W^{(q)\prime}(a)}. \quad (7)$$

**Remark 5.** Note that, by definition, the function $H$ is strictly positive, strictly decreasing and satisfies

$$\lim_{a \to \infty} H(a) = 0, \quad \lim_{a \to 0} H(a) = 1 - \frac{q[W^{(q)}(0)]^2}{W^{(q)\prime}(0+)}.$$

Therefore, the function $H$ has an inverse from $(0, 1 - q/(q + \Pi(0,\infty)))$ onto $(0,\infty)$ when $\sigma = 0$ and $\Pi(0,\infty) < \infty$, and from $(0,1)$ onto $(0,\infty)$ otherwise.

Similarly, taking $\kappa_b := \inf\{t > 0 : Y_t \in (b,\infty)\}$, with $b > 0$, we know from Proposition 2 in Pistorius (2004) that

$$\mathbb{E}_x\left[e^{-q\kappa_b}\right] = \frac{Z^{(q)}(x)}{Z^{(q)}(b)}, \quad x \leq b. \quad (8)$$

In addition, we know from (Avram et al. 2007, page 167) that

$$\mathbb{E}_x\left[\int_{[0,\kappa_b]} e^{-qt}\, dR_t^0\right] = -\overline{Z}^{(q)}(x) + \Phi(q)^{-1} Z^{(q)}(x) - \frac{\psi'(0+)}{q}$$

$$+ \left(\overline{Z}^{(q)}(b) - \Phi(q)^{-1} Z^{(q)}(b) + \frac{\psi'(0+)}{q}\right) \frac{Z^{(q)}(x)}{Z^{(q)}(b)}$$

$$= -k^{(q)}(x) + \frac{Z^{(q)}(x)}{Z^{(q)}(b)} k^{(q)}(b), \quad x \leq b, \tag{9}$$

where

$$k^{(q)}(x) := \overline{Z}^{(q)}(x) + \frac{\psi'(0+)}{q}. \tag{10}$$

### 3.2. Optimal Dividends without Transaction Cost and with Capital Injection

When $\delta = 0$, Equation (2) becomes

$$v_\Lambda^\pi(x) := v_{0,\Lambda}^\pi(x) = \mathbb{E}_x\left[\int_0^\infty e^{-qt}\, dL_t^\pi - \Lambda \int_0^\infty e^{-qt}\, dR_t^\pi\right],$$

for any initial capital $x \geq 0$ and admissible policy $\pi = \{L^\pi, R^\pi\} \in \Theta$. Consider the strategy $\pi_{a,0} = \{L^{a,0}, R^{a,0}\}$, which consists in setting reflecting barriers at $a$ and $0$, respectively. The controlled risk process $X^{\pi_{a,0}} = X - L^{a,0} + R^{a,0}$ is a doubly reflected spectrally negative Lévy process and was studied by Avram et al. (2007). Intuitively, the process behaves similar to a Lévy process when it is inside $[0, a]$, but when it tries to cross above the level $a$ or below the level $0$ it is forced to stay inside $[0, a]$. Using Theorem 1 from Avram et al. (2007), we have that for $a > 0$ and $x \in [0, a]$,

$$\mathbb{E}_x\left[\int_0^\infty e^{-qt}\, dL_t^{a,0}\right] = \frac{Z^{(q)}(x)}{qW^{(q)}(a)}, \tag{11}$$

$$\mathbb{E}_x\left[\int_0^\infty e^{-qt}\, dR_t^{a,0}\right] = \frac{Z^{(q)}(a)}{qW^{(q)}(a)} Z^{(q)}(x) - k^{(q)}(x). \tag{12}$$

Note that the expression in Equation (12) is finite under our assumption that $\psi'(0+) > -\infty$. Using the expressions above, we can see that, for $\Lambda \geq 1$,

$$v_\Lambda^a(x) := v_\Lambda^{\pi_{a,0}}(x) = \begin{cases} Z^{(q)}(x)\zeta_\Lambda(a) + \Lambda k^{(q)}(x), & \text{if } 0 \leq x \leq a, \\ x - a + v_\Lambda^a(a), & \text{if } x > a, \end{cases} \tag{13}$$

where

$$\zeta_\Lambda(a) := \frac{1 - \Lambda Z^{(q)}(a)}{qW^{(q)}(a)}, \quad a > 0. \tag{14}$$

Equation (13) suggests that, to find the best barrier strategy we should maximize the function $\zeta_\Lambda$. Thus, we can define the candidate for the optimal barrier by

$$a_\Lambda = \sup\{a \geq 0 : \zeta_\Lambda(a) \geq \zeta_\Lambda(x), \text{ for all } x \geq 0\}. \tag{15}$$

**Remark 6.** *Note that $\zeta_\Lambda : (0, \infty) \longrightarrow (-\infty, 0)$ and satisfies*

$$\lim_{a \to 0} \zeta_\Lambda(a) = -\frac{\Lambda - 1}{qW^{(q)}(0)} \quad \text{and} \quad \lim_{a \to \infty} \zeta_\Lambda(a) = -\frac{\Lambda}{\Phi(q)}.$$

Here, in case that $X$ is of unbounded variation, the first equality is understood to be $-\infty$. The barrier level $a_\Lambda$, given in Equation (15), corresponds with the level defined in Avram et al. (2007). Using the definition of the function $H$, we have that

$$\frac{d\zeta_\Lambda(a)}{da} = \frac{\Lambda W^{(q)\prime}(a)}{q[W^{(q)}(a)]^2}(H(a) - 1/\Lambda).$$

Since $H$ is strictly decreasing, $\zeta_\Lambda$ has a unique maximum at $a_\Lambda$ that is either a critical point, which is a solution of $H(a) = \frac{1}{\Lambda}$, or $0$ if the right-hand derivative of $\zeta_\Lambda$ is negative at $0$. Therefore, by Remark 5,

$$a_\Lambda = \begin{cases} 0, & \text{if } \sigma = 0,\ \Pi(0,\infty) < \infty \text{ and } \Lambda < 1 + \frac{q}{\Pi(0,\infty)}, \\ H^{-1}(1/\Lambda), & \text{otherwise}. \end{cases} \qquad (16)$$

In addition, note that $\zeta_\Lambda$ is strictly increasing on $(0, a_\Lambda)$ and strictly decreasing on $(a_\Lambda, \infty)$.

Hence, from Avram et al. (2007), we know that the value function in Equation (3) and the optimal strategy are given by $V_\Lambda := V_{0,\Lambda} = v_\Lambda^{a_\Lambda}$ and $\pi_{0,a_\Lambda}$, where $v_\Lambda^{a_\Lambda}$ and $a_\Lambda$ are as in Equations (13) and (16), respectively.

**Remark 7.** *Note that the optimal barrier $a_\Lambda \to \infty$ as $\Lambda \to \infty$.*

## 4. Capital Injection and Fixed Transaction Cost

In this section, we solve the problem in Equation (3) in the presence of a fixed transaction cost $\delta > 0$. We consider strategies where the capital injection policy is $R^0$, given in Equation (6), and the dividend strategy is the so-called reflected $(c_1, c_2)$-policy, defined below.

### 4.1. Value Function of Reflected $(c_1, c_2)$-Policies

Let $(c_1, c_2)$ be a pair such that $0 \leq c_1 < c_2$. In this subsection, we define the reflected $(c_1, c_2)$-policy, denoted by $\pi_{(c_1,c_2),0}$, and under which we construct the controlled process. Let $Y = X + R^0$ be the Lévy process reflected from below $0$, so we set

$$X_t^{(c_1,c_2),0} = Y_t, \quad \text{for } t \leq T_1^{c_1,c_2},$$

where $T_1^{c_1,c_2} = \inf\{t > 0 : Y_t > c_2\}$. The process then jumps downward by $Y_{T_1^{c_1,c_2}} - c_1$ so that $X_{T_1^{c_1,c_2}}^{(c_1,c_2),0} = c_1$. Now, for $T_1^{c_1,c_2} \leq t < T_2^{c_1,c_2} = \inf\{t > T_1^{c_1,c_2} : X_t^{(c_1,c_2),0} > c_2\}$, $X^{(c_1,c_2),0}$ is the reflected process from below at $0$ of $X_t + (c_1 - X_{T_1^{c_1,c_2}})$, and $X_{T_2^{c_1,c_2}}^{(c_1,c_2),0} = c_1$. By repeating this procedure, we can construct the process inductively. The process $X^{(c_1,c_2),0}$ clearly admits the decomposition

$$X_t^{(c_1,c_2),0} = X_t - L_t^{(c_1,c_2),0} + R_t^{(c_1,c_2),0}, \quad t \geq 0,$$

where $L^{(c_1,c_2),0}$ and $R^{(c_1,c_2),0}$ are the cumulative amounts of dividend payments and capital injection, respectively.

Let us compute the expected NPV of dividends with transaction costs for this strategy. For this purpose, we denote

$$f_{c_1,c_2}(x) = \mathbb{E}_x\left[\int_0^\infty e^{-qt}\, d\left(L_t^{(c_1,c_2),0} - \delta \sum_{0 \leq s \leq t} 1_{\{\Delta L_s^{(c_1,c_2),0} > 0\}}\right)\right].$$

If $x < c_2$, by the Strong Markov Property and Equation (8), we obtain that

$$f_{c_1,c_2}(x) = \mathbb{E}_x\left[e^{-qT_1^{c_1,c_2}}\right] f_{c_1,c_2}(c_2) = \frac{Z^{(q)}(x)}{Z^{(q)}(c_2)} f_{c_1,c_2}(c_2). \tag{17}$$

When $x \geq c_2$, an amount $x - c_1$ is paid as dividends and a transaction cost $\delta$ is incurred immediately, so by using Equation (17) we obtain

$$f_{c_1,c_2}(x) = x - c_1 - \delta + f_{c_1,c_2}(c_1) = x - c_1 - \delta + \frac{Z^{(q)}(c_1)}{Z^{(q)}(c_2)} f_{c_1,c_2}(c_2).$$

Hence, taking $x = c_2$, and solving for $f_{c_1,c_2}(c_2)$ we get

$$f_{c_1,c_2}(c_2) = (c_2 - c_1 - \delta) \frac{Z^{(q)}(c_2)}{Z^{(q)}(c_2) - Z^{(q)}(c_1)}.$$

Using the aforementioned expression in Equation (17), we have for $x < c_2$,

$$f_{c_1,c_2}(x) = (c_2 - c_1 - \delta) \frac{Z^{(q)}(x)}{Z^{(q)}(c_2) - Z^{(q)}(c_1)}. \tag{18}$$

Now, let us calculate the expected NPV of the injected capital denoted by

$$g_{c_1,c_2}(x) = \mathbb{E}_x\left[\int_0^\infty e^{-qt}\, dR_t^{(c_1,c_2),0}\right].$$

Again, by the Strong Markov Property, noting that $T_1^{c_1,c_2} = \inf\{t > 0 : Y_t \in (c_2, \infty)\}$ and Equations (8)–(9), we have for $x \geq 0$

$$g_{c_1,c_2}(x) = \mathbb{E}_x\left[\int_{[0,T_1^{c_1,c_2}]} e^{-qt}\, dR_t^0\right] + \mathbb{E}_x\left[e^{-qT_1^{c_1,c_2}}\right] g_{c_1,c_2}(c_1)$$

$$= -k^{(q)}(x) + k^{(q)}(c_2)\frac{Z^{(q)}(x)}{Z^{(q)}(c_2)} + \frac{Z^{(q)}(x)}{Z^{(q)}(c_2)} g_{c_1,c_2}(c_1).$$

Thus, setting $x = c_1$ and solving for $g_{c_1,c_2}(c_1)$, we obtain

$$g_{c_1,c_2}(c_1) = \left(-k^{(q)}(c_1) + k^{(q)}(c_2)\frac{Z^{(q)}(c_1)}{Z^{(q)}(c_2)}\right)\frac{Z^{(q)}(c_2)}{Z^{(q)}(c_2) - Z^{(q)}(c_1)}$$

$$= \left(-\overline{Z}^{(q)}(c_1) + \overline{Z}^{(q)}(c_2)\frac{Z^{(q)}(c_1)}{Z^{(q)}(c_2)}\right)\frac{Z^{(q)}(c_2)}{Z^{(q)}(c_2) - Z^{(q)}(c_1)} - \frac{\psi'(0+)}{q}.$$

Putting the pieces together, we obtain

$$g_{c_1,c_2}(x) = -k^{(q)}(x) + k^{(q)}(c_2)\frac{Z^{(q)}(x)}{Z^{(q)}(c_2)}$$

$$+ \left(\left(-\overline{Z}^{(q)}(c_1) + \overline{Z}^{(q)}(c_2)\frac{Z^{(q)}(c_1)}{Z^{(q)}(c_2)}\right)\frac{Z^{(q)}(c_2)}{Z^{(q)}(c_2) - Z^{(q)}(c_1)} - \frac{\psi'(0+)}{q}\right)\frac{Z^{(q)}(x)}{Z^{(q)}(c_2)}$$

$$= -k^{(q)}(x) + \overline{Z}^{(q)}(c_2)\frac{Z^{(q)}(x)}{Z^{(q)}(c_2)} + \left(-\overline{Z}^{(q)}(c_1) + \overline{Z}^{(q)}(c_2)\frac{Z^{(q)}(c_1)}{Z^{(q)}(c_2)}\right)\frac{Z^{(q)}(x)}{Z^{(q)}(c_2) - Z^{(q)}(c_1)}$$

$$= Z^{(q)}(x)\left(\frac{\overline{Z}^{(q)}(c_2) - \overline{Z}^{(q)}(c_1)}{Z^{(q)}(c_2) - Z^{(q)}(c_1)}\right) - k^{(q)}(x).$$

Hence, we have the following result.

**Lemma 1.** *The expected NPV associated with a reflected $(c_1, c_2)$-policy is given by*

$$v_{\delta,\Lambda}^{c_1,c_2}(x) := v_{\delta,\Lambda}^{\pi_{(c_1,c_2),0}}(x) = \begin{cases} Z^{(q)}(x) G_\Lambda(c_1, c_2) + \Lambda k^{(q)}(x), & \text{if } x \leq c_2, \\ x - c_1 - \delta + v_{\delta,\Lambda}^{c_1,c_2}(c_1), & \text{if } x > c_2, \end{cases}$$

*where*

$$G_\Lambda(c_1, c_2) := \frac{c_2 - c_1 - \delta - \Lambda\left(\overline{Z}^{(q)}(c_2) - \overline{Z}^{(q)}(c_1)\right)}{Z^{(q)}(c_2) - Z^{(q)}(c_1)}, \quad \text{for all } c_2 > c_1 \geq 0. \tag{19}$$

**Remark 8.** *Note that $G_\Lambda$ is $C^2$ on $\mathcal{A} := \{(c_1, c_2) \in \mathbb{R}_+^2 : c_1 < c_2\}$, and*

$$\lim_{c_2 \downarrow c_1} G_\Lambda(c_1, c_2) = -\infty, \text{ for } c_1 \geq 0 \text{ fixed},$$

$$\lim_{|c_1| + |c_2| \to \infty} G_\Lambda(c_1, c_2) = \lim_{c_2 \to \infty} G_\Lambda(c_1, c_2) = -\frac{\Lambda}{\Phi(q)}.$$

### 4.2. Choice of Optimal Thresholds

To choose the optimal thresholds among reflected policies, we maximize the function $G_\Lambda$.

**Proposition 1.** *The function $G_\Lambda$, defined in Equation (19), attains its maximum on $\mathcal{A}$.*

**Proof.** Let $c_1 \geq 0$ be fixed. The first derivative of $G_\Lambda$ with respect to $c_2$ is given by

$$\partial_{c_2} G_\Lambda(c_1, c_2) = \frac{q F_\Lambda(c_1, c_2) W^{(q)}(c_2)}{(Z^{(q)}(c_2) - Z^{(q)}(c_1))^2}, \tag{20}$$

where

$$F_\Lambda(c_1, c_2) := \frac{(Z^{(q)}(c_2) - Z^{(q)}(c_1))}{q W^{(q)}(c_2)} (1 - \Lambda Z^{(q)}(c_2)) - \left(c_2 - c_1 - \delta - \Lambda\left(\overline{Z}^{(q)}(c_2) - \overline{Z}^{(q)}(c_1)\right)\right)$$

$$= -\Lambda \left[\frac{(Z^{(q)}(c_2))^2}{q W^{(q)}(c_2)} - \overline{Z}^{(q)}(c_2) - \left(\frac{(Z^{(q)}(c_2))}{q W^{(q)}(c_2)} Z^{(q)}(c_1) - \overline{Z}^{(q)}(c_1)\right)\right]$$

$$+ \frac{Z^{(q)}(c_2) - Z^{(q)}(c_1)}{q W^{(q)}(c_2)} - (c_2 - c_1 - \delta). \tag{21}$$

On the other hand, taking $a = c_2$ in Equation (12), we see

$$\frac{[Z^{(q)}(c_2)]^2}{q W^{(q)}(c_2)} - k^{(q)}(c_2) \geq 0 \quad \text{and} \quad \frac{Z^{(q)}(c_2)}{q W^{(q)}(c_2)} Z^{(q)}(c_1) - k^{(q)}(c_1) \geq 0.$$

Then, using Equation (10), we have

$$F_\Lambda(c_1, c_2) < \frac{Z^{(q)}(c_2)}{q W^{(q)}(c_2)} + \Lambda\left[\frac{Z^{(q)}(c_2)}{q W^{(q)}(c_2)} Z^{(q)}(c_1) - k^{(q)}(c_1)\right] - (c_2 - c_1 - \delta). \tag{22}$$

Therefore, since $\lim_{c_2 \to \infty} \frac{Z^{(q)}(c_2)}{q W^{(q)}(c_2)} = \frac{1}{\Phi(q)}$ (see Remark 3), the right-hand side of the aforementioned inequality goes to $-\infty$ as $c_2$ goes to $\infty$, which implies

$$\partial_{c_2} G_\Lambda(c_1, c_2) < 0, \quad \text{for } c_2 \text{ large enough}. \tag{23}$$

From here and Remark 8, we obtain that there exists $c^* \in (c_1, \infty)$ (that depends on $c_1$) such that

$$G_\Lambda(c_1, c_2) \leq G_\Lambda(c_1, c^*), \quad \text{for all } c_2 > c_1.$$

Taking $d^*(c_1) := \sup\{c^* > c_1 : G_\Lambda(c_1, c_2) \leq G_\Lambda(c_1, c^*) \text{ for all } c_2 > c_1\}$, with $c_1 \geq 0$, we see $d^*(c_1) < \infty$ for each $c_1 \geq 0$, since Equation (23) holds. From Equation (20) and the fact that $\partial_{c_2} G_\Lambda(c_1, d^*(c_1)) = 0$, it follows that $F_\Lambda(c_1, d^*(c_1)) = 0$ for $c_1 \geq 0$. Then, by the definitions of $F_\Lambda$ and $\zeta_\Lambda$—see Equations (21) and (14), respectively—we get

$$G_\Lambda(c_1, d^*(c_1)) = \frac{d^*(c_1) - c_1 - \delta - \Lambda(\overline{Z}^{(q)}(d^*(c_1)) - \overline{Z}^{(q)}(c_1))}{Z^{(q)}(d^*(c_1)) - Z^{(q)}(c_1)} = \zeta_\Lambda(d^*(c_1)), \quad \text{for each } c_1 \geq 0.$$

Now, let us take $\tilde{c}_1 > a_\Lambda$ (where $a_\Lambda$ is defined in Equation (14)). Then, using the fact that $\zeta_\Lambda$ is strictly decreasing in $(a_\Lambda, \infty)$ (see Remark 6), we have that for any $c_2 > c_1 > d^*(\tilde{c}_1)$ it holds that $d^*(\tilde{c}_1) < d^*(c_1)$ and

$$G_\Lambda(c_1, c_2) \leq G_\Lambda(c_1, d^*(c_1)) = \zeta_\Lambda(d^*(c_1)) < \zeta_\Lambda(d^*(\tilde{c}_1)) = G_\Lambda(\tilde{c}_1, d^*(\tilde{c}_1)).$$

This implies that the maximum of the function $G_\Lambda$ has to be achieved on the set

$$\{(c_1, c_2) \in \mathbb{R}_+^2 : c_1 < c_2 \text{ and } c_1 \in [0, \tilde{c}_1]\}.$$

Finally, from Equation (22), we obtain

$$F_\Lambda(c_1, c_2) < \frac{Z^{(q)}(c_2)}{qW^{(q)}(c_2)} + \Lambda \sup_{c_1 \in [0,\tilde{c}_1]} \left[\frac{Z^{(q)}(c_2)}{qW^{(q)}(c_2)} Z^{(q)}(c_1) - k^{(q)}(c_1)\right] - (c_2 - \tilde{c}_1 - \delta), \quad \text{for } c_1 \in [0, \tilde{c}_1].$$

Hence, for any $c_1 \in [0, \tilde{c}_1]$, we can find $\tilde{c}_2 > \tilde{c}_1$ such that

$$\partial_{c_2} G_\Lambda(c_1, c_2)(c_1, c_2) < 0, \quad \text{for any } 0 \leq c_1 \leq \tilde{c}_1 \text{ and } 0 \leq c_2 \leq \tilde{c}_2.$$

Therefore, the function $G_\Lambda$ attains its maximum on the set

$$\{(c_1, c_2) \in [0, \tilde{c}_1] \times [0, \tilde{c}_2] : c_1 < c_2\} \subset \mathcal{A}. \quad \square$$

Note that by Proposition 1 the set $\mathcal{B} \subset \mathcal{A}$ defined as

$$\mathcal{B} := \{(c_1^*, c_2^*) \in \mathcal{A} : G_\Lambda(c_1^*, c_2^*) \geq G_\Lambda(c_1, c_2) \text{ for all } (c_1, c_2) \in \mathcal{A}\},$$

is not empty. Moreover, since $G_\Lambda \in C^1(\mathcal{A})$ and using Equation (14), it follows that

$$\partial_{c_1} G_\Lambda(c_1^*, c_2^*) = \frac{qW^{(q)}(c_1^*)}{Z^{(q)}(c_2^*) - Z^{(q)}(c_1^*)} (G_\Lambda(c_1^*, c_2^*) - \zeta_\Lambda(c_1^*)) \leq 0, \quad \text{for } (c_1^*, c_2^*) \in \mathcal{B}, \quad (24)$$

with equality if $c_1 > 0$, and

$$\partial_{c_2} G_\Lambda(c_1^*, c_2^*) = -\frac{qW^{(q)}(c_2^*)}{Z^{(q)}(c_2^*) - Z^{(q)}(c_1^*)} (G_\Lambda(c_1^*, c_2^*) - \zeta_\Lambda(c_2^*)) = 0, \quad \text{for } (c_1^*, c_2^*) \in \mathcal{B}. \quad (25)$$

**Proposition 2.** *There exists a unique pair $(c_1^\Lambda, c_2^\Lambda)$ in $\mathcal{B}$. Furthermore, $0 \le c_1^\Lambda \le a_\Lambda < c_2^\Lambda < \infty$, with $a_\Lambda$ defined in Equation (16), and the value function associated with the $(c_1^\Lambda, c_2^\Lambda)$-policy is*

$$v_{\delta,\Lambda}^{c_1^\Lambda, c_2^\Lambda}(x) = \begin{cases} Z^{(q)}(x)\zeta_\Lambda(c_2^\Lambda) + \Lambda k^{(q)}(x), & \text{if } x \le c_2^\Lambda, \\ x - c_2^\Lambda + v_{\delta,\Lambda}^{c_1^\Lambda, c_2^\Lambda}(c_2^\Lambda), & \text{if } x > c_2^\Lambda. \end{cases} \quad (26)$$

**Proof.** Let $M$ be the maximum value of $G_\Lambda$ in $\mathcal{B}$; therefore, for any $(c_1^*, c_2^*) \in \mathcal{B}$, we have that $\zeta_\Lambda(c_2^*) = M$ by Equation (25). From Remark 6, we know that $\zeta_\Lambda$ is strictly increasing on $(0, a_\Lambda)$ and strictly decreasing on $(a_\Lambda, \infty)$. If $\zeta_\Lambda(0) \ge M$, $\zeta_\Lambda$ attains $M$ at a unique $c_2^\Lambda > a_\Lambda$ and therefore $(0, c_2^\Lambda)$ is the only point that satisfies Equation (24). On the other hand, if $\zeta_\Lambda(0) < M$, $\zeta_\Lambda$ can only attain the value $M$ at a unique $c_1^\Lambda < a_\Lambda$ and a unique $c_2^\Lambda > a_\Lambda$. Hence, $(c_1^\Lambda, c_2^\Lambda)$ is the only point that satisfies Equations (24) and (25), that is, the only existing point in $\mathcal{B}$. Now, from Lemma 1 and using that $G_\Lambda(c_1^\Lambda, c_2^\Lambda) = \zeta_\Lambda(c_2^\Lambda)$, we obtain the first part of Equation (26). For the second part, let $x > c_2^\Lambda$, then

$$v_{\delta,\Lambda}^{c_1^\Lambda, c_2^\Lambda}(x) = x - c_1^\Lambda - \delta + v_{\delta,\Lambda}^{c_1^\Lambda, c_2^\Lambda}(c_1^\Lambda) = x - c_2^\Lambda + c_2^\Lambda - c_1^\Lambda - \delta + v_{\delta,\Lambda}^{c_1^\Lambda, c_2^\Lambda}(c_1^\Lambda) = x - c_2^\Lambda + v_{\delta,\Lambda}^{c_1^\Lambda, c_2^\Lambda}(c_2^\Lambda). \quad \square$$

The following properties of $v_{\delta,\Lambda}^{c_1^\Lambda, c_2^\Lambda}$ are used below in the verification theorem.

**Remark 9.** *From Equations (10) and (26), we note*

$$v_{\delta,\Lambda}^{c_1^\Lambda, c_2^\Lambda}(x) \ge \frac{\Lambda \psi'(0+)}{q} + Z^{(q)}(c_2^\Lambda)\zeta(c_2^\Lambda), \quad \text{for } x > 0.$$

**Remark 10** (Continuity/smoothness at zero). *Note that for $x < 0$, $v_{\delta,\Lambda}^{c_1^\Lambda, c_2^\Lambda}(x) = v_{\delta,\Lambda}^{c_1^\Lambda, c_2^\Lambda}(0) + \Lambda x$. Therefore,*

(i) $v_{\delta,\Lambda}^{c_1^\Lambda, c_2^\Lambda}$ *is continuous at zero.*
(ii) *For the case of unbounded variation, we have that*

$$v_{\delta,\Lambda}^{c_1^\Lambda, c_2^\Lambda'}(0+) = qW^{(q)}(0+)\zeta_\Lambda(c_2^\Lambda) + \Lambda = \Lambda = v_{\delta,\Lambda}^{c_1^\Lambda, c_2^\Lambda'}(0-).$$

### 4.3. Verification

Let us denote by $v_{\delta,\Lambda}$ the function given in Equation (26), which is the optimal value function among reflected policies. We now prove some properties of this function.

**Lemma 2.** *The function $v_{\delta,\Lambda}$ is $C^2((0, \infty) \setminus \{c_2^\Lambda\})$ and $C^1(0, \infty)$.*

**Proof.** By Assumption 1, we have that, for each $q \ge 0$, the function $W^{(q)}$ is continuously differentiable on $(0, \infty)$. This implies, by Equation (26), that $v_{\delta,\Lambda}$ is $C^2((0, \infty) \setminus \{c_2^\Lambda\})$. On the other hand, using Equation (26), we have that for $x \le c_2^\Lambda$,

$$v_{\delta,\Lambda}'(x) = qW^{(q)}(x)\zeta_\Lambda(c_2^\Lambda) + \Lambda Z^{(q)}(x) = qW^{(q)}(x)\left(\frac{1 - \Lambda Z^{(q)}(c_2^\Lambda)}{qW^{(q)}(c_2^\Lambda)}\right) + \Lambda Z^{(q)}(x).$$

This implies that $v_{\delta,\Lambda}'(c_2^\Lambda-) = 1$. For $x > c_2^\Lambda$, we obtain by Equation (26) that

$$v_{\delta,\Lambda}'(c_2^\Lambda+) = 1 = v_{\delta,\Lambda}'(c_2^\Lambda-),$$

which implies the result. $\square$

Let $\mathcal{L}$ be the operator defined as follows,

$$\mathcal{L}F(x) := \gamma F'(x) + \frac{\sigma^2}{2}F''(x) + \int_{(0,\infty)} (F(x-z) - F(x) + F'(x)z\mathbf{1}_{\{0<z\leq 1\}})\Pi(dz), \quad x > 0,$$

where $x \in \mathbb{R}$ and $F$ is a function on $\mathbb{R}$ such that $\mathcal{L}F(x)$ is well defined.

**Proposition 3.**

1. $(\mathcal{L} - q)v_{\delta,\Lambda}(x) = 0$ for $x < c_2^\Lambda$.
2. $(\mathcal{L} - q)v_{\delta,\Lambda}(x) \leq 0$ for $x > c_2^\Lambda$.

**Proof.**

1. By the proof of Theorem 2.1 in Bayraktar et al. (2014a), we have that for $0 < x < c_2^\Lambda$,

$$(\mathcal{L} - q)\left(\overline{Z}^{(q)}(x) + \frac{\psi'(0+)}{q}\right) = 0 \quad \text{and} \quad (\mathcal{L} - q)Z^{(q)}(x) = 0.$$

This implies that for $0 < x < c_2^\Lambda$,

$$(\mathcal{L} - q)v_{\delta,\Lambda}(x) = 0.$$

2. We note that $v_{\delta,\Lambda}(y) = u_\Lambda^{c_2^\Lambda}(y)$ for all $y \geq 0$, where $u_\Lambda^a$ is the barrier strategy at the level $a$ for the dividend problem with capital injection given by Equation (13). Therefore,

    (i) If we take $y \leq x$, and $c_2^\Lambda \leq x$, we obtain

$$u_\Lambda^x(y) = Z^{(q)}(y)\zeta_\Lambda(x) + \Lambda\left(\overline{Z}^{(q)}(y) + \frac{\psi'(0+)}{q}\right).$$

Recall the functions $\zeta_\Lambda$ and $H$ are as in Equations (7) and (14), respectively. Then,

$$\lim_{y\uparrow x} \frac{d^2 u_\Lambda^x}{dy^2}(y) = \Lambda q W^{(q)}(x) + q W^{(q)'}(x)\zeta_\Lambda(x)$$

$$= \frac{W^{(q)'}(x)}{W^{(q)}(x)}\left(1 - \Lambda\left(Z^{(q)}(x) - qW^{(q)}(x)\frac{W^{(q)}(x)}{W^{(q)'}(x)}\right)\right)$$

$$= \frac{W^{(q)'}(x)}{W^{(q)}(x)}(1 - \Lambda H(x))$$

$$= -q W^{(q)}(x)\zeta'_\Lambda(x).$$

By Proposition 2, we know that $a_\Lambda < c_2^\Lambda \leq x$. Then, $\lim_{y\uparrow x}\frac{d^2 u_\Lambda^x}{dy^2}(y) \geq 0 = \frac{d^2 u_\Lambda^{c_2^\Lambda}}{dx^2}(x)$, since $\zeta'_\Lambda(x) < 0$ by Remark 6.

    (ii) We have for $y \in [0, c_2^\Lambda]$,

$$\frac{du_\Lambda^{c_2^\Lambda}}{dy}(y) = \Lambda Z^{(q)}(y) + q W^{(q)}(y)\zeta_\Lambda(c_2^\Lambda) \geq \Lambda Z^{(q)}(y) + q W^{(q)}(y)\zeta_\Lambda(x) = \frac{du_\Lambda^x}{dy}(y),$$

which comes from the fact that for $x \geq c_2^\Lambda > a_\Lambda$, then $\zeta_\Lambda(c_2^\Lambda) \geq \zeta_\Lambda(x)$ by Remark 2. On the other hand, for $y \in (c_2^\Lambda, x]$, we have, using the fact that $\zeta_\Lambda(y) \geq \zeta_\Lambda(x)$,

$$\frac{du_\Lambda^x}{dy}(y) = \Lambda Z^{(q)}(y) + qW^{(q)}(y)\zeta_\Lambda(x)$$

$$\leq \Lambda Z^{(q)}(y) + qW^{(q)}(y)\zeta_\Lambda(y)$$

$$= \Lambda Z^{(q)}(y) + qW^{(q)}(y)\frac{1 - \Lambda Z^{(q)}(y)}{qW^{(q)}(y)} = 1 = \frac{du_\Lambda^{c_2^\Lambda}}{dy}(y).$$

(iii) We note that

$$u_\Lambda^x(c_2^\Lambda) = \Lambda\left(\overline{Z}^{(q)}(c_2^\Lambda) + \frac{\psi'(0+)}{q}\right) + Z^{(q)}(c_2^\Lambda)\zeta_\Lambda(x)$$

$$\leq \Lambda\left(\overline{Z}^{(q)}(c_2^\Lambda) + \frac{\psi'(0+)}{q}\right) + Z^{(q)}(c_2^\Lambda)\zeta_\Lambda(c_2^\Lambda) = u_\Lambda^{c_2^\Lambda}(c_2^\Lambda).$$

This and Point (ii) imply that $(u_\Lambda^{c_2^\Lambda} - u_\Lambda^x)(x) \geq 0$.
(iv) We have

$$\frac{du_\Lambda^{c_2^\Lambda}}{dx}(x) = 1 = \lim_{y \to x} \frac{du_\Lambda^x}{dy}(y).$$

Thus, by similar arguments to those in the proof of Theorem 2 in Loeffen (2008), we obtain the result. □

**Lemma 3.**

1. For $x > 0$, we have that $v'_{\delta,\Lambda}(x) \leq \Lambda$.
2. For $x \geq y \geq 0$, we have that $v_{\delta,\Lambda}(x) - v_{\delta,\Lambda}(y) \geq x - y - \delta$.

**Proof.** 1. By Equation (5) together with Equation (26), we note that for $x \leq c_2^\Lambda$,

$$v'_{\delta,\Lambda}(x) = \Lambda\left(Z^{(q)}(x) - \frac{W^{(q)}(x)}{W^{(q)}(c_2^\Lambda)}Z^{(q)}(c_2^\Lambda)\right) + \frac{W^{(q)}(x)}{W^{(q)}(c_2^\Lambda)}$$

$$= \Lambda \mathbb{E}_x\left[e^{-q\tau_0^-}\mathbf{1}_{\{\tau_0^- < \tau_{c_2^\Lambda}^+\}}\right] + \mathbb{E}_x\left[e^{-q\tau_{c_2^\Lambda}^+}\mathbf{1}_{\{\tau_{c_2^\Lambda}^+ < \tau_0^-\}}\right]$$

$$\leq \Lambda\left(\mathbb{E}_x\left[e^{-q\tau_0^-}\mathbf{1}_{\{\tau_0^- < \tau_{c_2^\Lambda}^+\}}\right] + \mathbb{E}_x\left[e^{-q\tau_{c_2^\Lambda}^+}\mathbf{1}_{\{\tau_{c_2^\Lambda}^+ < \tau_0^-\}}\right]\right)$$

$$\leq \Lambda\left(\mathbb{P}_x\left[\tau_0^- < \tau_{c_2^\Lambda}^+\right] + \mathbb{P}_x\left[\tau_{c_2^\Lambda}^+ < \tau_0^-\right]\right) = \Lambda.$$

On the other hand, $v'_{\delta,\Lambda}(x) = 1 \leq \Lambda$ for $x > c_2^\Lambda$.
2. Let us consider $c_2^\Lambda \geq x \geq y$. We note that

$$v_{\delta,\Lambda}(x) - v_{\delta,\Lambda}(y) = \Lambda\left(\overline{Z}^{(q)}(x) - \overline{Z}^{(q)}(y)\right) + \left(Z^{(q)}(x) - Z^{(q)}(y)\right)\zeta_\Lambda(c_2^\Lambda)$$

$$= \Lambda\left(\overline{Z}^{(q)}(x) - \overline{Z}^{(q)}(y)\right) + (Z^{(q)}(x) - Z^{(q)}(y))G_\Lambda(c_1^\Lambda, c_2^\Lambda)$$

$$\geq \Lambda\left(\overline{Z}^{(q)}(x) - \overline{Z}^{(q)}(y)\right) + (Z^{(q)}(x) - Z^{(q)}(y))G_\Lambda(x,y)$$

$$= \Lambda\left(\overline{Z}^{(q)}(x) - \overline{Z}^{(q)}(y)\right) + (Z^{(q)}(x) - Z^{(q)}(y))\frac{x - y - \delta - \Lambda\left(\overline{Z}^{(q)}(x) - \overline{Z}^{(q)}(y)\right)}{Z^{(q)}(x) - Z^{(q)}(y)}$$

$$= x - y - \delta. \tag{27}$$

Now, suppose that $x \geq y \geq c_2^\Lambda$, then using Equation (26) we obtain

$$v_{\delta,\Lambda}(x) - v_{\delta,\Lambda}(y) = x - y \geq x - y - \delta.$$

Finally, for the case $x \geq c_2^\Lambda \geq y$, by Equation (27), we have

$$v_{\delta,\Lambda}(x) - v_{\delta,\Lambda}(y) = x - c_2^\Lambda + v_{\delta,\Lambda}(c_2^\Lambda) - v_{\delta,\Lambda}(y) \geq x - c_2^\Lambda + (c_2^\Lambda - y - \delta) = x - y - \delta. \quad \square$$

Now, we proceed to the verification theorem that proves the optimality of the $(c_1^\Lambda, c_2^\Lambda)$-policy.

**Theorem 1** (Verification Theorem). *Let $V_{\delta,\Lambda}$, $v_{\delta,\Lambda}$ be as in Equations (3) and (26), respectively. Then, $v_{\delta,\Lambda}(x) = V_{\delta,\Lambda}(x)$ for all $x \geq 0$. Hence, the $(c_1^\Lambda, c_2^\Lambda)$-policy is optimal.*

**Proof.** By the definition of $V_{\delta,\Lambda}$, $v_{\delta,\Lambda}(x) \leq V_{\delta,\Lambda}(x)$ for all $x \geq 0$. Let us verify that $v_{\delta,\Lambda}(x) \geq v_{\delta,\Lambda}^\pi(x)$ for all admissible $\pi \in \Theta_\delta$ and for all $x \geq 0$. Recall that $v_{\delta,\Lambda}^\pi$ is defined in Equation (2). Take $\pi = \{L^\pi, R^\pi\} \in \Theta_\delta$ fixed and let $(T_n)_{n \in \mathbb{N}}$ be the sequence of stopping times where $T_n := \inf\{t > 0 : X_t^\pi > n\}$. Since $X^\pi = X - L^\pi + R^\pi$, with $X$ being a spectrally negative Lévy process, it is a semi-martingale and $v_{\delta,\Lambda}$ is sufficiently smooth on $(0, \infty)$ by Lemma 2, and continuous (respectively, continuously differentiable) at zero for the case of bounded variation (respectively, unbounded variation) by Remark 10, we can use the change of variables/Meyer-Itô's formula (cf. Theorems II.31 and II.32 of Protter (2005)) on the stopped process $(e^{-q(t \wedge T_n)} v_{\delta,\Lambda}(X_{t \wedge T_n}^\pi); t \geq 0)$ to deduce under $\mathbb{P}_x$ that

$$e^{-q(t \wedge T_n)} v_{\delta,\Lambda}(X_{t \wedge T_n}^\pi) - v_{\delta,\Lambda}(x) = \int_0^{t \wedge T_n} e^{-qs}(\mathcal{L} - q)v_{\delta,\Lambda}(X_{s-}^\pi)ds + M_{t \wedge T_n} + J_{t \wedge T_n} \tag{28}$$

$$+ \int_{[0, t \wedge T_n]} e^{-qs} v_{\delta,\Lambda}'(X_{s-}^\pi) dR_s^{\pi,c},$$

where $M$ is a local martingale with $M_0 = 0$, $R^{\pi,c}$ is the continuous part of $R^\pi$, and $J$ is a jump process, which is given by

$$J_t = \sum_{0 \leq s \leq t} e^{-qs}\left(v_{\delta,\Lambda}(X_{s-}^\pi + \Delta[X + R^\pi]_s) - v_{\delta,\Lambda}(X_{s-}^\pi + \Delta X_s)\right) 1_{\{\Delta[X + R^\pi]_s \neq 0\}}$$

$$+ \sum_{0 \leq s \leq t} e^{-qs}\left(v_{\delta,\Lambda}(X_{s-}^\pi + \Delta[X + R^\pi]_s - \Delta L_s^\pi) - v_{\delta,\Lambda}(X_{s-}^\pi + \Delta[X + R^\pi]_s)\right) 1_{\{\Delta L_s^\pi \neq 0\}}, \quad \text{for } t \geq 0.$$

On the other hand, by Part (1) of Lemma 3, we obtain that

$$\int_{[0, t \wedge T_n]} e^{-qs} v_{\delta,\Lambda}'(X_{s-}^\pi) dR_s^{\pi,c}$$

$$+ \sum_{0 \leq s \leq t \wedge T_n} e^{-qs} [v_{\delta,\Lambda}(X_{s-}^\pi + \Delta[X + R^\pi]_s) - v_{\delta,\Lambda}(X_{s-}^\pi + \Delta X_s)] 1_{\{\Delta[X + R^\pi]_s \neq 0\}}$$

$$\leq \Lambda \int_{[0, t \wedge T_n]} e^{-qs} dR_s^{\pi,c} + \Lambda \sum_{0 \leq s \leq t \wedge T_n} e^{-qs} \Delta R_s^\pi = \Lambda \int_{[0, t \wedge T_n]} e^{-qs} dR_s^\pi.$$

Similarly, by Part (2) of Lemma 3,

$$\sum_{0 \leq s \leq t \wedge T_n} e^{-qs}[v_{\delta,\Lambda}(X^\pi_{s-} + \Delta[X+R^\pi]_s - \Delta L^\pi_s) - v_{\delta,\Lambda}(X^\pi_{s-} + \Delta[X+R^\pi]_s)]1_{\{\Delta L^\pi_s \neq 0\}}$$

$$\leq -\sum_{0 \leq s \leq t \wedge T_n} e^{-qs} \Delta L^\pi_s + \delta \sum_{0 \leq s \leq t \wedge T_n} e^{-qs} 1_{\{\Delta L^\pi_s > 0\}}$$

$$= -\int_{[0,t \wedge T_n]} e^{-qs} \, d\left( L^\pi_s - \delta \sum_{0 \leq u \leq s} 1_{\{\Delta L^\pi_u > 0\}} \right).$$

Hence, from Equation (28), we derive that

$$v_{\delta,\Lambda}(x) \geq -\int_0^{t \wedge T_n} e^{-qs} (\mathcal{L} - q) v_{\delta,\Lambda}(X^\pi_{s-}) ds - \Lambda \int_{[0,t \wedge T_n]} e^{-qs} \, dR^\pi_s$$

$$+ \int_{[0,t \wedge T_n]} e^{-qs} \, d\left( L^\pi_s - \delta \sum_{0 \leq u \leq s} 1_{\{\Delta L^\pi_u > 0\}} \right) - M_{t \wedge T_n} + e^{-q(t \wedge T_n)} v_{\delta,\Lambda}(X^\pi_{t \wedge T_n}).$$

Using Proposition 3 along with Point 3 in the proof of Lemma 6 in Loeffen (2009), and that $X^\pi_{s-} \geq 0$ a.s. for $s \geq 0$, we observe that

$$v_{\delta,\Lambda}(x) \geq \int_{[0,t \wedge T_n]} e^{-qs} \, d\left( L^\pi_s - \delta \sum_{0 \leq u \leq s} 1_{\{\Delta L^\pi_u > 0\}} \right) - \Lambda \int_{[0,t \wedge T_n]} dR^\pi_s - M_{t \wedge T_n} + e^{-q(t \wedge T_n)} v_{\delta,\Lambda}(X^\pi_{t \wedge T_n})$$

$$\geq \int_{[0,t \wedge T_n]} e^{-qs} \, d\left( L^\pi_s - \delta \sum_{0 \leq u \leq s} 1_{\{\Delta L^\pi_u > 0\}} \right) - \Lambda \int_{[0,t \wedge T_n]} dR^\pi_s - M_{t \wedge T_n}$$

$$+ e^{-q(t \wedge T_n)} \left( \frac{\Lambda \psi'(0+)}{q} + Z^{(q)}(c_2^\Lambda) \left( \zeta(c_2^\Lambda) \right) \right), \quad (29)$$

where the last inequality follows from Remark 9. In addition, by the compensation formula (see, e.g., Corollary 4.6 of Kyprianou (2014)), $(M_{t \wedge T_n} : t \geq 0)$ is a zero-mean $\mathbb{P}_x$-martingale. Now, taking expected value in Equation (29) and letting $(t \wedge T_n) \nearrow \infty$ $\mathbb{P}_x$-a.s., the monotone convergence theorem, applied separately for $\mathbb{E}_x \left[ \int_{[0,t \wedge T_n]} e^{-qs} \, d\left( L^\pi_s - \delta \sum_{0 \leq u \leq s} 1_{\{\Delta L^\pi_u > 0\}} \right) \right]$ and $\mathbb{E}_x \left( \Lambda \int_{[0,t \wedge T_n]} e^{-qs} \, dR^\pi_s \right)$, gives

$$v_{\delta,\Lambda}(x) \geq \mathbb{E}_x \left( \int_{[0,\infty)} e^{-qs} \, d\left( L^\pi_s - \delta \sum_{0 \leq u \leq s} 1_{\{\Delta L^\pi_u > 0\}} \right) - \Lambda \int_{[0,\infty)} e^{-qs} \, dR^\pi_s \right) = v^\pi_{\delta,\Lambda}(x).$$

This completes the proof. □

## 5. Optimal Dividends with Capital Injection Constraint

In this section, we are interested in maximizing the expected NPV of the dividend strategy subject to a constraint in the expected present value of the injected capital. Specifically, we aim to solve

$$V_\delta(x, K) := \sup_{\pi \in \Theta_\delta} \mathbb{E}_x \left[ \int_0^\infty e^{-qt} \, d\left( L^\pi_t - \delta \sum_{0 \leq s < t} 1_{\{\Delta L^\pi_s > 0\}} \right) \right] \quad \text{s.t.} \quad \mathbb{E}_x \left[ \int_0^\infty e^{-qt} \, dR^\pi_t \right] \leq K, \quad (30)$$

for any $x \geq 0$ and $K \geq 0$. Strategies $\pi$ that do not satisfy the capital injection constraint are called infeasible. Recall that the insurance company has to inject capital to ensure the non-negativity of the risk process. Therefore, small values of $K$ require very low dividend payments to keep the risk process non-negative, or would even make the problem infeasible. In the latter case, we define the value function as $-\infty$.

To solve this problem, we use the solution of the optimal dividend problem with capital injection found in the section above. Thus, for $\Lambda \geq 0$, we define the function

$$v_{\delta,\Lambda}^\pi(x, K) := v_{\delta,\Lambda}^\pi(x) + \Lambda K,$$

with $v_{\delta,\Lambda}^\pi$ as in Equation (2). It is easy to check that $V_\delta(x, K) = \sup_{\pi \in \Theta_\delta} \inf_{\Lambda \geq 0} v_{\delta,\Lambda}^\pi(x, K)$ since for infeasible strategies $\inf_{\Lambda \geq 0} v_{\delta,\Lambda}^\pi(x, K) = -\infty$. By interchanging the sup with the inf we obtain an upper bound for $V_\delta(x, K)$, the so-called weak duality. Hence, the dual problem of Equation (30) is defined as

$$V_\delta^D(x, K) := \inf_{\Lambda \geq 0} \sup_{\pi \in \Theta_\delta} v_{\delta,\Lambda}^\pi(x, K) = \inf_{\Lambda \geq 0} \left\{ \Lambda K + \sup_{\pi \in \Theta_\delta} v_{\delta,\Lambda}^\pi(x) \right\} = \inf_{\Lambda \geq 1} \left\{ \Lambda K + V_{\delta,\Lambda}(x) \right\}, \quad (31)$$

with $V_{\delta,\Lambda}$ given in Equation (3). The last equality in Equation (31) is true, since $V_{\delta,\Lambda}(x)$ is infinite for any $\Lambda < 1$; see Remark 2. The main goal is to prove that $V_\delta^D(x, K) \leq V_\delta(x, K)$.

### 5.1. No Transaction Cost

In this subsection, we consider the problem in Equation (30) without transaction cost, i.e., $\delta = 0$. For this case, we denote $V(x, K) := V_0(x, K)$ and $V^D(x, K) := V_0^D(x, K)$. From Section 3.2, recall that for each $\Lambda \geq 1$, the optimal strategy is a barrier strategy, which is determined by $a_\Lambda$ defined in Equation (16), and its NPV satisfies $V_\Lambda = v_\Lambda^{a_\Lambda}$, where $v_\Lambda^{a_\Lambda}$ is as in Equation (13). Given a barrier strategy at $a > 0$ and $x \in [0, a]$, the expected NPV of the injected capital is given by the function

$$\Psi_x(a) := \mathbb{E}_x \left[ \int_0^\infty e^{-qt} \, dR_t^{a,0} \right] = \frac{Z^{(q)}(a)}{qW^{(q)}(a)} Z^{(q)}(x) - k^{(q)}(x), \quad (32)$$

with $k^{(q)}$ as in Equation (10). Clearly, if $x > a$, then $\Psi_x(a) = \Psi_a(a)$. We also define

$$\underline{K}_x := \lim_{a \to \infty} \Psi_x(a). \quad (33)$$

Using Equation (12) and the properties of scale functions (see Remark 3 (3)),

$$\underline{K}_x = -k^{(q)}(x) + \frac{Z^{(q)}(x)}{\Phi(q)}.$$

Note that $\underline{K}_x$ is the expected present value of the injected capital for the pay-nothing strategy $\pi_{PN} := \{0, R^0\}$. Therefore, letting $a \to \infty$ in Equation (13), it can be verified

$$v_\Lambda^{\pi_{PN}}(x, K) = \Lambda(K - \underline{K}_x).$$

Hence, if $K \geq \underline{K}_x$, then for any $x \geq 0$,

$$V(x, K) = \sup_{\pi \in \Theta} \inf_{\Lambda \geq 0} v_\Lambda^\pi(x, K) \geq \inf_{\Lambda \geq 0} v_\Lambda^{\pi_{PN}}(x, K) = 0. \quad (34)$$

Conversely, if $K < \underline{K}_x$, the problem in Equation (30) is infeasible, which is verified below.

**Lemma 4.** *If $K < \underline{K}_x$, then $V(x, K) = -\infty$.*

**Proof.** First, by Remark 7 and Equation (11), it is easy to verify that

$$\lim_{\Lambda \to \infty} \mathbb{E}_x \left[ \int_0^\infty e^{-qt} \, dL_t^{a_\Lambda,0} \right] = 0, \quad \text{for } x \geq 0. \quad (35)$$

Then,
$$V^D(x, K) = \inf_{\Lambda \geq 1} \{\Lambda K + V_\Lambda(x)\}$$
$$= \inf_{\Lambda \geq 1} \left\{ \mathbb{E}_x \left[ \int_0^\infty e^{-qt} \, dL_t^{a_\Lambda, 0} \right] + \Lambda(K - \Psi_x(a_\Lambda)) \right\}$$
$$\leq \lim_{\Lambda \to \infty} \left\{ \mathbb{E}_x \left[ \int_0^\infty e^{-qt} \, dL_t^{a_\Lambda, 0} \right] + \Lambda(K - \Psi_x(a_\Lambda)) \right\} = -\infty.$$

Now, since $V(x, K) \leq V^D(x, K)$ for any $x \geq 0$, $K \geq 0$, we have the result. □

The next lemma allows us to prove that, when $K = \underline{K}_x$, Equation (34) holds with equality, and it is used to prove the main result of this subsection.

**Lemma 5.** *Let $x \geq 0$ be fixed. The function $\Psi_x$ is strictly decreasing on $(0, \infty)$.*

**Proof.** First, consider the case when $x < a$. Then, by Remark 4 (i), we have that $\dfrac{qW^{(q)}(a)}{Z^{(q)}(a)}$ is strictly increasing and the lemma is obtained. Now, when $x \geq a > 0$, a simple calculation shows that

$$\frac{d\Psi_a(a)}{da} = -\frac{Z^{(q)}(a)\left(W^{(q)\prime}(a)Z^{(q)}(a) - q[W^{(q)}(a)]^2\right)}{q[W^{(q)}(a)]^2} = -\frac{Z^{(q)}(a)W^{(q)\prime}(a)}{q[W^{(q)}(a)]^2} H(a),$$

which is strictly negative, by Remarks 3 and 5. From here, we conclude the assertion of the lemma. □

**Lemma 6.** *If $K = \underline{K}_x$, then $V(x, K) = 0$ and the optimal strategy is the pay-nothing strategy $\pi_{PN}$.*

**Proof.** By Equation (34), we know that $V(x, K) \geq 0$. On the other hand, from Lemma 5 and Equation (33), we have that $\Lambda(K - \Psi_x(a_\Lambda)) \leq 0$ for all $\Lambda \geq 0$. Then, using Equations (33) and (35)

$$V^D(x, K) = \inf_{\Lambda \geq 1} \{\Lambda K + V_\Lambda(x)\} = \inf_{\Lambda \geq 1} \left\{ \mathbb{E}_x \left[ \int_0^\infty e^{-qt} \, dL_t^{a_\Lambda, 0} \right] + \Lambda(K - \Psi_x(a_\Lambda)) \right\}$$
$$\leq \lim_{\Lambda \to \infty} \mathbb{E}_x \left[ \int_0^\infty e^{-qt} \, dL_t^{a_\Lambda, 0} \right] = 0. \quad \square$$

Now, we define
$$\overline{K} := \lim_{a \to 0} \Psi_a(a).$$

Using Equation (12), we have that $\overline{K} = \infty$ when the risk process has unbounded variation. Otherwise, by Remark 3 (2),
$$\overline{K} = \frac{c - \psi'(0+)}{q}, \tag{36}$$

and $\overline{K}$ corresponds to the expected NPV of the injected capital for the strategy $\pi_{0,0}$ (see Equation (4.5) in Avram et al. (2007)).

**Lemma 7.** *Assume that the risk process $X$ has bounded variation. If $K \geq \overline{K}$, then $V(x, K) = K + V_1(x)$, with $V_1(x) = x + \dfrac{\psi'(0+)}{q}$.*

**Proof.** If the Lévy measure is finite, by Equation (16), we have that $a_1 = 0$. The same is true for the infinite Lévy measure case since $H^{-1}(1) = 0$ by Remark 5. Using Equation (13) and Remark 6, we obtain

$$V_1(x) = v_1^0(x) = x + \frac{c}{q} - \overline{K} = x + \frac{\psi'(0+)}{q}, \quad \text{for } x \geq 0. \tag{37}$$

Now, by Equations (31), (36) and (37) and the weak duality, we get

$$V(x,K) \leq V^D(x,K) \leq K + v_1^0(x) = K + V_1(x).$$

Since $K \geq \overline{K}$, $\pi_{0,0}$ is a feasible strategy. Then, using Equation (11), it yields,

$$V(x,K) \geq \inf_{\Lambda \geq 1} \{v_\Lambda^0(x) + \Lambda K\} = x + K - \frac{c - \psi'(0+)}{q} + \frac{c}{q} = K + V_1(x).$$

Therefore, $V(x,K) = K + V_1(x)$. □

We are now ready for the main result of this subsection.

**Theorem 2.** *Assume $\delta = 0$ and let $V$ and $V^D$ as in Equation (30) and Equation (31), respectively, then $V = V^D$. Furthermore, if $x$ and $K$ are such that $K \in (\underline{K}_x, \overline{K})$, then*

$$V(x,K) = \Lambda^* K + V_{\Lambda^*}(x) = \mathbb{E}_x \left[ \int_0^\infty e^{-qt} \, dL_t^{a^*,0} \right], \tag{38}$$

*where $a^* = \Psi_x^{-1}(K)$, and $\Lambda^* = \dfrac{1}{H(a^*)}$.*

**Proof.** Lemmas 4, 6 and 7 show imply that Equation (38) holds when $x$ and $K$ are such that $K \in [0, \underline{K}_x] \cup [\overline{K}, \infty)$. Assume now that $K \in (\underline{K}_x, \overline{K})$, then by Lemma 5 the function $\Psi_x$ is injective, so there exists a unique $a^* > 0$ such that $\Psi_x(a^*) = K$. Note that from Equation (16), we have that there exists a unique $\Lambda^*$ such that $a_{\Lambda^*} = a^*$. Then,

$$V^D(x,K) \leq \Lambda^* K + V_{\Lambda^*}(x)$$

$$= \Lambda^* K + \mathbb{E}_x \left[ \int_0^\infty e^{-qt} \, dL_t^{a^*,0} \right] - \Lambda^* \Psi_x(a^*)$$

$$= \mathbb{E}_x \left[ \int_0^\infty e^{-qt} \, dL_t^{a^*,0} \right].$$

Meanwhile, since the strategy $\pi_{a^*,0}$ is feasible, we see

$$V(x,K) \geq \inf_{\Lambda \geq 1} \{v_\Lambda^{\pi_{a^*,0}}(x) + \Lambda K\} = \inf_{\Lambda \geq 1} \left\{ \mathbb{E}_x \left[ \int_0^\infty e^{-qt} \, dL_t^{a^*,0} \right] + \Lambda(K - \Psi_x(a^*)) \right\}$$

$$= \mathbb{E}_x \left[ \int_0^\infty e^{-qt} \, dL_t^{a^*,0} \right].$$

This implies that $V^D(x,K) \leq V(x,K)$. Finally, the weak duality gives Equation (46). □

### 5.2. With Transaction Cost

Now, we consider the problem given in Equation (30) with transaction cost $\delta > 0$. From the previous section, we know that optimal strategies are $(c_1^\Lambda, c_2^\Lambda)$-reflected strategies with $(c_1^\Lambda, c_2^\Lambda)$ given in Proposition 2.

**Proposition 4.** *The curve $\Lambda \mapsto (c_1^\Lambda, c_2^\Lambda)$ is continuous and unbounded, for $\Lambda \in [1, \infty)$.*

**Proof.** From Remark 7 and the fact that $a_\Lambda < c_2^\Lambda$ (by Proposition 2), we know that $c_2^\Lambda \to \infty$ as $\Lambda \to \infty$, so the curve is unbounded. To show the continuity of the curve, we consider two cases and use the implicit function theorem. To this end, suppose first $c_1^\Lambda = 0$. Defining $f(\Lambda, c_2) := G_\Lambda(0, c_2) - \zeta_\Lambda(c_2)$, we have $f(\Lambda, c_2^\Lambda) = 0$. Then,

$$\frac{\partial f}{\partial c_2}(\Lambda, c_2^\Lambda) = \frac{\partial G_\Lambda}{\partial c_2}(0, c_2^\Lambda) - \zeta_\Lambda'(c_2^\Lambda) = -\zeta_\Lambda'(c_2^\Lambda) > 0,$$

since $c_2^\Lambda > a_\Lambda$. From here, we see that the conditions of the implicit function theorem are satisfied. Now, if $c_1^\Lambda > 0$, define the function $f(\Lambda, c_1, c_2) = (f_1(\Lambda, c_1, c_2), f_2(\Lambda, c_1, c_2))$ by

$$f_1(\Lambda, c_1, c_2) := G_\Lambda(c_1, c_2) - \zeta_\Lambda(c_1),$$
$$f_2(\Lambda, c_1, c_2) := G_\Lambda(c_1, c_2) - \zeta_\Lambda(c_2).$$

Then, $f(\Lambda, c_1^\Lambda, c_2^\Lambda) = (0, 0)$. Again, simple calculations show that the Jacobian determinant of this system of equations is $\zeta_\Lambda'(c_2^\Lambda) \zeta_\Lambda'(c_1^\Lambda) < 0$, since $c_1^\Lambda < a_\Lambda < c_2^\Lambda$. Therefore, the curve $\Lambda \mapsto (c_1^\Lambda, c_2^\Lambda)$ is continuous, for $\Lambda \in [1, \infty)$. □

Next, we analyze the level curves of the constraint. Let $\overline{\Psi}_x(c_1, c_2)$ be the expected present value of the injected capital under a $(c_1, c_2)$-reflected policy. Then, the calculations given in the proof of Lemma 1 show that

$$\overline{\Psi}_x(c_1, c_2) := \mathbb{E}_x \left[ \int_0^\infty e^{-qt} \, dR_t^{(c_1, c_2), 0} \right]$$

$$= \begin{cases} Z^{(q)}(x) \dfrac{\overline{Z}^{(q)}(c_2) - \overline{Z}^{(q)}(c_1)}{Z^{(q)}(c_2) - Z^{(q)}(c_1)} - k^{(q)}(x), & \text{if } 0 \le x \le c_2, \\ \dfrac{\overline{Z}^{(q)}(c_2) Z^{(q)}(c_1) - \overline{Z}^{(q)}(c_1) Z^{(q)}(c_2)}{Z^{(q)}(c_2) - Z^{(q)}(c_1)} - \dfrac{\psi'(0+)}{q}, & \text{if } x > c_2. \end{cases} \quad (39)$$

**Remark 11.** Note that $\lim_{c_1 \to c_2} \overline{\Psi}_x(c_1, c_2) = \Psi_x(c_2)$, where $\Psi_x$ is as in Equation (32).

The next lemmas describe some properties of $\overline{\Psi}_x(c_1, c_2)$.

**Lemma 8.** *Let $x \ge 0$ be fixed.*

1. *If $c_1 \ge 0$ is fixed, then the function $\overline{\Psi}_x(c_1, c_2)$, given in Equation (39), is strictly decreasing for all $c_2 > c_1$, and*

$$\lim_{c_2 \to \infty} \overline{\Psi}_x(c_1, c_2) = K_x, \quad (40)$$

*where $K_x$ is defined in Equation (33).*

2. *If $c_2 > 0$ is fixed, $\overline{\Psi}_x(c_1, c_2)$ is strictly decreasing for all $c_1 \in [0, c_2)$.*

**Proof.** Let $c_1 \ge 0$ be fixed. First, assume that $c_2 \ge x$. To show that $\overline{\Psi}_x(c_1, c_2)$ is strictly decreasing, it is sufficient to verify that

$$\frac{\overline{Z}^{(q)}(c_2) - \overline{Z}^{(q)}(c_1)}{Z^{(q)}(c_2) - Z^{(q)}(c_1)}, \quad (41)$$

is strictly decreasing, which is true if

$$\frac{\partial}{\partial c_2} \left[ \frac{\overline{Z}^{(q)}(c_2) - \overline{Z}^{(q)}(c_1)}{Z^{(q)}(c_2) - Z^{(q)}(c_1)} \right] = \frac{Z^{(q)}(c_2)}{Z^{(q)}(c_2) - Z^{(q)}(c_1)} - \frac{qW^{(q)}(c_2)(\overline{Z}^{(q)}(c_2) - \overline{Z}^{(q)}(c_1))}{[Z^{(q)}(c_2) - Z^{(q)}(c_1)]^2} < 0. \quad (42)$$

Since $Z^{(q)}$ is a strictly log-convex function on $[0,\infty)$ by Remark 4 (i),

$$\frac{qW^{(q)}(\eta)}{Z^{(q)}(\eta)} < \frac{qW^{(q)}(\varsigma)}{Z^{(q)}(\varsigma)}, \text{ for } \eta \text{ and } \varsigma \text{ such that } \eta < \varsigma.$$

Taking $\varsigma = c_2$ in the inequality above and integrating between $c_1$ and $c_2$, it follows that

$$Z^{(q)}(c_2) < \frac{qW^{(q)}(c_2)[\overline{Z}^{(q)}(c_2) - \overline{Z}^{(q)}(c_1)]}{Z^{(q)}(c_2) - Z^{(q)}(c_1)}. \tag{43}$$

Then, Equation (43) yields Equation (42) and hence Equation (41) is strictly decreasing. For the case $x > c_2$, it can be verified that

$$\frac{\partial}{\partial c_2}\left[Z^{(q)}(c_2)\frac{\overline{Z}^{(q)}(c_2) - \overline{Z}^{(q)}(c_1)}{Z^{(q)}(c_2) - Z^{(q)}(c_1)} - \overline{Z}^{(q)}(c_2)\right]$$

$$= \frac{Z^{(q)}(c_1)}{Z^{(q)}(c_2) - Z^{(q)}(c_1)}\left[Z^{(q)}(c_2) - \frac{qW^{(q)}(c_2)[\overline{Z}^{(q)}(c_2) - \overline{Z}^{(q)}(c_1)]}{Z^{(q)}(c_2) - Z^{(q)}(c_1)}\right]. \tag{44}$$

Then, using Equations (43) and (44), we obtain that $\overline{\Psi}_x(c_1, c_2)$ is strictly decreasing for all $c_2 \in (c_1, x)$. Similarly, we obtain Point 2 of the lemma. Now, by L'Hôpital's rule together with Exercise 8.5 (i) in Kyprianou (2014), it is not difficult to see that Equation (40) holds for any $c_1 \geq 0$. □

Note that Equation (34) still holds if $K \geq \underline{K}_x$. On the other hand, using that $c_2^\Lambda \to \infty$ as $\Lambda \to \infty$ together with Equation (18) we have that

$$\lim_{\Lambda \to \infty} \mathbb{E}_x\left[\int_0^\infty e^{-qt} dL_t^{(c_1^\Lambda, c_2^\Lambda),0}\right] = \lim_{\Lambda \to \infty}(c_2^\Lambda - c_1^\Lambda - \delta)\frac{Z^{(q)}(x)}{Z^{(q)}(c_2^\Lambda) - Z^{(q)}(c_1^\Lambda)} = 0, \tag{45}$$

by Remark 3 (3).

**Remark 12.** *Using the same arguments as in Lemma 7, we have that $c_1^1 = a_1 = 0 < c_2^1$ for bounded and unbounded variation processes. Similarly, if $x$ and $K$ are such that $K \geq \overline{\Psi}_x(0, c_2^1) =: \overline{K}_x$, then $V_\delta(x, K) = V_{\delta,1}(x) + K$. Note also that $\overline{K}_x < \overline{K}$.*

**Lemma 9.** *Let $x \geq 0$ be fixed. Then, for each $K \in (\underline{K}_x, \overline{K}_x)$, there exist $\underline{c} \leq \overline{c}$ such that the level curve $L_K(\overline{\Psi}_x) := \{(c_1, c_2) : \overline{\Psi}_x(c_1, c_2) = K\}$ is continuous, contained in the set $[0, \underline{c}] \times [\underline{c}, \overline{c}]$ and contains the points $(0, \overline{c})$ and $(\underline{c}, \overline{c})$.*

**Proof.** The continuity of the level curve is obtained as an immediate consequence of the continuity of $\overline{\Psi}_x$. Observe that, by Lemma 5, we know the existence of $\underline{c} > 0$ such that $\Psi_x(\underline{c}) = K$. Meanwhile, from Lemma 8, we have that there exists $\overline{c} \in [\underline{c}, \infty)$ such that $\overline{\Psi}_x(0, \overline{c}) = K$. Now, the fact that the level curve $L_K(\overline{\Psi}_x)$ is contained in $[0, \underline{c}] \times [\underline{c}, \overline{c}]$ is a consequence of Remark 11 and Lemma 8. □

**Remark 13.** *Lemmas 4 and 9 yield that the parametric curve $\Lambda \mapsto (c_1^\Lambda, c_2^\Lambda)$ and the level curve $L_K(\overline{\Psi}_x)$ must intersect, i.e., there exists $\Lambda^*$ such that $\overline{\Psi}_x(c_1^{\Lambda^*}, c_2^{\Lambda^*}) = K$, for $K \in (\underline{K}_x, \overline{K}_x]$.*

By similar arguments as in the proof of Theorem 2, by Remarks 12 and 13, and using Equation (45), we get the following result, whose proof is omitted.

**Theorem 3.** *Assume $\delta > 0$ and let $V_\delta$ and $V_\delta^D$ as in Equations (30) and (31), respectively, then $V_\delta = V_\delta^D$. Furthermore, if $x, K$ are such that*

1. $K < \underline{K}_x$, then $V_\delta(x, K) = -\infty$;

2. $K = \underline{K}_x$, then $V_\delta(x, K) = 0$;
3. $K \geq \overline{K}_x$, then $V_\delta(x, K) = V_{\delta,1}(x) + K$; and
4. $K \in (\underline{K}_x, \overline{K}_x)$, then there exists $\Lambda^* \geq 1$ such that

$$V_\delta(x, K) = \Lambda^* K + V_{\delta,\Lambda^*}(x) = \mathbb{E}_x \left[ \int_0^\infty e^{-qt} \, d \left( L_t^{(c_1^{\Lambda^*}, c_2^{\Lambda^*}), 0} - \delta \sum_{0 \leq s < t} \mathbf{1}_{\left\{ \Delta L_s^{(c_1^{\Lambda^*}, c_2^{\Lambda^*}), 0} > 0 \right\}} \right) \right]. \quad (46)$$

## 6. Numerical Examples

In this section, we confirm the obtained results by a sequence of numerical examples. Here, we assume that $X$ is of the form

$$X_t - X_0 = t + 0.5 B_t - \sum_{n=1}^{N_t} Z_n, \quad 0 \leq t < \infty,$$

where $B = \{B_t : t \geq 0\}$, $N = \{N_t : t \geq 0\}$, and $Z = \{Z_n\}_{n \geq 1}$ are a standard Brownian motion, a Poisson process with arrival rate $\lambda = 0.4$, and an i.i.d. sequence of random variables with distribution Gamma (1,2), respectively, which are assumed mutually independent. Since there is no closed form for the scale function $W^{(q)}$ associated with $X$, we use a numerical algorithm presented in Surya (2008) in order to approximate the inverse Laplace transform of Equation (4). Similarly, we approximate the derivatives of the scale functions and use the trapezoidal rule to calculate its integrals.

We first consider the case without transaction cost presented in Section 3.2. In Figure 1 (left), we plot the function $x \mapsto V_\Lambda(x) + \Lambda K$ for various values of $\Lambda$ and a fixed value of $K$. For $x \geq x_0$, where $x_0$ is such that $\underline{K}_{x_0} = K$, its minimum over the considered values of $\Lambda$ provides (an approximation of) $V(x, K)$, indicated by the solid red line in the plot. Since the process has unbounded variation, then $\overline{K} = \infty$. In Figure 1 (right), we plot, for $x > x_0$, the Lagrange multiplier $\Lambda^*$ given in Theorem 2. We observe that $\Lambda^*$ goes to infinity as $x \downarrow x_0$ and remains always above 1.

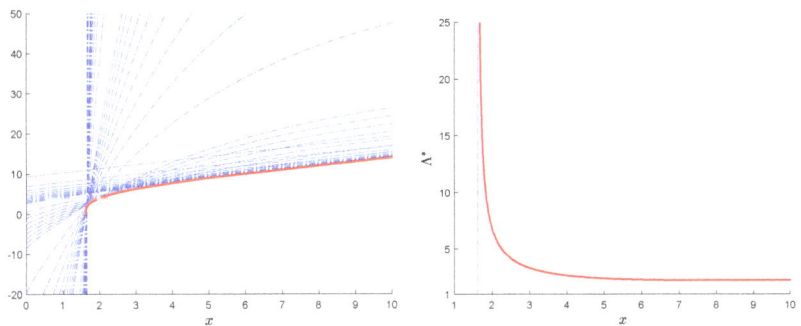

**Figure 1.** (Left) Plots of $x \mapsto V_\Lambda(x) + \Lambda K$ for $\Lambda = 1, 1.1, \ldots, 2, 3, \ldots, 10, 20, \ldots, 100, 200, \ldots, 1000, 2000, \ldots, 10,000, 20,000$ (dotted) for the case $K = 2.7$. The minimum of $V_\Lambda(x) + \Lambda K$ over $\Lambda$ is shown in solid bold-face red line. (**Right**) Plot of the Lagrange multiplier $\Lambda^*$ for $x > x_0$, where $x_0$ is such that $\underline{K}_{x_0} = K$.

In Figure 2, we show the values of $V(x, K)$ and Lagrange multiplier $\Lambda^*$ as functions of $(x, K)$. It is confirmed that $V(x, K)$ increases as $x$ and $K$ increase, while $\Lambda^*$ increases as $x$ and $K$ decrease.

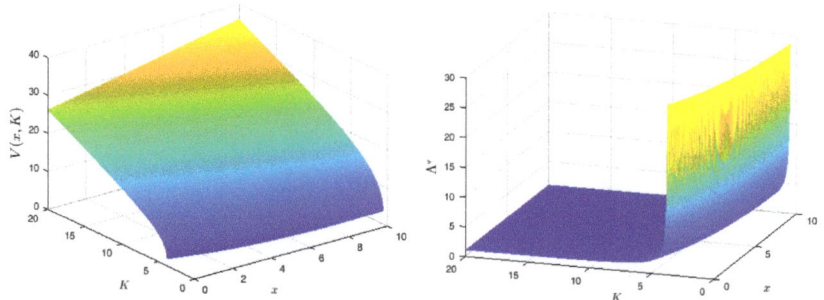

**Figure 2.** Plots of $V(x,K)$ (**left**); and the Lagrange multiplier $\Lambda^*$ (**right**) as functions of $x$ and $K$.

We now move to the case with transaction cost. First, we illustrate the results shown in Section 4. In Figure 3 (left), we plot the function $x \mapsto \zeta_\Lambda(x)$ for the values of $\Lambda = 1, \ldots, 9$. We also plot its maximum value attained at $a_\Lambda$ and the value attained at the corresponding optimal values $(c_1^\Lambda, c_2^\Lambda)$ with transaction cost $\delta = 0.05$. Note that, when $\Lambda = 1$, $a_\Lambda = c_1^\Lambda = 0$ and for the other values of $\Lambda$, $\zeta_\Lambda(c_1^\Lambda) = \zeta_\Lambda(c_2^\Lambda) < \zeta_\Lambda(a_\Lambda)$. In Figure 3 (right), we plot the optimal thresholds $a_\Lambda, c_1^\Lambda$ and $, c_2^\Lambda$ as function of $\Lambda$.

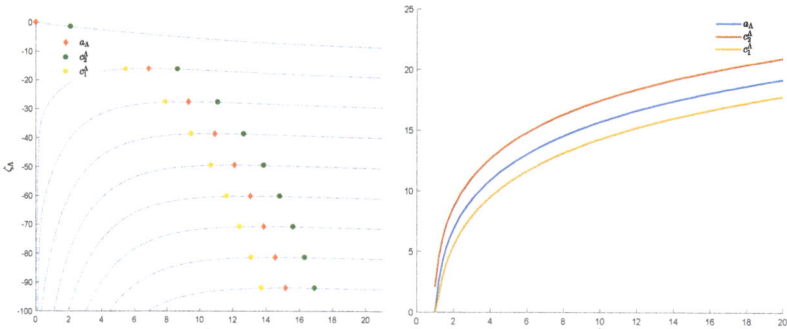

**Figure 3.** (**Left**) Plots of $x \mapsto \zeta_\Lambda(x)$ for $\Lambda = 1, \ldots, 9$ and the corresponding values of $a_\Lambda, c_1^\Lambda$ and $, c_2^\Lambda$ for $\delta = 0.05$. (**Right**) Plots of the functions $\Lambda \mapsto a_\Lambda, c_1^\Lambda$ and $, c_2^\Lambda$.

In Figure 4, we illustrate the findings of Section 5.2. This figure is analogous to Figure 1 but with transaction cost $\delta$ as above. It can be seen that the change in the function $V_\delta(x,K)$ is relatively very small, but the change in the optimal Lagrange multiplier $\Lambda^*$ is significant, being smaller in the case of transaction cost. A similar figure as Figure 2 in the case of transaction cost is omitted since both have the same shape.

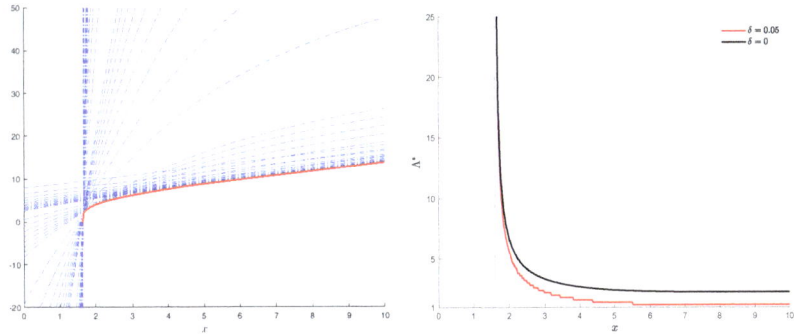

**Figure 4.** (**Left**) Plots of $x \mapsto V_\Lambda(x) + \Lambda K$ for $\Lambda = 1, 1.1, \ldots, 2, 3, \ldots, 10$, $20, \ldots, 100, 200, \ldots, 1000, 2000, \ldots, 10{,}000, 20{,}000$ (dotted) for the case $K = 2.7$. The minimum of $V_{\delta,\Lambda}(x) + \Lambda K$ over $\Lambda$ is plotted in solid bold-face red line. (**Right**) Plots of the Lagrange multipliers $\Lambda^*$ for $x > x_0$, where $x_0$ is such that $\underline{K}_{x_0} = K$ with $\delta = 0$ and $\delta = 0.05$.

## 7. Concluding Remarks

In this study, we proved that the optimal strategy for the bail-out dividend problem with fixed transaction costs is given by a reflected $(c_1, c_2)$-policy. We also characterized the optimal thresholds and gave a semi-explicit form for the value function in terms of the scale functions. In addition, we used the previous results to solve the constrained dividend maximization problem with the restriction that the expected present value of the capital injected is bounded by a given constant. The solution of the constrained problem can provide the insurance company with a guideline to maximize the profits of the shareholders taking into account the risk of bail-out losses.

It is a legitimate and interesting question whether the optimal strategy and the associated value function with transaction costs (i.e., $\delta > 0$) converge to the corresponding optimal strategy and its value function without transaction cost as $\delta \downarrow 0$. Although we conjecture that indeed this is the case, further investigation is needed.

Another interesting generalization would involve considering fixed and proportional costs for the capital injection as well. We conjecture that in this case the optimal strategy would consist in a double band strategy, that is, a band strategy similar to the dividend payment strategy given in this paper, and a band strategy for the capital injection, which consists in pushing the process to a positive level each time the surplus process tries to cross below 0. We leave this problem as an opportunity for future research.

**Author Contributions:** All authors equally contributed to this paper.

**Funding:** This research received no external funding.

**Acknowledgments:** The authors would like to thank the anonymous reviewers for their comments and suggestions, which helped to improve significantly the quality of this paper. They would also like to thank Florin Avram who encouraged them to contribute to this special issue of *Risks*.

**Conflicts of Interest:** The authors declare no conflict of interest.

## References

Avram, Florin, Zbigniew Palmowski, and Martin Pistorius. 2007. On the optimal dividend problem for a spectrally negative Lévy process. *The Annals of Applied Probability* 17: 156–80. [CrossRef]

Bayraktar, Erhan, Andreas E. Kyprianou, and Kazutoshi Yamazaki. 2014a. On optimal dividends in the dual problem. *ASTIN Bulletin* 43: 359–72. [CrossRef]

Bayraktar, Erhan, Andreas E. Kyprianou, and Kazutoshi Yamazaki. 2014b. Optimal dividends in the dual model under transaction costs. *Insurance: Mathematics and Economics* 54: 133–43. [CrossRef]

Bertoin, Jean. 1998. *Lévy Processes*. Cambridge: Cambridge University Press.

Chan, Terence, Andreas E. Kyprianou, and Mladen Savov. 2011. Smoothness of scale functions for spectrally negative Lévy processes. *Probability Theory and Related Fields* 150: 691–708. [CrossRef]

Hernández, Camilo, Mauricio Junca, and Harold Moreno-Franco. 2018. A time of ruin constrained optimal dividend problem for spectrally one-sided Lévy processes. *Insurance: Mathematics and Economics* 79: 57–68. [CrossRef]

Kuznetsov, Alexey, Andreas E. Kyprianou, and Victor Rivero. 2013. The theory of scale functions for spectrally negative Lévy processes. *Levy Matters II* 2061: 97–186.

Kyprianou, Andreas E. 2014. *Fluctuations of Lévy Processes with Applications*. Universitext. Berlin and Heidelberg: Springer.

Loeffen, Ronnie. 2008. On the optimality of the barrier strategy in de Finetti's problem for spectrally negative Lévy processes. *The Annals of Applied Probability* 18: 1669–80. [CrossRef]

Loeffen, Ronnie L. 2009. An optimal dividends problem with transaction costs for spectrally negative Lévy processes. *Insurance: Mathematics and Economics* 45: 41–48. [CrossRef]

Pistorius, Martijn R. 2004. On exit and ergodicity of the spectrally one-sided Lévy process reflected at its infimum. *Journal of Theoretical Probability* 17: 183–220. [CrossRef]

Protter, Philip. 2005. *Stochastic Integration and Differential Equations*, 2nd ed. Berlin: Springer.

Surya, Budhig A. 2008. Evaluating scale functions of spectrally negative Lévy processes. *Journal of Applied Probability* 45: 135–49. [CrossRef]

© 2019 by the authors. Licensee MDPI, Basel, Switzerland. This article is an open access article distributed under the terms and conditions of the Creative Commons Attribution (CC BY) license (http://creativecommons.org/licenses/by/4.0/).

Article

# The Løkka–Zervos Alternative for a Cramér–Lundberg Process with Exponential Jumps

**Florin Avram [1,*], Dan Goreac [2,3] and Jean-François Renaud [4]**

1. Laboratoire de Mathématiques Appliquées, Université de Pau, 64012 Pau, France
2. School of Mathematics and Statistics, Shandong University, Weihai 264209, China; dan.goreac@u-pem.fr
3. LAMA, Univ Gustave Eiffel, UPEM, Univ Paris Est Creteil, CNRS, F-77447 Marne-la-Vallée, France
4. Département de Mathématiques, Université du Québec à Montréal (UQAM), Montréal, QC H2X 3Y7, Canada; renaud.jf@uqam.ca
* Correspondence: florin.avram@orange.fr or florin.avram@univ-pau.fr

Received: 28 October 2019; Accepted: 2 December 2019; Published: 10 December 2019

**Abstract:** In this paper, we study a stochastic control problem faced by an insurance company allowed to pay out dividends and make capital injections. As in (Løkka and Zervos (2008); Lindensjö and Lindskog (2019)), for a Brownian motion risk process, and in Zhu and Yang (2016), for diffusion processes, we will show that the so-called Løkka–Zervos alternative also holds true in the case of a Cramér–Lundberg risk process with exponential claims. More specifically, we show that: if the cost of capital injections is *low*, then according to a double-barrier strategy, it is optimal to pay dividends and inject capital, meaning ruin never occurs; and if the cost of capital injections is *high*, then according to a single-barrier strategy, it is optimal to pay dividends and never inject capital, meaning ruin occurs at the first passage below zero.

**Keywords:** stochastic control; optimal dividends; capital injections; bankruptcy; barrier strategies; reflection and absorption; scale functions

---

## 1. Introduction

Risk theory initially revolved around minimizing the probability of ruin. However, shareholders are more interested in maximizing the value of the company than minimizing risks. Therefore, (de Finetti 1957) suggested finding the optimal dividend policies which maximize the expected value of the sum of discounted future dividend payments up to the time of ruin; see also (Miller and Modigliani 1961). Another interesting objective, as suggested by (Shreve et al. 1984), is to maximize the expected discounted cumulative dividends while redressing the reserves by injecting capital each time it becomes necessary.

This note is motivated by subsequent results obtained by (Løkka and Zervos 2008; Lindensjö and Lindskog 2019) for a Brownian motion with drift, and by (Zhu and Yang 2016) for diffusions. Their results state that, depending on the size of transaction costs, one of the following strategies is optimal:

1. if the cost of capital injections is *low*, then according to a double-barrier strategy, it is optimal to pay dividends and to inject capital, meaning ruin never occurs;
2. if the cost of capital injections is *high*, then according to a single-barrier strategy, it is optimal to pay dividends and never inject capital, meaning ruin occurs at the first passage below zero.

### 1.1. The Model

In what follows, we will use the following notation: the law of a Markov process $X$ when starting from $X_0 = x$ will be denoted by $\mathbb{P}_x$, and the corresponding expectation by $\mathbb{E}_x$. We write $\mathbb{P}$ and $\mathbb{E}$ when $x = 0$.

To fix ideas, let us start with the Cramér–Lundberg risk model for $t \geq 0$ (see, for example, Dufresne and Gerber 1991; Albrecher and Asmussen 2010):

$$X_t = x + ct - S_t, \quad \text{where } S_t = \sum_{i=1}^{N_t} C_i. \tag{1}$$

Here, $x \geq 0$ is the initial surplus, $c \geq 0$ is the linear premium rate, and $\{C_i, i = 1, 2, \ldots\}$ are independent and identically distributed random variables, with distribution function $F$ and mean $m_1 = \int_0^\infty z F(dz)$ representing non-negative jumps/claims. The inter-arrival times between these jumps are independent and exponentially distributed with mean $1/\lambda$, and $N_t$ denotes the time-$t$ value of the associated Poisson process counting the arrivals of claims on the interval $[0, t]$. We will assume the positive profit condition $p := c - \lambda m_1 > 0$.

The process given in (1) is a particular case of a spectrally negative Lévy process (SNLP), that is, a Lévy process without positive jumps, where in this case there is also a finite mean. More precisely, such a process is defined by adding a Brownian perturbation to (1), and by assuming that $S_t$ is a subordinator with a $\sigma$-finite Lévy measure $\Pi(dx)$, having possibly infinite activity near the origin, that is, $\Pi(0, \infty) = \infty$. For a SNLP, the positive profit condition becomes $p = c - \int_0^\infty x \Pi(dx) > 0$. Note that for the SNLP given in (1), we have $\Pi(dx) = \lambda F(dx)$ so $\Pi(0, \infty) = \lambda$. See, for example, (Bertoin 1998) for more details.

The main result of our paper assumes that the claim sizes/jumps are exponentially distributed with mean $1/\mu$, that is, that $F(z) = 1 - e^{-\mu z}$ when $z > 0$. However, as most of our intermediate results hold for a general SNLP, they will be stated in this more general context. Unfortunately, one key fact below holds only for a Cramér–Lundberg process with exponential jumps. Consequently, in the general SNLP case, the Løkka–Zervos alternative is still an open problem.

Recall that a SNLP $X$ is characterized by its Laplace exponent defined by $\psi(\theta) = \ln \mathbb{E}\left[e^{\theta X_1}\right]$. For the Cramér-Lundberg process $X$ given in (1), we have

$$\psi(\theta) = c\theta + \int_0^\infty \left(e^{-\theta z} - 1\right) \lambda F(dz)$$

and, in the case of exponential jumps, we further have

$$\psi(\theta) = \theta \left(c - \frac{\lambda}{\mu + \theta}\right). \tag{2}$$

### 1.2. The Problem

For the stochastic control problem considered in this paper, an admissible strategy is represented by a pair $(C, D)$ composed of a non-decreasing, left-continuous, and adapted stochastic process $D = \{D_t, t \geq 0\}$ and $C = \{C_t, t \geq 0\}$, where $D_t$ represents the cumulative amount of dividends paid up to time $t$, while $C_t$ represents the cumulative amount of capital injections made up to time $t$. We assume $D_0 = 0$ and $C_0 = 0$. For a given strategy $(C, D)$, the corresponding controlled surplus process $U = \{U_t, t \geq 0\}$ is defined by $U_t = X_t - D_t + C_t$. Define also $\tau = \inf\{t > 0 : U_t < 0\}$.

For a given initial surplus $x \geq 0$, let $\mathcal{A}(x)$ be the corresponding set of admissible strategies. Also, let $q > 0$ be the discounting rate and let $k > 1$ be the proportional cost of injecting capital. The objective is to maximize the value of a strategy using the following objective function:

$$J(x, C, D) = \mathbb{E}_x \left[\int_0^\tau e^{-qt} \left(dD_t - k dC_t\right)\right], \tag{3}$$

that is, the goal is to find the optimal value function

$$V_k(x) = \sup_{(C,D) \in \mathcal{A}(x)} J(x, C, D).$$

For a general Markov process $X$, our problem amounts to solving (in a viscosity sense) the following Hamilton–Jacobi–Bellman (HJB) equation:

$$\begin{cases} \max\{(\mathcal{L}-q)V(x), 1-V'(x), V'(x)-k, -V(x)\} \leq 0, & x \geq 0 \\ \max\{(\mathcal{L}-q)V(x), V'(x)-k, -V(x)\} \leq 0, & x < 0 \end{cases} \quad (4)$$

where $\mathcal{L}$ is the infinitesimal generator associated with the underlying uncontrolled process $X$ (see also (Zhu and Yang 2016, sct. 3.6) for the case of diffusions). For the Cramér-Lundberg process with exponential jumps, the operator is

$$\mathcal{L}V(x) = cV'(x) + \lambda\mu \int_0^\infty (V(x-z) - V(x))e^{-\mu z} dz. \quad (5)$$

The second part of (4) is associated to the possibility of modifying the surplus by a lump sum dividend payment (see (6) below), and the third part to capital injections.

In the cases already studied, the Løkka–Zervos alternative reduces to the following dilemma: shall we declare bankruptcy at level 0, or shall we use capital injections to maintain the surplus positive?

The classical problems studied by (de Finetti 1957; Shreve et al. 1984) are revisited in Section 2 and new results are obtained. In Section 3, we prove that the Løkka–Zervos alternative holds for a Cramér–Lundberg model with exponential jumps.

## 2. The Classical Dividend Problems for SNLPs

In this section, we review de Finetti's, as well as Shreve, Lehoczky, and Gaver's optimal dividend problems for general spectrally negative Lévy processes. As is well-known, the value functions can be expressed in terms of scale functions (see, for example, Avram et al. 2004, 2019; Bertoin 1998; Kyprianou 2014).

### 2.1. De Finetti's Problem

De Finetti's problem corresponds to the case where $k = \infty$, implying that $C \equiv 0$, that is, capital injections cannot be profitable. In this case, the controlled process is ruined as soon as it goes below zero. For this problem, the optimal value function will be denoted by $V^{dF}$.

It is well-known that for this problem, constant barrier strategies are very important. For $b \geq 0$, the (horizontal) barrier strategy at level $b$ is the strategy with a cumulative amount of dividends paid until time $t > 0$ given by $D_t^b = \left(\sup_{0 < s \leq t} X_s - b\right)_+$. If $X_0 = x > b$, then $D_{0+}^b = x - b$ (a lump sum payment is made). For such a strategy, the value function is such that $J(x, 0, D^b) = V^b(x)$, where

$$V^b(x) := \mathbb{E}_x\left[\int_0^{\tau^b} e^{-qt} dD_t^b\right],$$

where $\tau^b$ is the time of ruin for the controlled process $U_t^b = X_t - D_t^b$. In this case, $\mathbb{P}_x\left(\tau^b < \infty\right) = 1$.

It is well-known that, for a SNLP (see, for example, Avram et al. 2007),

$$V^b(x) = \begin{cases} \frac{W_q(x)}{W_q'(b)}, & x \leq b, \\ x - b + \frac{W_q(b)}{W_q'(b)}, & x > b, \end{cases} \quad (6)$$

where the $q$-scale function $W_q$ (Bertoin 1998) is given through its Laplace transform:

$$\int_0^\infty e^{-\theta x} W_q(x) dx = \frac{1}{\psi(\theta) - q}, \quad (7)$$

for all $\theta > \Phi(q) = \sup\{s \geq 0 : \psi(s) = q\}$.

It is known (see Theorem 1.1 in (Loeffen and Renaud 2010)) that if the tail of the jump distribution is log-convex, then an optimal dividend policy is formed by the barrier strategy at level $b^*$, where $b^*$ is the last maximum of the *barrier function*

$$H^{dF}(b) = \frac{1}{W'_q(b)}, \quad b > 0. \tag{8}$$

In this case, the optimal value function $V^{dF}$ is given by $V^{b^*}$.

Consequently, for a Cramér–Lundberg risk process with exponentially distributed claims, the optimal value function $V^{dF}$ is equal to the value function of a barrier strategy. More precisely, for $X$ given in (1) with exponential jumps, we have

$$W_q(x) = \frac{A_+ e^{\rho_+ x} - A_- e^{\rho_- x}}{c(\rho_+ - \rho_-)}, \quad x \geq 0,$$

where

$$\rho_\pm = \frac{1}{2c}\left(-(\mu c - \lambda - q) \pm \sqrt{(\mu c - \lambda - q)^2 + 4\mu q c}\right)$$

are such that $\rho_- \leq 0 \leq \rho_+ = \Phi(q)$, and where $A_\pm = \mu + \rho_\pm$. In this case, the barrier function $H^{dF}$ has a unique maximum at level

$$b^* = \begin{cases} \frac{1}{\rho_+ - \rho_-} \log\left(\frac{\rho_-^2(\mu+\rho_-)}{\rho_+^2(\mu+\rho_+)}\right) & \text{if } (q+\lambda)^2 - c\lambda\mu < 0, \\ 0 & \text{if } (q+\lambda)^2 - c\lambda\mu \geq 0, \end{cases}$$

and we have $V^{dF} = V^{b^*}$.

### 2.2. Shreve, Lehoczky, and Gaver's Problem

In Shreve, Lehoczky, and Gaver's problem, capital is injected as soon as it is necessary—that is, to keep the controlled process non-negative, so $\tau = \infty$. In this case, the controlled process is never ruined—zero acts as a (lower) reflecting barrier. For this problem, the optimal value function will be denoted by $V_k^{SLG}$.

It is well-known that for this problem, double-barrier strategies play an important role. For $b \geq 0$, a double-barrier strategy with an upper barrier at level $b$ is such that $0 \leq U_t^{0,b} = X_t - D_t^b + C_t^0 \leq b$ for all $t \geq 0$. As capital injections are now considered, the process $D^b$ here is different from the one in de Finetti's problem; see (Avram et al. 2007) for details. For such a strategy, the value function is such that $J(x, C^0, D^b) = V_k^{0,b}(x)$, where

$$V_k^{0,b}(x) := \mathbb{E}_x\left[\int_0^\infty e^{-qt}\left(dD_t^b - k dC_t^0\right)\right],$$

and the optimal value function is such that $V_k^{SLG}(x) = \sup_{b \geq 0} V_k^{0,b}(x)$.

It is well-known that for a SNLP,

$$V_k^{0,b}(x) = \begin{cases} k\left(\overline{Z}_q(x) + \frac{p}{q}\right) + Z_q(x) H_k^{SLG}(b), & x \leq b, \\ x - b + V_k^{0,b}(b), & x > b, \end{cases} \tag{9}$$

where

$$Z_q(x) = 1 + q\int_0^x W_q(y)dy \quad \text{and} \quad \overline{Z}_q(x) = \int_0^x Z_q(y)dy, \tag{10}$$

and where, for $x \geq 0$, the *barrier function* is defined by

$$H_k^{SLG}(b) = \frac{1 - kZ_q(b)}{qW_q(b)}, \quad b > 0. \tag{11}$$

The next proposition—namely, Proposition 1—contains new results, as well as results taken from Lemma 2 of (Avram et al. 2007). In particular, we provide a new relationship (see (13)) between the value functions of de Finetti's and Shreve, Lehoczky, and Gaver's problems.

The main object in this next proposition is the function $k_f \colon [0, \infty) \to [k_0, \infty)$ defined by

$$k_f(b) := \frac{W_q'(b)}{Z_q(b)W_q'(b) - qW_q^2(b)}, \tag{12}$$

where

$$k_0 := k_f(0_+) = \frac{W_q'(0_+)}{W_q'(0_+) - qW_q^2(0_+))} = \begin{cases} 1, & \text{if } X \text{ is of unbounded variation,} \\ 1 + \frac{q}{\Pi(0, \infty)}, & \text{if } X \text{ is of bounded variation.} \end{cases}$$

The function $k_f$ is increasing, thanks to (Avram et al. 2004, Theorem 1)[1]. Indeed, it is known that

$$\mathbb{E}_x\left[e^{-q\tau^b}\right] = Z_q(x) - q\frac{W_q(b)}{W_q'(b)}W_q(x),$$

so $k_f(b) = \frac{1}{\mathbb{E}_b[e^{-q\tau^b}]}$. The statement follows from the fact that the map $b \mapsto \mathbb{E}_b\left[e^{-q\tau^b}\right]$ is decreasing.

The monotonicity allows us to re-parametrize the problem in terms of the optimal barrier, $b_k$ associated to a fixed cost, $k$.

**Proposition 1.** *Assume X is a SNLP. We have the following results:*

(a) *For fixed $x$ and $b$, the function $k \mapsto V_k^{SLG}(x)$ is non-increasing.*
(b) *For $k = k_f(b)$, the value function defined in (9) can be written as follows:*

$$V_{k_f(b)}^{0,b}(x) = k_f(b)\left[\overline{Z}_q(x) + \frac{p}{q} - Z_q(x)V^b(b)\right]$$

$$= k_f(b)\left[Z_q^{(1)}(x) + Z_q(x)\left(\frac{p}{q} - V^b(b)\right)\right]. \tag{13}$$

*where $V^b(x)$ is defined in (6) and*

$$Z_q^{(1)}(x) := \int_0^x (Z_q(y) - pW_q(y))\,dy. \tag{14}$$

(c) *For fixed $k$, the barrier function $H_k^{SLG}$ has a unique point of maximum $b_k \geq 0$. It is decreasing, and thus $b_k = 0$ if, and only if $k \in (1, k_0]$. Finally, if $b_k > 0$, then $k = k_f(b_k)$.*

**Remark 1.** *Note that*

$$Z_q^{(1)}(x) = \left.\frac{\partial Z_{q,\theta}(x)}{\partial \theta}\right|_{\theta=0}$$

---

[1] Some papers refer to this as the log-convexity of $Z_q(x)$.

where $Z_{q,\theta}(x) = (\psi(\theta) - q)\int_0^\infty e^{-\theta y} W_q(x+y)dy$. The function $Z_{q,\theta}$ was introduced simultaneously in (Avram et al. 2015; Ivanovs and Palmowski 2012) (but was already present implicitly in (Avram et al. 2004, Theorem 1), where it was presented as an Esscher transform of $Z_q(x)$). It was first used as a generating function for Gerber–Shiu penalty functions induced by polynomial rewards $1, x, x^2$, which were denoted respectively by $Z_q, Z_q^{(1)}, Z_q^{(2)}, \ldots$, and started also being used intensively in exponential Parisian ruin problems following the work of (Albrecher et al. 2016). See (Avram et al. 2019) for more information.

**Proof.** (a) The result follows from the fact that $k \mapsto V_k^{0,b}(x)$ is decreasing (by definition) and because $V_k^{SLG}(x)$ is obtained by a maximization of $V_k^{0,b}(x)$ over all barrier levels $b$ (chosen independently of $k$).
(b) Recalling (9), we need to show that

$$-H_{k_f(b)}^{SLG}(b) = k_f(b)V^b(b). \tag{15}$$

Indeed, it is easy to check that the equality

$$\frac{kZ_q(b) - 1}{qW_q(b)} = k\frac{W_q(b)}{W_q'(b)}$$

holds for $k = k_f(b)$.

(c) It is well-known (see Avram et al. 2007, Lemma 2) that $H_k^{SLG}$ is an increasing-decreasing function in $b$, with a unique maximum $b_k \geq 0$. For the sake of completeness, let us reproduce this proof. The derivative of the barrier function (11) satisfies

$$q\frac{H'W_q^2}{W_q'}(b) = f(b) := k\frac{\Delta_q^{(ZW)}(b)}{W_q'(b)} - 1 = k\,\mathbb{E}_b[e^{-q\tau^b}] - 1 = \frac{k}{k_f(b)} - 1, \tag{16}$$

where

$$\Delta_q^{(ZW)}(b) := Z^{(q)}(b)W_q'(b) - \left(Z^{(q)}\right)'(b)W_q(b). \tag{17}$$

Therefore, the sign of the derivative of the barrier function (11) coincides with that of $f$. Clearly, the latter function $f$ is decreasing in $b$ from $\lim_{b \to 0} f(b) = \frac{k}{k_0} - 1$ to $-1$. □

**Remark 2.** *In conclusion, if $k \leq k_0$, then the barrier function $H_k^{SLG}$ reaches its unique maximum at $b_k = 0$ and, if $k > k_0$, then $b_k$ is such that $k = k_f(b_k)$.*

**Remark 3.** *The previous proposition suggests the following new (and short) proof of the Løkka–Zervos alternative in the Brownian motion case. It is easy to verify that*

$$Z_q^{(1)}(x) + Z_q(x)\left(\frac{p}{q} - V^b(b)\right) = Z_q^{(1)}(x) = \frac{\sigma^2}{2}W_q(b),$$

*which yields $V_{k_f(b^*)}^{0,b^*}(x) = V^{dF}(x)$, where $b^*$ denotes the optimal barrier level in de Finetti's problem, as defined in Section 2.1. Then, use the monotonicity of $V_k^{SLG}(x)$ in $k$.*

Similar computations below will establish the Løkka–Zervos alternative in the Cramér-Lundberg case with exponential jumps.

## 3. The Løkka–Zervos Alternative for a Cramér–Lundberg Model with Exponential Jumps

Here is our main result.

**Theorem 1.** *For a Cramér–Lundberg process with exponentially distributed jumps, the Løkka–Zervos alternative holds with two regimes separated by the threshold $k = k_f(b^*)$—that is, for all $x \geq 0$,*

$$V_k^{SLG}(x) \geq V^{dF}(x) \quad \text{if, and only if} \quad k \leq k_f(b^*).$$

**Proof.** By Proposition 1, we know that for fixed $x$ and $b$, the function $k \mapsto V_k^{SLG}(x) = \sup_{b \geq 0} V_k^{0,b}(x)$ is non-increasing. One deduces that, for all $x \geq 0$,

$$V_k^{SLG}(x) \geq V_{k_f(b^*)}^{0,b^*}(x) \quad \text{if and only if} \quad k \leq k_f(b^*).$$

Therefore, the Løkka–Zervos alternative follows from Lemma 1, given below, where it is proved that

$$V_{k_f(b^*)}^{0,b^*}(x) = V^{dF}(x). \tag{18}$$

□

Recall that we assume $q > 0$.

**Lemma 1.** *For a Cramér–Lundberg process with exponentially distributed jumps, we have:*

(a) $Z_q(x) + \mu Z_q^{(1)}(x) = c W_q(x)$, for all $x > 0$;

(b) $\dfrac{1}{k_f(b^*)} = \dfrac{c W_q'(b^*)}{\mu}$;

(c) $V_{k_f(b^*)}^{0,b^*}(x) = V^{dF}(x)$, for all $x > 0$.

**Proof.** (a) This can be verified by taking Laplace transforms and using (2). Letting $\widehat{\overline{F}}(s)$ denote the Laplace transform of the tail distribution function $\overline{F}(z) = 1 - F(z)$, it amounts to checking

$$\frac{c}{\psi(s) - q} = \frac{c - \lambda \widehat{\overline{F}}(s)}{\psi(s) - q} + \mu \frac{c - \lambda \widehat{\overline{F}}(s) - p}{s(\psi(s) - q)} \iff \lambda \widehat{\overline{F}}(s) = \mu \lambda \frac{\mu^{-1} - \widehat{\overline{F}}(s)}{s},$$

which holds true for exponential jumps.

(b) Manipulating the Kolmogorov IDE for $Z_q$, we can reduce it to

$$c Z_q''(x) + (c\mu - \lambda - q) Z_q'(x) - q \mu Z_q(x) = q W_q'(x) \left( c + (c\mu - \lambda - q) \frac{W_q(x)}{W_q'(x)} - \mu \frac{Z_q(x)}{W_q'(x)} \right) = 0.$$

At $x = b^*$, using the fact that

$$V^{dF}(b^*) = \frac{p}{q} - \frac{1}{\mu} \tag{19}$$

(see, for example, Equation (5.24) in Gerber et al. 2006) together with simple algebraic manipulations yields

$$\frac{Z_q(b^*)}{W_q'(b^*)} = q \left( \frac{p}{q} - \frac{1}{\mu} \right)^2 + \frac{c}{\mu}.$$

This is equivalent to the result.

(c) From (13) and part (a), we get

$$V_{k_f(b^*)}^{0,b^*}(x) = \frac{k_f(b^*)}{\mu} \left( \mu Z_q^{(1)}(x) + Z_q(x) \right) = \frac{c k_f(b^*)}{\mu} W_q(x).$$

Then the result follows from (b), as well as the fact that $V^{dF} = V^{b^*}$. □

## 4. Conclusions and Conjecture

We believe it is important to study the Løkka–Zervos alternative for spectrally negative Lévy processes and for spectrally negative additive Markov processes, both practically and methodologically. We conjecture that in those more general cases, more than two regimes will be involved, giving rise to *Løkka–Zervos alternatives*.

**Author Contributions:** The authors have contributed equally to this work.

**Funding:** J.-F.R. is acknowledging financial support from the Natural Sciences and Engineering Research Council of Canada (NSERC).

**Conflicts of Interest:** The authors declare no conflict of interest.

## References

Albrecher, Hansjörg, and Sören Asmussen. 2010. *Ruin Probabilities*. Singapore: World Scientific, vol. 14.

Albrecher, Hansjörg, Jevgenijs Ivanovs, and Xiaowen Zhou. 2016. Exit identities for Lévy processes observed at Poisson arrival times. *Bernoulli* 22: 1364–82. [CrossRef]

Avram, Florin, Andreas Kyprianou, and Martijn Pistorius. 2004. Exit problems for spectrally negative Lévy processes and applications to (Canadized) Russian options. *The Annals of Applied Probability* 14: 215–38.

Avram, Florin, Danijel Grahovac, and Ceren Vardar-Acar. 2019. The W, Z scale functions kit for first passage problems of spectrally negative Lévy processes, and applications to the optimization of dividends. *arXiv* arXiv:1706.06841.

Avram, Florin, Zbigniew Palmowski, and Martijn R. Pistorius. 2007. On the optimal dividend problem for a spectrally negative Lévy process. *The Annals of Applied Probability* 17: 156–80. [CrossRef]

Avram, Florin, Zbigniew Palmowski, and Martijn R. Pistorius. 2015. On Gerber–Shiu functions and optimal dividend distribution for a Lévy risk process in the presence of a penalty function. *The Annals of Applied Probability* 25: 1868–935. [CrossRef]

Bertoin, Jean. 1998. *Lévy Processes*. Cambridge: Cambridge University Press, vol. 121.

de Finetti, Bruno. 1957. Su un'impostazione alternativa della teoria collettiva del rischio. Paper presented at Transactions of the XVth International Congress of Actuaries, Sydney, Australia, October 21–27, vol. 2, pp. 433–43.

Dufresne, Francois, and Hans U. Gerber. 1991. Risk theory for the compound Poisson process that is perturbed by diffusion. *Insurance: Mathematics and Economics* 10: 51–59. [CrossRef]

Gerber, Hans U., Elias S. W. Shiu, and Nathaniel Smith. 2006. Maximizing dividends without bankruptcy. *ASTIN Bulletin* 36: 5–23 [CrossRef]

Ivanovs, Jevgenijs, and Zbigniew Palmowski. 2012. Occupation densities in solving exit problems for Markov additive processes and their reflections. *Stochastic Processes and Their Applications* 122: 3342–60. [CrossRef]

Kyprianou, Andreas. 2014. *Fluctuations of Lévy Processes with Applications: Introductory Lectures*. Berlin: Springer.

Lindensjö, Kristoffer, and Filip Lindskog. 2019. Optimal dividends and capital injection under dividend restrictions. *arXiv* arXiv:1902.06294.

Loeffen, Ronnie L., and Jean-François Renaud. 2010. De Finetti's optimal dividends problem with an affine penalty function at ruin. *Insurance: Mathematics and Economics* 46: 98–108. [CrossRef]

Løkka, Arne, and Mihail Zervos. 2008. Optimal dividend and issuance of equity policies in the presence of proportional costs. *Insurance: Mathematics and Economics* 42: 954–61. [CrossRef]

Miller, Merton H., and Franco Modigliani. 1961. Dividend policy, growth, and the valuation of shares. *Journal of Business* 34: 411–33. [CrossRef]

Shreve, Steven E., John P. Lehoczky, and Donald P. Gaver. 1984. Optimal consumption for general diffusions with absorbing and reflecting barriers. *SIAM Journal on Control and Optimization* 22: 55–75. [CrossRef]

Zhu, Jinxia, and Hailiang Yang. 2016. Optimal capital injection and dividend distribution for growth restricted diffusion models with bankruptcy. *Insurance: Mathematics and Economics* 70: 259–71. [CrossRef]

 © 2019 by the authors. Licensee MDPI, Basel, Switzerland. This article is an open access article distributed under the terms and conditions of the Creative Commons Attribution (CC BY) license (http://creativecommons.org/licenses/by/4.0/).

Article
# Potential Densities for Taxed Spectrally Negative Lévy Risk Processes

Wenyuan Wang [1] and Xiaowen Zhou [2,*]

[1] School of Mathematical Sciences, Xiamen University, Xiamen 361005, China
[2] Department of Mathematics and Statistics, Concordia University, Montreal, QC H3G 1M8, Canada
* Correspondence: xiaowen.zhou@concordia.ca; Tel.: +1-514-8482424 (ext. 3220)

Received: 29 May 2019; Accepted: 17 July 2019; Published: 2 August 2019

**Abstract:** This paper revisits the spectrally negative Lévy risk process embedded with the general tax structure introduced in Kyprianou and Zhou (2009). A joint Laplace transform is found concerning the first down-crossing time below level 0. The potential density is also obtained for the taxed Lévy risk process killed upon leaving $[0, b]$. The results are expressed using scale functions.

**Keywords:** spectrally negative Lévy process; general tax structure; first crossing time; joint Laplace transform; potential measure

## 1. Introduction

The study of loss-carry-forward tax was initiated in Albrecher and Hipp (2007) under the framework of the classical compound Poisson risk model, in which the taxation is imposed at a fixed rate as long as the surplus process of the company stays at the running supremum. In particular, criteria were obtained for the optimal taxation level that maximizes the expected (discounted) accumulated tax payments, and an interesting simple relationship was recovered between the ruin probabilities for scenarios with and without tax. Later, the underlying risk process was generalized to the spectrally negative Lévy process in Albrecher et al. (2008), followed by further generalizations in terms of the tax structures in Kyprianou and Zhou (2009) and Kyprianou and Ott (2012), where new identities were derived for the two-sided exit problem and a generalized version of the Gerber-Shiu function as well as the the net present value of tax payments until ruin. On the other hand, Wei (2009) studied the asymptotic formulas of the ruin probability for the classical compound Poisson risk process with constant credit interest rate and surplus-dependent tax rate. In the mean time, a similar problem for spectrally negative Lévy risk process with periodic tax was investigated in Hao and Tang (2009). In addition, Renaud (2009) obtained the explicit expressions for arbitrary moments of the accumulated discounted tax payments. Moreover, the Gerber-Shiu functions were presented in Wei et al. (2010) and Cheung and Landriault (2012) for Markov-modulated risk models with constant tax rate and for classical compound Poisson risk model with surplus-dependent premium and tax rate, respectively. Several two-sided exit problems for the time-homogeneous diffusion risk processes with surplus-dependent tax rate were investigated in Li et al. (2013). Concerning the related optimization problem, identification of the optimal taxation strategy that maximizes the expected (discounted) accumulated tax pay-out until ruin was addressed in Wang and Hu (2012). More recently, the two-sided exit problem in terms of the linear draw-down time for a spectrally negative Lévy risk process with loss-carry-forward tax was solved in Avram et al. (2017). Another periodic taxation different from that in Hao and Tang (2009) was introduced in Zhang et al. (2017), where the Gerber-Shiu function for the spectrally negative Lévy risk process was studied.

In the present paper, we continue the study of the general tax structure introduced in Kyprianou and Zhou (2009) for the spectrally negative Lévy risk process. Our main goal is to identify the potential density of the taxed Lévy risk process killed upon leaving interval $[0, b]$. To our best knowledge, such a potential density has not been obtained in the literature for taxed processes. In addition, we also find a joint Laplace transform concerning the first down-crossing time of level 0. To this end, we adopt an excursion theory approach, and the aforementioned results are expressed using scale functions for spectrally negative Lévy processes.

This paper is arranged as follows. In Section 2, some preliminary results concerning the spectrally negative Lévy processes and the taxed Lévy risk models with general tax structure are reviewed. The main results on the taxed spectrally negative Lévy risk processes with proofs are included in Section 3.

## 2. Mathematical Presentation of the Problem

Under probability laws $\{\mathbb{P}_x; x \in \mathbb{R}\}$ and with natural filtration $\{\mathcal{F}_t; t \geq 0\}$, let $X = \{X(t); t \geq 0\}$ be a spectrally negative Lévy process with the usual exclusion of pure increasing linear drift and the negative of a subordinator. Denote by $\{\overline{X}(t) := \sup_{0 \leq s \leq t} X(s); t \geq 0\}$ the running supremum process of $X$. We assume that, in the case of no control, the risk process evolves according to the law of $\{X(t); t \geq 0\}$.

Define a measurable function $\gamma : [0, +\infty) \to [0, 1)$ satisfying

$$\int_0^{+\infty} (1 - \gamma(z)) \, dz = +\infty, \tag{1}$$

and

$$\overline{\gamma}_x(z) := x + \int_x^z (1 - \gamma(w)) dw$$

for $z \geq x \geq 0$. Intuitively, when an insurer is in a profitable situation at time $t$, the proportion of the insurer's income that is paid out as tax at time $t$ is equal to $\gamma(\overline{X}(t))$, where the insurer is in a profitable situation at time $t$ if $X(t) = \overline{X}(t)$. Therefore, the cumulative tax until time $t$ is given by

$$\int_0^t \gamma(\overline{X}(s)) \, d\overline{X}(s),$$

which is well defined because $\{\overline{X}(t); t \geq 0\}$ is a process of bounded variation (see also, Kyprianou and Ott 2012), and the controlled aggregate surplus process is also well defined by

$$U(t) = X(t) - \int_0^t \gamma(\overline{X}(s)) \, d\overline{X}(s). \tag{2}$$

The taxed risk process given by Equation (2) was first investigated in Kyprianou and Zhou (2009) for $[0, 1)$-valued function $\gamma$ satisfying Equation (1).

The objective of the present paper is to find the expressions for the joint Laplace transform of the ruin time and the position at ruin, conditioning on the event of ruin occurring before the up-crossing time of $b$

$$\mathbb{E}_x \left( e^{-q\sigma_0^- + \theta U(\sigma_0^-)}; \sigma_0^- < \sigma_b^+ \right) \tag{3}$$

and the $q$-potential measuring up to the exit time of $[0, b]$

$$\int_0^\infty e^{-qt} \mathbb{P}_x(U(t) \in du, t < \sigma_b^+ \wedge \sigma_0^-) dt \tag{4}$$

with $b \geq x \geq 0$ and $q, \theta > 0$, where $\sigma_0^-$ and $\sigma_b^+$ are the first down-crossing time of level 0 (or, ruin time) and the first up-crossing time of level $b$ defined, respectively, by

$$\sigma_0^- := \inf\{t \geq 0 : U(t) < 0\} \text{ and } \sigma_b^+ := \inf\{t \geq 0 : U(t) > b\}$$

with the usual convention $\inf \emptyset := \infty$.

The $q$-potential measure defined in Equation (4) plays an essential role in the theory of temporally homogeneous Markov processes. For example, the $q$-potential measure has turned out to be helpful in solving the occupation time involved problems (see Landriault et al. 2011 and Li and Zhou 2014), and the fluctuation problems for Lévy processes that are observed at Poisson arrival epochs (see Albrecher et al. 2016). In addition, the fluctuation quantity of Equation (3) involving the first passage times has been extensively investigated for various kinds of stochastic processes. Because $\sigma_0^-$ can be viewed as the ruin time of a risk process, the study of Equation (3) finds interesting applications in risk theory. Identifying the joint distribution of the ruin time and the deficit at the ruin time, which is the key topic of the well-known Gerber-Shiu function theory, has attracted broad attention in the community of actuarial sciences (see Gerber and Shiu 1998 and the special issue launched in the journal "Insurance: Mathematics and Economics" themed on Gerber-Shiu functions in 2010).

To keep our paper self-contained, we briefly introduce some basic facts of the spectrally negative Lévy process. Let the Laplace exponent of $X$ be defined by

$$\psi(\theta) := \log \mathbb{E}_x[e^{\theta(X_1 - x)}],$$

which is known to be finite for all $\theta \in [0, \infty)$, and is strictly convex and infinitely differentiable. As defined in Bertoin (1996), for each $q \geq 0$ the scale function $W^{(q)} : [0, \infty) \to [0, \infty)$ is the unique strictly increasing and continuous function with Laplace transform

$$\int_0^\infty e^{-\theta x} W^{(q)}(x) dx = \frac{1}{\psi(\theta) - q}, \quad \theta > \Phi(q),$$

where $\Phi(q)$ is the largest solution of the equation $\psi(\theta) = q$. For convenience, we extend the domain of $W^{(q)}$ to the whole real line by setting $W^{(q)}(x) = 0$ for all $x < 0$. In particular, write $W = W^{(0)}$ for simplicity. When $X$ has sample paths of unbounded variation, or when $X$ has sample paths of bounded variation and the Lévy measure has no atoms, then the scale function $W^{(q)}$ is continuously differentiable over $(0, \infty)$. Interested readers are referred to Chan et al. (2011) for more detailed discussions on the smoothness of scale functions. In this paper, we only need the scale function $W^{(q)}$ to be right-differentiable over $(0, \infty)$, which is readily satisfied since $W^{(q)}$ is both right- and left-differentiable over $(0, \infty)$ (see, for example, p. 291 of Pistorius 2007 and Lemma 1 of Pistorius 2004). In the sequel, denote by $W_+^{(q)'}(x)$ the right-derivative of $W^{(q)}$ in $x$.

Further, define

$$Z^{(q)}(x) = 1 + q \int_0^x W^{(q)}(z) dz, \quad q \geq 0, x \geq 0,$$

and

$$Z^{(q)}(x, \theta) = e^{\theta x} \left(1 + (q - \psi(\theta)) \int_0^x e^{-\theta z} W^{(q)}(z) dz\right), \quad q, \theta \geq 0, x \geq 0,$$

with $Z^{(q)}(x) = 1$ and $Z^{(q)}(x, \theta) = e^{\theta x}$ for $x < 0$.

We also briefly recall concepts in excursion theory for the reflected process $\{\overline{X}(t) - X(t); t \geq 0\}$, and we refer to Bertoin (1996) and Kyprianou (2014) for more details. For $x \in \mathbb{R}$, the process $\{L(t) := \overline{X}(t) - x; t \geq 0\}$ serves as a local time at 0 for the Markov process $\{\overline{X}(t) - X(t); t \geq 0\}$ under $\mathbb{P}_x$. Define the corresponding inverse local time as

$$L^{-1}(t) := \inf\{s \geq 0 : L(s) > t\} = \sup\{s \geq 0 : L(s) \leq t\}.$$

Let $L^{-1}(t-) := \lim_{s \uparrow t} L^{-1}(s)$. The Poisson point process of excursions indexed by this local time is denoted by $\{(t, e_t); t \geq 0\}$

$$e_t(s) := X(L^{-1}(t)) - X(L^{-1}(t-) + s), \quad s \in (0, L^{-1}(t) - L^{-1}(t-)],$$

whenever $L^{-1}(t) - L^{-1}(t-) > 0$. For the case of $L^{-1}(t) - L^{-1}(t-) = 0$, define $e_t = \Upsilon$ with $\Upsilon$ being an additional isolated point. Accordingly, we denote a generic excursion as $\varepsilon(\cdot)$ (or, $\varepsilon$ for short) belonging to the space $\mathcal{E}$ of canonical excursions. The intensity measure of the Poisson point process $\{(t, \varepsilon_t); t \geq 0\}$ is given by $dt \times dn$ where $n$ is a $\sigma$-finite measure on the space $\mathcal{E}$. The lifetime of a canonical excursion $\varepsilon$ is denoted by $\zeta$, and its excursion height is denoted by $\bar{\varepsilon} = \sup_{t \in [0,\zeta]} \varepsilon(t)$. The first passage time of a canonical excursion $\varepsilon$ is defined by

$$\rho_b^+ = \rho_b^+(\varepsilon) := \inf\{t \in [0, \zeta] : \varepsilon(t) > b\}$$

with the convention $\inf \emptyset := \zeta$.

Denote by $\varepsilon_g$ the excursion (away from 0) with left-end point $g$ for the reflected process $\{\overline{X}(t) - X(t); t \geq 0\}$, and by $\zeta_g$ and $\bar{\varepsilon}_g$ denote its lifetime and excursion height, respectively; see Section IV.4 of Bertoin (1996).

## 3. Main Results

For the process $X$, define its first down-crossing time of level 0 and up-crossing time of level $b$, respectively, by

$$\tau_0^- := \inf\{t \geq 0 : X(t) < 0\} \quad \text{and} \quad \tau_b^+ := \inf\{t \geq 0 : X(t) > b\}.$$

From Kyprianou (2014), the resolvent measure corresponding to $X$ is absolutely continuous with respect to the Lebesgue measure and has a version of density given by

$$q \int_0^\infty e^{-qt} \mathbb{E}_x(f(X(t)); t < \tau_0^- \wedge \tau_b^+) \, dt$$

$$= q \int_0^b f(y) \left( \frac{W^{(q)}(x)}{W^{(q)}(b)} W^{(q)}(b-y) - W^{(q)}(x-y) \right) dy, \quad x \in [0, b]. \tag{5}$$

In preparation for showing the main results, we first present the following Lemma 1 which gives the joint Laplace transform of $\rho_z^+$ and the overshoot at $\rho_z^+$ of a canonical excursion $\varepsilon$ with respect to the excursion measure $n$, which is a $\sigma$-finite measure on the space $\mathcal{E}$ of canonical excursions (see Section 2).

**Lemma 1.** *For any $q, z > 0$, we have*

$$n\left(e^{-q\rho_z^+ + \theta(z-\varepsilon(\rho_z^+))}; \bar{\varepsilon} > z\right)$$
$$= \frac{W_+^{(q)'}(z)}{W^{(q)}(z)} Z^{(q)}(z, \theta) - \theta Z^{(q)}(z, \theta) - (q - \psi(\theta)) W^{(q)}(z). \quad (6)$$

*In particular*

$$n\left(e^{-q\rho_z^+}; \bar{\varepsilon} > z\right) = \frac{W_+^{(q)'}(z)}{W^{(q)}(z)} Z^{(q)}(z) - q W^{(q)}(z),$$

*and*

$$n\left(e^{\theta(z-\varepsilon(\rho_z^+))}; \bar{\varepsilon} > z\right) = \frac{W_+'(z)}{W(z)} Z(z, \theta) - \theta Z(z, \theta) + \psi(\theta) W(z).$$

**Proof.** Taking use of the first result in Proposition 2 of Pistorius (2007), we can prove the desired results following the arguments in Lemma 2.2 of Kyprianou and Zhou (2009). □

Proposition 1 gives the joint Laplace transform of $\sigma_0^-$ and the position of the process $U$ at $\sigma_0^-$. It is similar to Theorem 1.3 of Kyprianou and Zhou (2009) where the Lévy measure is involved in the expression. The following joint Laplace transform is expressed in terms of scale functions.

**Proposition 1.** *For any $q, \theta > 0$ and $0 \leq x \leq b$ we have*

$$\mathbb{E}_x\left(e^{-q\sigma_0^- + \theta U(\sigma_0^-)}; \sigma_0^- < \sigma_b^+\right)$$
$$= \int_x^b \frac{1}{(1 - \gamma(\bar{\gamma}_x^{-1}(z)))} \exp\left(-\int_x^z \frac{W_+^{(q)'}(w)}{(1 - \gamma(\bar{\gamma}_x^{-1}(w))) W^{(q)}(w)} dw\right)$$
$$\times \left(\frac{W_+^{(q)'}(z)}{W^{(q)}(z)} Z^{(q)}(z, \theta) - \theta Z^{(q)}(z, \theta) - (q - \psi(\theta)) W^{(q)}(z)\right) dz. \quad (7)$$

**Proof.** By Theorems 1.1 in Kyprianou and Zhou (2009) (with minor adaptation), for $0 \leq x \leq a$, one has

$$\mathbb{E}_x\left(e^{-q\sigma_a^+}; \sigma_a^+ < \sigma_0^-\right) = \exp\left(-\int_x^a \frac{W_+^{(q)'}(w)}{(1 - \gamma(\bar{\gamma}_x^{-1}(w))) W^{(q)}(w)} dw\right). \quad (8)$$

For $0 \leq x \leq b$, by Equation (8) and the compensation formula in excursion theory we have

$$\mathbb{E}_x\left(e^{-q\sigma_0^- + \theta U(\sigma_0^-)}; \sigma_0^- < \sigma_b^+\right)$$

$$= \mathbb{E}_x\left(\sum_g e^{-qg} \prod_{r<g} \mathbf{1}_{\{\bar{\varepsilon}_r \leq \overline{\gamma}_x(x+L(r)), \overline{\gamma}_x(x+L(g)) \leq b\}}\right.$$

$$\left. \times e^{-q\rho_{\overline{\gamma}_x(x+L(g))}^+(\varepsilon_g) + \theta\left(\overline{\gamma}_x(x+L(g)) - \varepsilon_g(\rho_{\overline{\gamma}_x(x+L(g))}^+(\varepsilon_g))\right)} \mathbf{1}_{\{\bar{\varepsilon}_g > \overline{\gamma}_x(x+L(g))\}}\right)$$

$$= \mathbb{E}_x\left(\int_0^\infty e^{-qt} \prod_{r<t} \mathbf{1}_{\{\bar{\varepsilon}_r \leq \overline{\gamma}_x(x+L(r)), \overline{\gamma}_x(x+L(t)) \leq b\}}\right.$$

$$\left. \times \int_{\mathcal{E}} e^{-q\rho_{\overline{\gamma}_x(x+L(t))}^+ + \theta\left(\overline{\gamma}_x(x+L(t)) - \varepsilon(\rho_{\overline{\gamma}_x(x+L(t))}^+)\right)} \mathbf{1}_{\{\bar{\varepsilon} > \overline{\gamma}_x(x+L(t))\}} n(d\varepsilon) dL(t)\right)$$

$$= \mathbb{E}_x\left(\int_0^{\overline{\gamma}_x^{-1}(b)-x} e^{-qL^{-1}(t-)} \prod_{r<L^{-1}(t-)} \mathbf{1}_{\{\bar{\varepsilon}_r \leq \overline{\gamma}_x(x+L(r))\}}\right.$$

$$\left. \times n\left(e^{-q\rho_{\overline{\gamma}_x(x+t)}^+ + \theta(\overline{\gamma}_x(x+t) - \varepsilon(\rho_{\overline{\gamma}_x(x+t)}^+))} \mathbf{1}_{\{\bar{\varepsilon} > \overline{\gamma}_x(x+t)\}}\right) dt\right)$$

$$= \int_0^{\overline{\gamma}_x^{-1}(b)-x} \exp\left(-\int_x^{\overline{\gamma}_x(x+t)} \frac{W_+^{(q)'}(w)}{(1-\gamma(\overline{\gamma}_x^{-1}(w)))W^{(q)}(w)} dw\right)$$

$$\times n\left(e^{-q\rho_{\overline{\gamma}_x(x+t)}^+ + \theta(\overline{\gamma}_x(x+t) - \varepsilon(\rho_{\overline{\gamma}_x(x+t)}^+))}; \bar{\varepsilon} > \overline{\gamma}_x(x+t)\right) dt$$

$$= \int_x^b \exp\left(-\int_x^s \frac{W_+^{(q)'}(w)}{(1-\gamma(\overline{\gamma}_x^{-1}(w)))W^{(q)}(w)} dw\right) \frac{n\left(e^{-q\rho_s^+ + \theta(s - \varepsilon(\rho_s^+))}; \bar{\varepsilon} > s\right)}{1 - \gamma(\overline{\gamma}_x^{-1}(s))} ds,$$

which together with Equation (6) yields Equation (7). □

**Remark 1.** Let $\gamma \equiv 0$ in Equation (7). Then $U(t) = X(t)$ for $t \geq 0$ and $\overline{\gamma}_x(z) \equiv z$ for $z \geq x$, and by Proposition 1 we have

$$\mathbb{E}_x(e^{-q\tau_0^- + \theta X(\tau_0^-)}; \tau_0^- < \tau_b^+)$$

$$= \int_x^b \frac{W^{(q)}(x)}{W^{(q)}(s)} \left(\frac{W_+^{(q)'}(s)}{W^{(q)}(s)} Z^{(q)}(s,\theta) - \theta Z^{(q)}(s,\theta) - (q - \psi(\theta))W^{(q)}(s)\right) ds$$

$$= -W^{(q)}(x) \int_x^b \frac{d}{ds}\left(\frac{Z^{(q)}(s,\theta)}{W^{(q)}(s)}\right) ds$$

$$= Z^{(q)}(x,\theta) - \frac{W^{(q)}(x)}{W^{(q)}(b)} Z^{(q)}(b,\theta),$$

which can be found in (8.12) (with an appropriate killing rate added) in Chapter 8 of Kyprianou (2014), or Albrecher et al. (2016).

Let $\gamma \equiv \alpha \in (0,1)$ or $\overline{\gamma}_x(z) = x + (1-\alpha)(z-x)$ in Equation (7), we have for $q, \theta > 0$ and $0 \le x \le b$

$$\mathbb{E}_x\left(e^{-q\sigma_0^- + \theta U(\sigma_0^-)}; \sigma_0^- < \sigma_b^+\right)$$

$$= \frac{1}{1-\alpha}\int_x^b \left(\frac{W^{(q)}(x)}{W^{(q)}(z)}\right)^{\frac{1}{1-\alpha}} \left(\frac{W_+^{(q)'}(z)}{W^{(q)}(z)} Z^{(q)}(z,\theta) - \theta Z^{(q)}(z,\theta) - (q-\psi(\theta))W^{(q)}(z)\right) dz.$$

Proposition 2 gives an expression of potential density for the process $U$.

**Proposition 2.** *The potential measure corresponding to $U$ is absolutely continuous with respect to the Lebesgue measure with density given by*

$$\int_0^\infty e^{-qt}\mathbb{P}_x(U(t) \in du, t < \sigma_b^+ \wedge \sigma_0^-) dt$$

$$= W^{(q)}(0)\frac{1}{1-\gamma(\overline{\gamma}_x^{-1}(u))}\exp\left(-\int_x^u \frac{W_+^{(q)'}(w)}{(1-\gamma(\overline{\gamma}_x^{-1}(w)))W^{(q)}(w)} dw\right)\mathbf{1}_{(x,b)}(u) du$$

$$+ \int_x^b \frac{1}{1-\gamma(\overline{\gamma}_x^{-1}(y))} \exp\left(-\int_x^y \frac{W_+^{(q)'}(w)}{(1-\gamma(\overline{\gamma}_x^{-1}(w)))W^{(q)}(w)} dw\right)$$

$$\times \left(W_+^{(q)'}(y-u) - \frac{W_+^{(q)'}(y)}{W^{(q)}(y)} W^{(q)}(y-u)\right)\mathbf{1}_{(0,y)}(u) dy du, \quad x, u \in [0,b], q > 0. \qquad (9)$$

**Proof.** Let $e_q$ be an exponentially distributed random variable independent of $X$ with mean $1/q$. For any continuous, non-negative and bounded function $f$, we have

$$\int_0^\infty q e^{-qt}\mathbb{E}_x(f(U(t)); t < \sigma_b^+ \wedge \sigma_0^-) dt$$

$$= \mathbb{E}_x\left(f(U(e_q))\mathbf{1}_{\{U(e_q) < \overline{U}(e_q), e_q < \sigma_b^+ \wedge \sigma_0^-\}}\right)$$

$$+ \mathbb{E}_x\left(\int_0^\infty q e^{-qt} f(U(t))\mathbf{1}_{\{U(t) = \overline{U}(t), t < \sigma_b^+ \wedge \sigma_0^-\}} dt\right). \qquad (10)$$

Note that $\int_0^t \mathbf{1}_{\{X(s) = \overline{X}(s)\}} ds = W^{(q)}(0)\overline{X}(t)$, see Corollary 6 in Chapter IV of Bertoin (1996). Recalling that $U(t) = \overline{U}(t)$ is equivalent to $X(t) = \overline{X}(t)$ which implies $t = L^{-1}(L(t))$, we have

$$\mathbb{E}_x\left(\int_0^\infty q e^{-qt} f(U(t))\mathbf{1}_{\{U(t) = \overline{U}(t), t < \sigma_b^+ \wedge \sigma_0^-\}} dt\right)$$

$$= \mathbb{E}_x\left(\int_0^\infty q e^{-qL^{-1}(L(t))} f(U(L^{-1}(L(t))))\mathbf{1}_{\{X(t) = \overline{X}(t), L^{-1}(L(t)) < \sigma_b^+ \wedge \sigma_0^-\}} dt\right)$$

$$= W(0)\mathbb{E}_x\left(\int_0^\infty q e^{-qL^{-1}(L(t))} f(U(L^{-1}(L(t))))\mathbf{1}_{\{L^{-1}(L(t)) < L^{-1}(\overline{\gamma}_x^{-1}(b)-x) \wedge \sigma_0^-\}} dL_t\right)$$

$$= qW(0)\int_0^{\overline{\gamma}_x^{-1}(b)-x} \exp\left(-\int_x^{\overline{\gamma}_x(x+t)} \frac{W_+^{(q)'}(w)}{(1-\gamma(\overline{\gamma}_x^{-1}(w)))W^{(q)}(w)} dw\right) f(\overline{\gamma}_x(x+t)) dt$$

$$= qW(0)\int_x^b \frac{1}{(1-\gamma(\overline{\gamma}_x^{-1}(y)))} \exp\left(-\int_x^y \frac{W_+^{(q)'}(w)}{(1-\gamma(\overline{\gamma}_x^{-1}(w)))W^{(q)}(w)} dw\right) f(y) dy, \qquad (11)$$

where Equation (8) is used in the last but one equation.

By the compensation formula in excursion theory and the memoryless property of the exponential random variable, one has

$$\mathbb{E}_x\left(f(U(e_q)); U(e_q) < \overline{U}(e_q), e_q < \sigma_b^+ \wedge \sigma_0^-\right)$$

$$= \mathbb{E}_x\left(\int_0^\infty \sum_g e^{-qg} \prod_{r<g} \mathbf{1}_{\{\bar{\varepsilon}_r \leq \bar{\gamma}_x(x+L(r)), \bar{\gamma}_x(x+L(g)) \leq b\}} f\left(\bar{\gamma}_x(x+L(g)) - \varepsilon_g(t-g)\right)\right.$$

$$\left. \times q e^{-q(t-g)} \mathbf{1}_{\{g<t<g+\zeta_g \wedge \rho^+_{\bar{\gamma}_x(x+L(g))}(\varepsilon_g)\}} dt\right)$$

$$= \mathbb{E}_x\left(\int_0^\infty e^{-qt} \prod_{r<t} \mathbf{1}_{\{\bar{\varepsilon}_r \leq \bar{\gamma}_x(x+L(r)), \bar{\gamma}_x(x+L(t)) \leq b\}}\right.$$

$$\left. \times \left(\int_{\mathcal{E}} \int_0^\infty q e^{-qs} f\left(\bar{\gamma}_x(x+L(t)) - \varepsilon(s)\right) \mathbf{1}_{\{s<\zeta \wedge \rho^+_{\bar{\gamma}_x(x+L(t))}\}} ds\, n(d\varepsilon)\right) dL(t)\right)$$

$$= q \int_0^{\bar{\gamma}_x^{-1}(b)-x} \exp\left(-\int_x^{\bar{\gamma}_x(x+t)} \frac{W_+^{(q)'}(w)}{(1-\gamma(\bar{\gamma}_x^{-1}(w)))W^{(q)}(w)} dw\right)$$

$$\times \int_0^\infty n\left(e^{-qs} f(\bar{\gamma}_x(x+t) - \varepsilon(s)) \mathbf{1}_{\{s<\zeta \wedge \rho^+_{\bar{\gamma}_x(x+t)}\}}\right) ds\, dt$$

$$= q \int_x^b \frac{1}{1-\gamma(\bar{\gamma}_x^{-1}(y))} \exp\left(-\int_x^y \frac{W_+^{(q)'}(w)}{(1-\gamma(\bar{\gamma}_x^{-1}(w)))W^{(q)}(w)} dw\right)$$

$$\times \int_0^\infty n\left(e^{-qs} f(y-\varepsilon(s)) \mathbf{1}_{\{s<\zeta \wedge \rho^+_y\}}\right) ds\, dy. \tag{12}$$

Applying the same arguments as in Equations (11) and (12), we have

$$\mathbb{E}_x\left(f(X(e_q))\mathbf{1}_{\{e_q<\tau_b^+ \wedge \tau_0^-\}}\right)$$

$$= \mathbb{E}_x\left(f(X(e_q))\mathbf{1}_{\{X(e_q)=\overline{X}(e_q), e_q<\tau_b^+ \wedge \tau_0^-\}}\right) + \mathbb{E}_x\left(f(X(e_q))\mathbf{1}_{\{X(e_q)<\overline{X}(e_q), e_q<\tau_b^+ \wedge \tau_0^-\}}\right)$$

$$= q \int_x^b \frac{W^{(q)}(x)}{W^{(q)}(y)} \left(W(0)f(y) + \int_0^\infty n\left(e^{-qs} f(y-\varepsilon(s)) \mathbf{1}_{\{s<\rho_y^+ \wedge \zeta\}}\right) ds\right) dy. \tag{13}$$

Equating the right hand sides of Equations (5) and (13) and then differentiating the resultant equation with respect to $b$ gives

$$\frac{W^{(q)}(x)}{W^{(q)}(b)} \left(W(0)f(b) + \int_0^\infty n\left(e^{-qs} f(b-\varepsilon(s)) \mathbf{1}_{\{s<\rho_b^+ \wedge \zeta\}}\right) ds\right)$$

$$= \frac{W^{(q)}(x)}{W^{(q)}(b)} \left(f(b)W(0) + \int_0^b f(y) \left(W_+^{(q)'}(b-y) - \frac{W_+^{(q)'}(b)}{W^{(q)}(b)} W^{(q)}(b-y)\right) dy\right),$$

or equivalently

$$\int_0^\infty n\left(e^{-qs} f(y-\varepsilon(s)) \mathbf{1}_{\{s<\rho_y^+ \wedge \zeta\}}\right) ds$$

$$= \int_0^y f(w) \left(W_+^{(q)'}(y-w) - \frac{W_+^{(q)'}(y)}{W^{(q)}(y)} W^{(q)}(y-w)\right) dw,$$

which together with Equation (12) yields

$$\mathbb{E}_x\left(f(U(e_q)); U(e_q) < \overline{U}(e_q), e_q < \sigma_b^+ \wedge \sigma_0^-\right)$$

$$= q \int_x^b \frac{1}{(1-\gamma(\overline{\gamma}_x^{-1}(y)))} \exp\left(-\int_x^y \frac{W_+^{(q)'}(w)}{(1-\gamma(\overline{\gamma}_x^{-1}(w)))W^{(q)}(w)} dw\right)$$

$$\times \int_0^y f(w) \left(W_+^{(q)'}(y-w) - \frac{W_+^{(q)'}(y)}{W^{(q)}(y)} W^{(q)}(y-w)\right) dw\, dy,$$

which combined with Equations (10) and (11) yields Equation (9). □

**Remark 2.** *Letting $\gamma \equiv 0$ in Equation (9), i.e., $U(t) = X(t)$ for $t \geq 0$ and $\overline{\gamma}_x(z) \equiv z$ for $z \geq x$, by Proposition 2 we have for $0 \leq x, u \leq b$*

$$\int_0^\infty e^{-qt} \mathbb{P}_x(X(t) \in du, t < \sigma_b^+ \wedge \sigma_0^-) dt$$

$$= \frac{W^{(q)}(0)W^{(q)}(x)\mathbf{1}_{(x,b)}(u)}{W^{(q)}(u)} du$$

$$+ \int_x^b \frac{W^{(q)}(x)}{W^{(q)}(y)} \left(W_+^{(q)'}(y-u) - \frac{W_+^{(q)'}(y)}{W^{(q)}(y)} W^{(q)}(y-u)\right) \mathbf{1}_{(0,y)}(u) dy\, du$$

$$= \frac{W^{(q)}(0)W^{(q)}(x)\mathbf{1}_{(x,b)}(u)}{W^{(q)}(u)} du + W^{(q)}(x) \int_x^b \frac{d}{dy}\left[\frac{W^{(q)}(y-u)}{W^{(q)}(y)}\right] \mathbf{1}_{(0,y)}(u) dy\, du$$

$$= W^{(q)}(x) \left(\frac{W^{(q)}(0)\mathbf{1}_{(x,b)}(u)}{W^{(q)}(u)} du + \int_x^b \frac{d}{dy}\left[\frac{W^{(q)}(y-u)}{W^{(q)}(y)}\right] dy\, du\mathbf{1}_{(0,x)}(u)\right.$$

$$\left.+ \int_u^b \frac{d}{dy}\left[\frac{W^{(q)}(y-u)}{W^{(q)}(y)}\right] dy\, du\mathbf{1}_{(x,b)}(u)\right)$$

$$= \frac{W^{(q)}(b-u)}{W^{(q)}(b)} W^{(q)}(x)\mathbf{1}_{(x,b)}(u) du + \left(\frac{W^{(q)}(b-u)}{W^{(q)}(b)} - \frac{W^{(q)}(x-u)}{W^{(q)}(x)}\right) W^{(q)}(x)\mathbf{1}_{(0,x)}(u) du$$

$$= \left(\frac{W^{(q)}(x)}{W^{(q)}(b)} W^{(q)}(b-u) - W^{(q)}(x-u)\right) \mathbf{1}_{(0,b)}(u) du,$$

*which recovers Equation (5).*
*Let $\gamma \equiv \alpha \in (0,1)$ or $\overline{\gamma}_x(z) = x + (1-\alpha)(z-x)$ in Equation (9), we have*

$$\int_0^\infty e^{-qt} \mathbb{P}_x(U(t) \in du, t < \sigma_b^+ \wedge \sigma_0^-) dt$$

$$= \frac{W^{(q)}(0)}{1-\alpha} \left(\frac{W^{(q)}(x)}{W^{(q)}(u)}\right)^{\frac{1}{1-\alpha}} \mathbf{1}_{(x,b)}(u) du + \frac{1}{1-\alpha} \int_x^b \left(\frac{W^{(q)}(x)}{W^{(q)}(y)}\right)^{\frac{1}{1-\alpha}}$$

$$\times \left(W_+^{(q)'}(y-u) - \frac{W_+^{(q)'}(y)}{W^{(q)}(y)} W^{(q)}(y-u)\right) \mathbf{1}_{(0,y)}(u) dy\, du, \quad x, u \in [0,b], \, q > 0.$$

**Author Contributions:** Conceptualization, X.Z.; methodology, W.W. and X.Z.; validation, W.W. and X.Z.; investigation, W.W. and X.Z.; writing—original draft preparation, W.W.; writing—review and editing, W.W. and X.Z.; supervision, X.Z.; project administration, X.Z.; funding acquisition, W.W. and X.Z.

**Funding:** This research was partly funded by the National Natural Science Foundation of China (Nos. 11601197; 11771018) and the Program for New Century Excellent Talents in Fujian Province University.

**Acknowledgments:** Wenyuan Wang thanks Concordia University where this paper was finished during his visit.

**Conflicts of Interest:** The author declare no conflict of interest.

## References

Albrecher, Hansjörg, Xiaowen Zhou, and Jean-Francois Renaud. 2008. A Lévy insurance risk process with tax. *Journal of Applied Probability* 45: 363–75. [CrossRef]

Albrecher, Hansjörg, and Christian Hipp. 2007. Lundberg's risk process with tax. *Blätter der DGVFM* 28: 13–28. [CrossRef]

Albrecher, Hansjörg, Jevgenijs Ivanovs, and Xiaowen Zhou. 2016. Exit identities for Lévy processes observed at Poisson arrival times. *Bernoulli* 22: 1364–82. [CrossRef]

Avram, Florin, Nhat Linh Vu, and Xiaowen Zhou. 2017. On taxed spectrally negative Lévy processes with draw-down stopping. *Insurance: Mathematics and Economics* 76: 69–74. [CrossRef]

Bertoin, Jean. 1996. *Lévy Process*. Cambridge: Cambridge University Press.

Chan, Terence, Andreas E. Kyprianou, and Mladen Savov. 2011. Smoothness of scale functions for spectrally negative Lévy processes. *Probability Theory and Related Fields* 150: 691–708. [CrossRef]

Cheung, Eric C. K., and David Landriault. 2012. On a risk model with surplus dependent premium and tax rates. *Methodology and Computing in Applied Probability* 14: 233–51. [CrossRef]

Gerber, Hans U., and Elias S. W. Shiu. 1998. On the time value of ruin. *North American Actuarial Journal* 2: 48–72. [CrossRef]

Hao, Xuemiao, and Qihe Tang. 2009. Asymptotic ruin probabilities of the Lévy insurance model under periodic taxation. *Astin Bulletin* 39: 479–94. [CrossRef]

Kyprianou, Andreas E. 2014. *Fluctuations of Lévy Processes with Applications*. Berlin/Heidelberg: Springer Science+Business Media.

Kyprianou, Andreas E., and Curdin Ott. 2012. Spectrally negative Lévy processes perturbed by functionals of their running supremum. *Journal of Applied Probability* 49: 1005–14. [CrossRef]

Kyprianou, Andreas E., and Xiaowen Zhou. 2009. General tax structures and the Lévy insurance risk model. *Journal of Applied Probability* 46: 1146–56. [CrossRef]

Landriault, David, Jean-François Renaud, and Xiaowen Zhou. 2011. Occupation times of spectrally negative Lévy processes with applications. *Stochastic Processes and Their Applications* 121: 2629–41. [CrossRef]

Li, Bin, Qihe Tang, and Xiaowen Zhou. 2013. A time-homogeneous diffusion model with tax. *Journal of Applied Probability* 50: 195–207. [CrossRef]

Li, Yingqiu, and Xiaowen Zhou. 2014. On pre-exit joint occupation times for spectrally negative Lévy processes. *Statistics and Probability Letters* 94: 48–55. [CrossRef]

Pistorius, Martijn R. 2004. On exit and ergodicity of the spectrally one-sided Lévy process reflected at its infimum. *Journal of Theoretical Probability* 17: 183–220. [CrossRef]

Pistorius, Martijn R. 2007. An excursion-theoretical approach to some boundary crossing problems and the Skorokhod embedding for reflected Lévy processes. In *Séminaire de Probabilités XL*. Berlin/Heidelberg: Springer, pp. 287–307.

Renaud, Jean-François. 2009. The distribution of tax payments in a Lévy insurance risk model with a surplus-dependent taxation structure. *Insurance: Mathematics and Economics* 45: 242–46. [CrossRef]

Wang, Wenyuan, and Yijun Hu. 2012. Optimal loss-carry-forward taxation for the Lévy risk model. *Insurance: Mathematics and Economics* 50: 121–30. [CrossRef]

Wei, Li. 2009. Ruin probability in the presence of interest earnings and tax payments. *Insurance: Mathematics and Economics* 45: 133–38. [CrossRef]

Wei, Jiaqin, Hailiang Yang, and Rongming Wang. 2010. On the markov-modulated insurance risk model with tax. *Blätter der DGVFM* 31: 65–78. [CrossRef]

Zhang, Zhimin, Eric CK Cheung, and Hailiang Yang. 2017. Lévy insurance risk process with Poissonian taxation. *Scandinavian Actuarial Journal* 2017: 51–87. [CrossRef]

 © 2019 by the authors. Licensee MDPI, Basel, Switzerland. This article is an open access article distributed under the terms and conditions of the Creative Commons Attribution (CC BY) license (http://creativecommons.org/licenses/by/4.0/).

Article
# Three Essays on Stopping

**Eberhard Mayerhofer**

Department of Mathematics and Statistics, University of Limerick, Limerick V94TP9X, Ireland;
eberhard.mayerhofer@ul.ie

Received: 27 September 2019; Accepted: 16 October 2019; Published: 18 October 2019

**Abstract:** First, we give a closed-form formula for first passage time of a reflected Brownian motion with drift. This corrects a formula by Perry et al. (2004). Second, we show that the maximum before a fixed drawdown is exponentially distributed for any drawdown, if and only if the diffusion characteristic $\mu/\sigma^2$ is constant. This complements the sufficient condition formulated by Lehoczky (1977). Third, we give an alternative proof for the fact that the maximum before a fixed drawdown is exponentially distributed for any spectrally negative Lévy process, a result due to Mijatović and Pistorius (2012). Our proof is similar, but simpler than Lehoczky (1977) or Landriault et al. (2017).

**Keywords:** reflected Brownian motion; linear diffusions; spectrally negative Lévy processes; drawdown

**MSC (2010):** 60J65; 60J75

---

## 1. Introduction

This paper comprises three essays on stopping.

In Section 2, we compute the Laplace transform of the first hitting time of a fixed upper barrier for a reflected Brownian motion with drift. This expands on and corrects a result by Perry et al. (2004).

In Section 3, we show, by using an intrinsic delay differential equation, that for a diffusion process, the maximum before a fixed drawdown threshold is generically exponentially distributed, only if the diffusion characteristic $\mu/\sigma^2$ is constant. This complements the sufficient condition formulated by Lehoczky (1977). By solving discrete delay differential equations, we further construct diffusions, where the exponential law only holds for specific drawdown sizes.

Section 4 uses Lehoczky (1977)'s argument to show that the maximum before a fixed drawdown threshold is exponentially distributed for any spectrally negative Lévy process, the parameter being the right-sided logarithmic derivative of the scale function. This yields an alternative proof to the original one in Mijatović and Pistorius (2012) and is also similar to the one in Landriault et al. (2017).

## 2. The First Hitting Time for a Reflected Brownian Motion With Drift

Let $X$ be a reflected Brownian motion on $[0,\infty)$, with drift $\mu$ and volatility $\sigma$. Then $X$ can be written as

$$X_t = x + \mu t + \sigma W_t + L_t,$$

where $W = (W_t)_{t\geq 0}$ is a standard Brownian motion, and $L = (L_t)_{t\geq 0}$ is an inon-decreasing process, such that the induced random measure $dL$ is supported on $\{X = 0\}$. Itô's formula implies that for any $f \in C_b^2([0,\infty))$ satisfying $f'(0+) = 0$, the process

$$f(X_t) - f(x) - \int_0^t \mathcal{A}_y f(X_s) ds$$

is a martingale, where $\mathcal{A}_y$ is the differential operator, defined by $\mathcal{A}_y f(y) = \frac{\sigma^2}{2} f''(y) + \mu f'(y)$.[1]
For $\delta \geq -x$, we define the first hitting time:

$$\tau_\delta := \inf\{t \geq 0 \mid X_t = \delta + x\}.$$

Since, before reaching the boundary 0, the process cannot be distinguished from a Brownian motion with drift, we may confine ourselves to computing $\tau_\delta$ for barriers $\delta + x$, where $\delta > 0$. Our aim is to compute the Laplace transform:

$$\Psi(\theta; \delta, x) := \mathbb{E}[e^{-\theta \tau_\delta} \mid X_0 = x], \quad \theta \geq 0.$$

**Theorem 1.** *For $\delta \geq 0$, the Laplace transform of the first hitting time of a reflected Brownian motion with drift $\mu$ and volatility $\sigma$ is given by*

$$\Psi(\theta; x, \delta) := e^{\frac{\delta \mu}{\sigma^2}} \frac{\sqrt{\mu^2 + 2\theta\sigma^2}\cosh\left(\frac{x\sqrt{\mu^2+2\theta\sigma^2}}{\sigma^2}\right) + \mu \sinh\left(\frac{x\sqrt{\mu^2+2\theta\sigma^2}}{\sigma^2}\right)}{\sqrt{\mu^2 + 2\theta\sigma^2}\cosh\left(\frac{(x+\delta)\sqrt{\mu^2+2\theta\sigma^2}}{\sigma^2}\right) + \mu \sinh\left(\frac{(x+\delta)\sqrt{\mu^2+2\theta\sigma^2}}{\sigma^2}\right)}. \tag{1}$$

**Proof.** Pick $\Phi \in C_c^\infty(\mathbb{R})$, such that $\Phi(\xi) = 1$ for $|\xi| \leq x + \delta$. Furthermore, let $\kappa \in \mathbb{R}$; then for any $\theta \geq 0$ and $t \geq 0$, the function

$$F(t, x) := e^{-\theta t} \Phi(x) \left( e^{-\kappa x} + \kappa x \right)$$

satisfies $f := F(t, \cdot) \in C_b^2$ and $f'(0) = 0$. According to the introductory notes of this section, the process $F(t, X_t) - \int_0^t \partial_s F(s, X_s) ds - \int_0^t \mathcal{A}_y F(s, X_s) ds$ is a uniformly bounded martingale; therefore, the stopped process

$$F(t, X_{t \wedge \tau_\delta}) - (e^{-\kappa x} + \kappa x) - \int_0^{t \wedge \tau_\delta} \partial_t F(s, X_s) ds - \int_0^{t \wedge \tau_\delta} \mathcal{A}_y F(s, X_s) ds$$

is also a true martingale, which starts at zero, $\mathbb{P}^x$- almost surely. Using the fact that $\Phi(X_{t \wedge \tau_\delta}) = 1$, we find that the stopped process satisfies for any $t \geq 0$,

$$e^{-\theta(t \wedge \tau_\delta)} \left( e^{-\kappa X_{t \wedge \tau_\delta}} + \kappa X_{t \wedge \tau_\delta} \right) - (e^{-\kappa x} + \kappa x) + \theta \int_0^{t \wedge \tau_\delta} e^{-\kappa X_s - \theta s} ds + \theta \kappa \int_0^{t \wedge \tau_\delta} e^{-\theta s} X_s ds$$

$$- \mu \int_0^{t \wedge \tau_\delta} \left( \kappa e^{-\theta s} - \kappa e^{-\kappa X_s - \theta s} \right) ds - \frac{\sigma^2 \kappa^2}{2} \int_0^{t \wedge \tau_\delta} e^{-\kappa X_s - \theta s} ds$$

$$= e^{-\theta(t \wedge \tau_\delta)} \left( e^{-\kappa X_{t \wedge \tau_\delta}} + \kappa X_{t \wedge \tau_\delta} \right) - (e^{-\kappa x} + \kappa x) + \theta \kappa \int_0^{t \wedge \tau_\delta} e^{-\theta s} X_s ds$$

$$- \frac{\mu \kappa}{\theta} \left( 1 - e^{-\theta(t \wedge \tau_\delta)} \right) - \left( \frac{\sigma^2 \kappa^2}{2} - \kappa \mu - \theta \right) \int_0^{t \wedge \tau_\delta} e^{-\kappa X_s - \theta s} ds.$$

Letting $t \to \infty$, we thus get by optional sampling,

$$(e^{-\kappa(x+\delta)} + \kappa(x+\delta)) \mathbb{E}[e^{-\theta \tau_\delta} \mid X_0 = x] - (e^{-\kappa x} + \kappa x) + \theta \kappa \mathbb{E}\left[\int_0^{\tau_\delta} e^{-\theta s} X_s ds \mid X_0 = x\right]$$

$$- \frac{\mu \kappa}{\theta}(1 - \mathbb{E}[e^{-\theta \tau_\delta} \mid X_0 = x]) - \left( \frac{\sigma^2 \kappa^2}{2} - \kappa \mu - \theta \right) \mathbb{E}\left[\int_0^{\tau_\delta} e^{-\kappa X_s - \theta s} ds \mid X_0 = x\right] = 0.$$

---

[1] In the language of linear diffusions Borodin and Salminen (2012), $X$ has infinitesimal generator $\mathcal{A}_y$ acting on the domain $\mathcal{D}(\mathcal{A}_y) = \{f \in C_b^2([0, \infty)) \mid f'(0+) = 0\}$.

For the two choices $\kappa \in \{\kappa_-, \kappa_+\}$, where

$$\kappa_\pm := \frac{\mu \pm \sqrt{\mu^2 + 2\theta\sigma^2}}{\sigma^2},$$

we thus obtain two equations, for two unknown moments,

$$\left(e^{-\kappa_\pm(x+\delta)} + \kappa_\pm(x+\delta) + \frac{\mu\kappa_\pm}{\theta}\right)\mathbb{E}[e^{-\theta\tau_\delta} \mid X_0 = x] + \theta\kappa_\pm\mathbb{E}\left[\int_0^\tau e^{-\theta s} X_s ds \mid X_0 = x\right]$$
$$= (e^{-\kappa_\pm x} + \kappa x) + \frac{\mu\kappa_\pm}{\theta}.$$

Solving this linear system for the involved moments yields the Laplace transform of $\tau_\delta$, Equation (1). □

**Remark 1.** *This result can also be obtained from a more general result for spectrally negative Lévy processes, reflected at an upper barrier (Avram et al. 2017, Proposition 4.B and Section 10.1). In fact, the distribution of $\tau^\delta$ is equal in distribution to the first hitting time 0 of the Brownian motion $X_t = \delta + \sigma B_t - \mu t$, starting at $\delta \geq 0$, reflected at $x + \delta > 0$. Its Laplace transform is therefore given by*

$$\psi_\theta^{[x+\delta]}(\delta) = e^{\frac{\mu\delta}{\sigma^2}} \frac{H(x)}{H(x+\delta)},$$

*where*

$$H(\xi) = \sqrt{2\theta\sigma^2 + \mu^2} \cosh\left(\frac{\xi\sqrt{2\theta\sigma^2 + \mu^2}}{\sigma^2}\right) + \mu \sinh\left(\frac{\xi\sqrt{2\theta\sigma^2 + \mu^2}}{\sigma^2}\right),$$

*(see Avram et al. (2017), Section 10.1).*

*Remarks On Perry et al. (2004)*

For another "sanity check" of Theorem 4, we compute the Laplace transform Equation (1) independently when $\mu = 0$ and $x = 0$. In this case, the reflected Brownian motion is equal to $|\sigma B|$ in law, where $B$ is a standard Brownian motion. But then $\tau_\delta$ is equal in distribution to

$$\tilde{\tau}_\delta := \inf\{s > 0 \mid B_s \in \{\pm\frac{\delta}{\sigma}\}\}.$$

Now, it is well known that the Laplace transform of $\tilde{\tau}_\delta$ is given by

$$\mathbb{E}[e^{-\tilde{\theta}\tau_\delta} \mid X_0 = x] = \frac{1}{\cosh(\frac{\delta}{\sigma}\sqrt{2\theta})}, \tag{2}$$

which indeed coincides with Equation (1) for $\mu \to 0$ and $x = 0$.

Perry et al. (2004), Formula (5.2), state a different Laplace transforms than our Theorem 4. Letting $\mu \to 0$ in Perry et al. (2004), Formula (5.2) indeed yields for $\sigma^2 = 1$ and $x = 0$,

$$\mathbb{E}[e^{-\theta\tau_\delta} \mid X_0 = x] = \frac{1}{\cosh(\delta\sqrt{\theta})},$$

which contradicts Equation (2). The proof of Perry et al. (2004), Lemma 5.1, cannot be rectified, however, by merely fixing the (obviously) missing factor of $1/2$ for $\alpha^2$ in the second line of their proof. Indeed, in the same line, they forget a factor $e^{-\kappa W(s)}$ in the second integrand; thus, by inserting special values of $\kappa$ into the process in line 2, one does not get rid of the local-time term, as claimed.

## 3. Diffusions with Exponentially Distributed Gains Before Fixed Drawdowns

Let $X$ be a diffusion process on the interval $[-a, \infty)$, satisfying the SDE

$$dX_t = \mu(X_t)dt + \sigma(X_t)dW_t, \quad X_0 = 0, \tag{3}$$

where $\mu(x)$ and $\sigma(x)$ are locally Lipschitz continuous functions of linear growth on $[-a, \infty)$, and $\sigma(x) > 0$ thereon.

For a threshold $0 < \delta \leq a$, we define $M^\delta$ as the maximum of $X$, prior to a drawdown of size $\delta$, that is

$$M^\delta = M(\tau^\delta), \quad \text{where} \quad M(t) := \max_{s \leq t} X_s, \quad \text{and} \quad \tau^\delta := \inf\{t > 0 \mid M_t - X_t = \delta\}.$$

We use the abbreviation $\Phi(x) := e^{-2\int_0^x \gamma(u)du}$, where $\gamma(x) = \mu(x)/\sigma^2(x)$. The following is due to Lehoczky (1977):

**Proposition 1.**

$$\log \mathbb{P}[M^\delta \geq \zeta] = -\int_0^\zeta \frac{\Phi(u)}{\int_{u-\delta}^u \Phi(s)ds} du, \quad \zeta \geq 0. \tag{4}$$

Caution is needed when interpreting the original paper Lehoczky (1977): Lehoczky uses the letter "a" for three different objects: The drift $\mu(x)$ is denoted as $a(x)$, while $-a$ is the left endpoint of the interval of the support of $X$; third, the threshold $\delta$ in his paper is also called $a$. An inspection of Lehozky's proof reveals that our more general version with $\delta \leq a$ holds.

In terms of diffusion characteristics, Lehoczky's result holds in a more general context. First, the assumption of locally Lipschitz coefficients are too strong, and can be relaxed. For example, we can relax to Hölder regularity of $\sigma(x)$ of order no worse than $1/2$, due to Yamada et al. (1971). In addition, we can allow reflecting or absorbing boundary conditions, thus include reflected diffusions. For instance, Proposition 1 holds for a Brownian motion with drift, starting at 0 and being reflected at $-a$, because the process $X$ cannot hit $-a$ before it reaches a strictly positive maximum, due to strict positive volatility $\sigma(0) > 0$.

From Equation (4), it can be seen that when $\mu/\sigma^2$ is constant, $M^\delta$ is exponentially distributed (the special case for for a Brownian motion with drift is due to Taylor (1975), and independently discovered by Golub et al. (2016)). Mijatović and Pistorius (2012) extended this result to spectrally negative Lévy processes: For those, $M^\delta$ is also exponentially distributed, with the parameter being the right-sided logarithmic derivative of the scale function, evaluated at the drawdown threshold.

This section characterizes the exponential law for diffusions:

**Theorem 2.** *The following are equivalent:*

1. $\mu(x)/\sigma^2(x)$ *is a constant on* $[-a, \infty)$.
2. *For each* $\delta > 0$, $M^\delta$ *is exponentially distributed.*

**Proof of the Theorem.** Sufficiency of the first condition for the second one follows directly from Proposition 1. Suppose, therefore, that for each $0 < \delta \leq a$, there exists $\Lambda(\delta) > 0$ such that $M^\delta$ is exponentially distributed with parameter $\Lambda(\delta)$. Then, due to Equation (4),

$$\int_0^\zeta \frac{\Phi(u)}{\int_{u-\delta}^u \Phi(s)ds} du = \Lambda(\delta)\zeta, \quad \zeta \geq 0, \quad \delta \leq a. \tag{5}$$

By this particular functional form, and, since $\mu/\sigma^2$ is continuous, it follows that the functions $\Lambda(\delta)$ and $\Phi(x)$ are continuously differentiable. By differentiating Equation (5) with respect to $\zeta$, we have

$$\Phi(\zeta) = \Lambda(\delta) \int_{\zeta-\delta}^{\zeta} \Phi(u)du, \quad \zeta \geq 0, \quad \delta \leq a, \tag{6}$$

and differentiating with respect to $\delta$ yields, in conjunction with the previous identity,

$$\frac{\Phi(\zeta-\delta)}{\Phi(\zeta)} = -\frac{\Lambda'(\delta)}{\Lambda^2(\delta)}, \quad \zeta \geq 0, \quad \delta \leq a.$$

Therefore, also

$$\frac{\Phi(\zeta)}{\Phi(\zeta+\delta)} = -\frac{\Lambda'(\delta)}{\Lambda^2(\delta)}, \quad \zeta \geq 0, \quad \delta \leq a,$$

and dividing the last two equations yields Lobacevsky's functional equation[2]

$$\Phi(\zeta-\delta)\Phi(\zeta+\delta) = \Phi(\zeta)^2, \quad \zeta \geq 0, \quad \delta \leq a, \tag{7}$$
$$\Phi(0) = 1.$$

Note, $\Phi$ is continuously differentiable, and strictly positive. Hence, by taking derivatives with respect to $\delta$, we get

$$\frac{\Phi'(\zeta-\delta)}{\Phi(\zeta-\delta)} = \frac{\Phi'(\zeta+\delta)}{\Phi(\zeta+\delta)},$$

and by setting $\zeta = \delta$, we thus have

$$\Phi'(2\zeta) = \alpha \Phi(2\zeta), \quad \Phi(0) = 1, \quad 0 < \zeta \leq a,$$

where $\alpha = \Phi'(0)/\Phi(0) \in \mathbb{R}$. We conclude that for some $\beta \in \mathbb{R}$,

$$\Phi(\zeta) = e^{\beta\zeta}, \quad 0 \leq \zeta \leq 2a. \tag{8}$$

By Equation (7), we can extend the exponential solution to $-a \leq \zeta < 0$: By setting $\zeta = 0$, we indeed have

$$\Phi(-\delta) = \frac{\Phi^2(0)}{\Phi(\delta)} = \frac{1}{e^{\beta\delta}} = e^{-\beta\delta}, \quad 0 < \delta \leq a.$$

Similarly, we can successively extend the validity of Equation (8) to the right, using the functional Equation (7). Now that $\Phi(\zeta) = e^{\beta\zeta}$ for all $\zeta \in [-a, \infty)$ we have, by taking the logarithmic derivative of $\Phi$, that $\mu(x)/\sigma^2(x)$ is indeed a constant on $[-a, \infty)$. □

Examples of processes for which the running maximum at drawdown is exponentially distributed are the following:

1. ($a = -\infty$): Brownian motion with drift $\sigma B_t + \mu t$.
2. ($a < \infty$): Reflected Brownian motion with drift, reflected at $-a$.
3. Similar examples as in 1 and 2 can be constructed, where $\mu(x)/\sigma^2(x)$ is constant. These include reflected diffusions.

However, there are processes that do not satisfy Theorem 2, even though they may exhibit exponentially distributed gains before $\delta$ drawdowns for specific choices of $\delta$. One can, for instance, let $\mu/\sigma^2$ be constant only on $[-1, \infty)$, and modify $\mu, \sigma^2$ on $[-2, -1)$ in such a way, that the SDE Equation (3) has unique global strong solution. Then, by Proposition 1, for any $\delta < 1$ the maximum at

---

[2] See (Aczél (1966) p. 82, Chapter 2 Equation (16)) and the references therein.

drawdown of size $\delta$ is exponentially distributed. It goes without saying, that there must exist $\delta > 1$, for which this is not the case.

Similar, but more sophisticated, examples can be constructed by solving delay differential equations for $\Phi(\cdot) = e^{-2\int_0^{\cdot} \mu(u)/\sigma^2(u)du}$, such that only for a specific threshold $\delta$, $M^\delta$ is exponentially distributed. Equation (6) reads in differential form:

$$\Phi'(\xi) = \Lambda(\delta)\left(\Phi(\xi) - \Phi(\xi - \delta)\right), \quad \xi \geq 0,$$

which is the simplest non-trivial (discrete) delay differential equation. To construct a diffusion process for which the maximum before a drawdown of size 1 is exponentially distributed with parameter one, we set $\Lambda(\delta) = \delta = 1$, and we choose a strictly positive continuous function $g(x)$ on $[-1,0]$ satisfying $g(0) = 1$. To obtain $\Phi$ on $[0,\infty)$, we solve

$$\Phi'(\xi) = \Phi(\xi) - \Phi(\xi - 1), \quad \xi \geq 0,$$

subject to $\Phi(\xi) = g(\xi)$ for $\xi \in [-1,0]$. This problem has a unique solution with exponential growth. However, if $g$ is not an exponentially linear function (that is, of the form $e^{\lambda x}$ for some $\lambda > 0$), then $\Phi$ is not, and therefore $\mu/\sigma^2$ is not constant. An underlying diffusion process $X$ with $M^1$ being exponentially distributed with parameter one can for instance be constructed, by solving SDE Equation (3), where $\sigma = 1$ and $\mu = -\frac{\Phi'(x)}{2\Phi(x)}$ on $[-1,\infty)$. Due to Theorem 2, $M^\delta$ is, in general, not exponentially distributed.

## 4. Lehoczky's Proof for Spectrally Negative Lévy Martingales

We study in this section the distribution of maximal gains[3] of processes, prior to the occurrence of a fixed loss $\delta > 0$. Golub et al. (2016, 2018) claim that for a Brownian motion (the toy model of a fair game), this gain is exponentially distributed, with parameter $\delta$; thus, on average, one gains $\delta$ before experiencing a loss of size $\delta$. This result is independent of the volatility of the Brownian motion. In private communication, Golub (2014) raised the question of whether similar scaling laws hold for other processes, e.g., other diffusion models, or processes with jumps. Such models are useful as benchmark models in the context of certain event-based high-frequency trading algorithms, where the Brownian motion is used as a proxy for an asset, and the location of the maximum suggests the beginning of a trend reversal.[4]

The conjecture that a fair game on average experiences the exact same gain as is lost later on may appear intuitive. And this is indeed the case for many continuous-time martingales, those who are time-changed Brownian motions, with a quadratic variation tending to infinity, along almost every path (because the timing is not relevant here). But it is not true for Lévy martingales, as can be seen from Theorem 4. Nevertheless, the (exponential) distribution of gains, not its parameter, is universal within the class of spectrally negative Lévy processes. Besides, the martingale property is not needed to arrive at this result.

After Theorem 4 was proved in the summer of 2019, F. Hubalek kindly pointed out that the result is, in identical form, preceded by Mijatović and Pistorius (2012). Our proof is, however, similar to the one of Lehoczky (1977), and is therefore an alternative, and simpler one. Finally, we also found a replication of Lehoczky's proof in Landriault et al. (2017), Lemma 3.1, however, this proof is also more difficult than ours due the more general discretization used therein.

---

[3] This random gain is called "overshoot" in Golub et al. (2016). In this section, we refrain from using this terminology due to its established meaning in the field of Lévy processes—it is the discrepancy between a certain threshold, and a jump processes' value, passing beyond that threshold.

[4] It goes without saying that the first time this maximum is attained is not a stopping time; otherwise, one could devise arbitrage strategies that short-sell the asset at the maximum.

We assume, that a Lévy process $X$ is given with downward jumps only but not equal to the negative of a Lévy subordinator and not being a deterministic drift[5]. Such a process is defined by its Lévy exponent

$$\Psi(\theta) := \frac{1}{t} \log \mathbb{E}[e^{\theta X_t}], \quad \theta > 0,$$

which is of the form

$$\Psi(\theta) = \mu\theta + \frac{\sigma^2\theta^2}{2} + \int_{(-\infty,0)} \left(e^{\theta\xi} - 1 - \theta\xi 1_{[-1,0)}(\xi)\right) \nu(d\xi), \quad \theta > 0,$$

with Lévy-Khintchine triplet $\mu \in \mathbb{R}$, $\sigma \in \mathbb{R}$ and a measure $\nu(d\xi)$ supported on $(-\infty, 0)$, integrating $\min(\xi^2, 1)$.

The scale function $W$ is the unique absolutely continuous function $[0, \infty) \to [0, \infty)$ with Laplace transform

$$\int_0^\infty e^{-\theta x} W(x) dx = \frac{1}{\Psi(\theta)}, \quad \theta > 0.$$

Since the processes lack positive jumps, they can only creep up. This assumption is essential to obtain exit probabilities from compact intervals and also for the main Theorem 4.

**Theorem 3.** *(Bertoin 1996, Theorem VII.8) Let $x, y > 0$, the probability that $X$ makes its first exit from $[-x, y]$ at $y$ is*

$$\mathbb{P}[\tau_y < \tau_{-x}] = \frac{W(x)}{W(x+y)}.$$

For a threshold $\delta > 0$, we define $M^\delta$ as the supremum of $X$, prior to a drawdown of size $\delta$, that is

$$M^\delta = M(\tau^\delta), \quad \text{where} \quad M(t) := \sup_{s \leq t} X_s, \quad \text{and} \quad \tau^\delta := \inf\{t > 0 \mid M_t - X_t \geq \delta\}.$$

We are ready to state and proof the main theorem:

**Theorem 4.** *For a spectrally negative Lévy process, the maximal gain $M^\delta$ before a $\delta$-loss is exponentially distributed with parameter equal to the logarithmic derivative of the scale function, that is,*

$$\mathbb{P}[M^\delta \geq \xi] = e^{-\frac{W'(\delta+)}{W(\delta)}\xi}.$$

**Proof of Theorem 4.** The proof is inspired by Golub et al. (2016), however, the exact same idea can be traced back to Lehoczky (1977) in the general context of univariate diffusions processes. Let $A_{k,n}$ be the event that $X$ reaches $k\xi/2^n$ before $-\delta + (k-1)/2^n\xi$ ($k = 1, \ldots, 2^n$). The set $\{M^\delta \geq \xi\}$ can be approximated by $\bigcap_{k=1}^{2^n} A_{k,n}$, which are decreasing for increasing $n$. In other words,

$$\{M^\delta \geq \xi\} = \bigcap_{n=1}^\infty \bigcap_{k=1}^{2^n} A_{k,n}.$$

Therefore,

$$\mathbb{P}[M^\delta \geq \xi] = \lim_{n \to \infty} \mathbb{P}\left[\bigcap_{k=1}^{2^n} A_{k,n}\right].$$

---

[5] This is the natural non-degeneracy condition of Bertoin (1996), Chapter VII to ensure that the process creeps up to any level.

Due to state-independence of the process (translation invariance) and the Markov property

$$\mathbb{P}\left[\bigcap_{k=1}^{2^n} A_{k,n}\right] = \mathbb{P}[A_{1,n}] \times \prod_{k=2}^{2^n} \mathbb{P}[A_{k,n} \mid A_{k-1,n}] = (\mathbb{P}[A_{1,n}])^{2^n} = \left(\frac{W(\delta)}{W(\delta + \xi/2^n)}\right)^{2^n},$$

where the last identity follows from Theorem 3. Since $W$ is differentiable from the right at $\delta$, applying L'Hospital's rule yields

$$\log \mathbb{P}[M^\delta \geq \xi] = \lim_{n \to \infty} \log(\mathbb{P}[A_{1,n}])^{2^n} = -\xi \frac{W'(\delta+)}{W(\delta)}.$$

□

**Remark 2.** *Theorem 4 implicitly requires right-differentiability of the scale functions, which is for free, because it can be rewritten as an integral of the tail of some finite measure, see (Bertoin (1996), Chapter VII). However, in many models, full $C^1$-regularity is guaranteed (cf. the characterization given by (Kuznetsov et al. (2012), Lemma 2.4)).*

*Examples*

The scale functions for the below processes are taken from review article of Hubalek and Kyprianou (2011). Throughout this section, $\mathcal{E}(\lambda)$ denotes the exponential distribution with parameter $\lambda > 0$.

**Example 1** (Compound Poisson Process). *Assume we have a compound Poisson process with negative exponentially distributed jumps,*

$$X_t = ct - \sum_{k=0}^{N_t^\lambda} \xi_k, \quad \xi_k \text{ i.i.d. and } \sim \mathcal{E}(\mu), \quad c - \lambda/\mu > 0.$$

We get

$$W(x) = \frac{1}{c}\left(1 + \frac{\lambda}{c\mu - \lambda}(1 - e^{-(\mu - \lambda/c)x})\right).$$

Clearly, $W \in C^1(0, \infty)$,

$$W'(x) = \frac{\lambda}{c^2} e^{-(\mu - \lambda/c)x}.$$

Therefore, by Theorem 4

$$M^\delta \sim \mathcal{E}\left(\frac{\lambda/c}{e^{\delta(\mu - \lambda/c)} - \frac{\lambda/c}{\mu - \lambda/c}}\right), \quad \lim_{\delta \downarrow 0} \mathbb{E}[M^\delta] = \lambda/c > 0, \quad \lim_{\delta \uparrow \infty} \mathbb{E}[M^\delta] = \mu - \lambda/c < \infty.$$

Unlike the previous example, the following two examples exhibit the same qualitative dependence on the threshold $\delta$, as the standard Brownian motion, where $M^\delta \sim \mathcal{E}(1/\delta)$: when $\delta \to 0$, the average maximum at drawdown of size $\delta$ tends to 0, and when $\delta \to \infty$, this average goes to infinity.

**Example 2** (Brownian motion with drift). *A Brownian motion with drift $\mu > 0$ and volatility $\sigma$,*

$$X_t = \mu t + \sigma B_t$$

*has scale function*

$$W(x) \sim e^{-\mu x/\sigma^2} \sinh(\sqrt{\mu} x/\sigma^2).$$

Hence,
$$\frac{W'(x)}{W(x)} = \frac{-\mu/\sigma^2 \sinh(\sqrt{\bar{\mu}}x/\sigma^2) + \sqrt{\bar{\mu}}/\sigma^2 \cosh(\sqrt{\bar{\mu}}x/\sigma^2)}{\sinh(\sqrt{\bar{\mu}}x/\sigma^2)}.$$

Therefore, by Theorem 4 (see, e.g., Golub et al. (2016)),
$$M^\delta \sim \mathcal{E}\left(\mu/\sigma^2 \left(\coth(\sqrt{\bar{\mu}}\delta/\sigma^2) - 1\right)\right).$$

**Example 3** (Caballero and Chaumont (2006)). *This is a Lévy process without diffusion component, defined by its Lévy measure*
$$\nu(d\xi) = \frac{e^{(\beta-1)\xi}}{(e^\xi - 1)^{\beta+1}}, \quad \xi < 0,$$

*where $\beta \in (1,2)$, and its Laplace exponent,*
$$\Psi(\theta) = \frac{\Gamma(\theta + \beta)}{\Gamma(\theta)\Gamma(\beta)}, \quad \theta > 0.$$

*The process exhibits Infinite variation jumps, and drifts to $-\infty$, because $\Psi'(0) < 0$. The scale function is*
$$W(x) = (1 - e^{-x})^{\beta-1}.$$

*Using Theorem 4, we thus get*
$$M^\delta \sim \mathcal{E}\left(\frac{\beta-1}{e^\delta-1}\right), \quad \mathbb{E}[M^\delta] = \frac{e^\delta - 1}{\beta - 1}.$$

The asymptotic behaviour of the logarithmic derivative of the scale function of a spectrally negative Lévy process can be characterized using the asymptotic behaviour of $W$ and $W'$, cf. (Kuznetsov et al. 2012, Chapter 3). For instance, $W(0) = W(0+) = 0$, if and only if the process is of infinite variation. In the case of finite variation, we can write the process as $\delta t - J_t$, where $J$ is a subordinator; and then $W(0) = 1/\delta > 0$. Furthermore, $W'(0+) = \infty$, if a diffusion component is present, or if the Lévy measure is infinite. These general findings are consistent with the three examples.

**Funding:** This research received no external funding.

**Acknowledgments:** I thank John Appleby, Florin Avram, Huayuan Dong, Friedrich Hubalek, Andreas Kyprianou and two anonymous referees for useful comments.

**Conflicts of Interest:** The author declares no conflict of interest.

## References

Aczél, János. 1966. *Lectures on Functional Equations and Their Applications*. Waltham: Academic Press, vol. 19.
Avram, Florin, Danijel Grahovac, and Ceren Vardar-Acar. 2017. The W, Z scale functions kit for first passage problems of spectrally negative Lévy processes, and applications to the optimization of dividends. *arXiv*. arXiv:1706.06841.
Bertoin, Jean. 1996. *Lévy Processes*. Cambridge: Cambridge University Press, vol. 121.
Borodin, Andrei. N., and Paavo Salminen. 2012. *Handbook of Brownian Motion-Facts and Formulae*. Basel: Birkhäuser.
Caballero, Maria Emilia, and Loïc Chaumont. 2006. Conditioned stable Lévy processes and the Lamperti representation. *Journal of Applied Probability* 43: 967–83. [CrossRef]
Golub, Anton. 2014. Flov Technologies, Zürich, Switzerland. Private communication.
Golub, Anton, Gregor Chliamovitch, Alexandre Dupuis, and Bastien Chopard. 2016. Multi-scale representation of high frequency market liquidity. *Algorithmic Finance* 5: 3–19. [CrossRef]
Golub, Anton, James B. Glattfelder, and Richard B. Olsen. 2018. The alpha engine: Designing an automated trading algorithm. In *High-Performance Computing in Finance*. London: Chapman and Hall/CRC, pp. 49–76.

Hubalek, Friedrich, and Andreas E. Kyprianou. 2011. Old and new examples of scale functions for spectrally negative Lévy processes. In *Seminar on Stochastic Analysis, Random Fields and Applications VI*. Basel: Springer, pp. 119–45.

Kuznetsov, Alexey., Andreas E. Kyprianou, and Victor Rivero. 2012. The theory of scale functions for spectrally negative Lévy processes. In *Lévy Matters II*. New York: Springer, pp. 97–186.

Landriault, David, Bin Li, and Hongzhong Zhang. 2017. On magnitude, asymptotics and duration of drawdowns for Lévy models. *Bernoulli* 23: 432–58. [CrossRef]

Lehoczky, John P. 1977. Formulas for stopped diffusion processes with stopping times based on the maximum. *The Annals of Probability* 5: 601–7. [CrossRef]

Mijatović, Aleksandar, and Martijn R. Pistorius. 2012. On the drawdown of completely asymmetric Lévy processes. *Stochastic Processes and their Applications* 122: 3812–36. [CrossRef]

Perry, David, Wolfgang Stadje, and Shelemyahu Zacks. 2004. The first rendezvous time of Brownian motion and compound Poisson-type processes. *Journal of Applied Probability* 41: 1059–70. [CrossRef]

Taylor, Howard M. 1975. A stopped Brownian motion formula. *The Annals of Probability* 3: 234–46. [CrossRef]

Yamada, Toshio, and Shinzo Watanabe. 1971. On the uniqueness of solutions of stochastic differential equations. *Journal of Mathematics of Kyoto University* 11: 155–67. [CrossRef]

© 2019 by the author. Licensee MDPI, Basel, Switzerland. This article is an open access article distributed under the terms and conditions of the Creative Commons Attribution (CC BY) license (http://creativecommons.org/licenses/by/4.0/).

Article

# Fluctuation Theory for Upwards Skip-Free Lévy Chains

Matija Vidmar

Department of Mathematics, Faculty of Mathematics and Physics, University of Ljubljana, 1000 Ljubljana, Slovenia; matija.vidmar@fmf.uni-lj.si

Received: 20 July 2018; Accepted: 16 September 2018; Published: 18 September 2018

**Abstract:** A fluctuation theory and, in particular, a theory of scale functions is developed for upwards skip-free Lévy chains, i.e., for right-continuous random walks embedded into continuous time as compound Poisson processes. This is done by analogy to the spectrally negative class of Lévy processes—several results, however, can be made more explicit/exhaustive in the compound Poisson setting. Importantly, the scale functions admit a linear recursion, of constant order when the support of the jump measure is bounded, by means of which they can be calculated—some examples are presented. An application to the modeling of an insurance company's aggregate capital process is briefly considered.

**Keywords:** Lévy processes; non-random overshoots; skip-free random walks; fluctuation theory; scale functions; capital surplus process

## 1. Introduction

It was shown in Vidmar (2015) that precisely two types of Lévy processes exhibit the property of non-random overshoots: those with no positive jumps a.s., and compound Poisson processes, whose jump chain is (for some $h > 0$) a random walk on $\mathbb{Z}_h := \{hk : k \in \mathbb{Z}\}$, skip-free to the right. The latter class was then referred to as "upwards skip-free Lévy chains". Also in the same paper it was remarked that this common property which the two classes share results in a more explicit fluctuation theory (including the Wiener-Hopf factorization) than for a general Lévy process, this being rarely the case (cf. (Kyprianou 2006, p. 172, sct. 6.5.4)).

Now, with reference to existing literature on fluctuation theory, the spectrally negative case (when there are no positive jumps, a.s.) is dealt with in detail in (Bertoin 1996, chp. VII); (Sato 1999, sct. 9.46) and especially (Kyprianou 2006, chp. 8). On the other hand, no equally exhaustive treatment of the right-continuous random walk seems to have been presented thus far, but see Brown et al. (2010); Marchal (2001); Quine (2004); (De Vylder and Goovaerts 1988, sct. 4); (Dickson and Waters 1991, sct. 7); (Doney 2007, sct. 9.3); (Spitzer 2001, passim).[1] In particular, no such exposition appears forthcoming for the continuous-time analogue of such random walks, wherein the connection and analogy to the spectrally negative class of Lévy processes becomes most transparent and direct.

In the present paper, we proceed to do just that, i.e., we develop, by analogy to the spectrally negative case, a complete fluctuation theory (including theory of scale functions) for upwards skip-free Lévy chains. Indeed, the transposition of the results from the spectrally negative to the skip-free setting is mostly straightforward. Over and above this, however, and beyond what is purely analogous to the exposition of the spectrally negative case, (i) further specifics of the reflected process (Theorem 1-1),

---

[1] However, such a treatment did eventually become available (several years after this manuscript was essentially completed, but before it was published), in the preprint Avram and Vidmar (2017).

of the excursions from the supremum (Theorem 1-3) and of the inverse of the local time at the maximum (Theorem 1-4) are identified, (ii) the class of subordinators that are the descending ladder heights processes of such upwards skip-free Lévy chains is precisely characterized (Theorem 4), and (iii) a linear recursion is presented which allows us to directly compute the families of scale functions (Equations (20), (21), Proposition 9 and Corollary 1).

Application-wise, note that the classical continuous-time Bienaymé-Galton-Watson branching process is associated with upwards skip-free Lévy chains via a suitable time change (Kyprianou 2006, sct. 1.3.4). Besides, our chains feature as a natural continuous-time approximation of the more subtle spectrally negative Lévy family, that, because of its overall tractability, has been used extensively in applied probability (in particular to model the risk process of an insurance company; see the papers Avram et al. (2007); Chiu and Yin (2005); Yang and Zhang (2001) among others). This approximation point of view is developed in Mijatović et al. (2014, 2015). Finally, focusing on the insurance context, the chains may be used directly to model the aggregate capital process of an insurance company, in what is a continuous-time embedding of the discrete-time compound binomial risk model (for which see Avram and Vidmar (2017); Bao and Liu (2012); Wat et al. (2018); Xiao and Guo (2007) and the references therein). We elaborate on this latter point of view in Section 5.

The organisation of the rest of this paper is as follows. Section 2 introduces the setting and notation. Then Section 3 develops the relevant fluctuation theory, in particular details of the Wiener-Hopf factorization. Section 4 deals with the two-sided exit problem and the accompanying families of scale functions. Finally, Section 5 closes with an application to the risk process of an insurance company.

## 2. Setting and Notation

Let $(\Omega, \mathcal{F}, \mathbb{F} = (\mathcal{F}_t)_{t \geq 0}, \mathsf{P})$ be a filtered probability space supporting a Lévy process (Kyprianou 2006, p. 2, Definition 1.1) $X$ ($X$ is assumed to be $\mathbb{F}$-adapted and to have independent increments relative to $\mathbb{F}$). The Lévy measure (Sato 1999, p. 38, Definition 8.2) of $X$ is denoted by $\lambda$. Next, recall from Vidmar (2015) (with $\mathrm{supp}(\nu)$ denoting the support (Kallenberg 1997, p. 9) of a measure $\nu$ defined on the Borel $\sigma$-field of some topological space):

**Definition 1** (Upwards skip-free Lévy chain). *$X$ is an upwards skip-free Lévy chain, if it is a compound Poisson process (Sato 1999, p. 18, Definition 4.2), viz. if $\mathsf{E}[e^{izX_t}] = e^{t \int (e^{izx} - 1)\lambda(dx)}$ for $z \in \mathbb{R}$ and $t \in [0, \infty)$, and if for some $h > 0$, $\mathrm{supp}(\lambda) \subset \mathbb{Z}_h$, whereas $\mathrm{supp}(\lambda|_{\mathcal{B}((0,\infty))}) = \{h\}$.*

**Remark 1.** *Of course to say that $X$ is a compound Poisson process means simply that it is a real-valued continuous-time Markov chain, vanishing a.s. at zero, with holding times exponentially distributed of rate $\lambda(\mathbb{R})$ and the law of the jumps given by $\lambda/\lambda(\mathbb{R})$ (Sato 1999, p. 18, Theorem 4.3).*

In the sequel, $X$ will be assumed throughout an upwards skip-free Lévy chain, with $\lambda(\{h\}) > 0$ ($h > 0$) and characteristic exponent $\Psi(p) = \int (e^{ipx} - 1)\lambda(dx)$ ($p \in \mathbb{R}$). In general, we insist on (i) every sample path of $X$ being càdlàg (i.e., right-continuous, admitting left limits) and (ii) $(\Omega, \mathcal{F}, \mathbb{F}, \mathsf{P})$ satisfying the standard assumptions (i.e., the $\sigma$-field $\mathcal{F}$ is P-complete, the filtration $\mathbb{F}$ is right-continuous and $\mathcal{F}_0$ contains all P-null sets). Nevertheless, we shall, sometimes and then only provisionally, relax assumption (ii), by transferring $X$ as the coordinate process onto the canonical space $\mathbb{D}_h := \{\omega \in \mathbb{Z}_h^{[0,\infty)} : \omega \text{ is càdlàg}\}$ of càdlàg paths, mapping $[0, \infty) \to \mathbb{Z}_h$, equipping $\mathbb{D}_h$ with the $\sigma$-algebra and natural filtration of evaluation maps; this, however, will always be made explicit. We allow $e_1$ to be exponentially distributed, mean one, and independent of $X$; then define $e_p := e_1/p$ ($p \in (0, \infty) \backslash \{1\}$).

Furthermore, for $x \in \mathbb{R}$, introduce $T_x := \inf\{t \geq 0 : X_t \geq x\}$, the first entrance time of $X$ into $[x, \infty)$. Please note that $T_x$ is an $\mathbb{F}$-stopping time (Kallenberg 1997, p. 101, Theorem 6.7). The supremum or maximum (respectively infimum or minimum) process of $X$ is denoted $\overline{X}_t := \sup\{X_s : s \in [0, t]\}$ (respectively $\underline{X}_t := \inf\{X_s : s \in [0, t]\}$) ($t \geq 0$). $\underline{X}_\infty := \inf\{X_s : s \in [0, \infty)\}$ is the overall infimum.

With regard to miscellaneous general notation we have:

1. The nonnegative, nonpositive, positive and negative real numbers are denoted by $\mathbb{R}_+ := \{x \in \mathbb{R} : x \geq 0\}$, $\mathbb{R}_- := \{x \in \mathbb{R} : x \leq 0\}$, $\mathbb{R}^+ := \mathbb{R}_+ \backslash \{0\}$ and $\mathbb{R}^- := \mathbb{R}_- \backslash \{0\}$, respectively. Then $\mathbb{Z}_+ := \mathbb{R}_+ \cap \mathbb{Z}$, $\mathbb{Z}_- := \mathbb{R}_- \cap \mathbb{Z}$, $\mathbb{Z}^+ := \mathbb{R}^+ \cap \mathbb{Z}$ and $\mathbb{Z}^- := \mathbb{R}^- \cap \mathbb{Z}$ are the nonnegative, nonpositive, positive and negative integers, respectively.

2. Similarly, for $h > 0$: $\mathbb{Z}_h^+ := \mathbb{Z}_h \cap \mathbb{R}_+$, $\mathbb{Z}_h^{++} := \mathbb{Z}_h \cap \mathbb{R}^+$, $\mathbb{Z}_h^- := \mathbb{Z}_h \cap \mathbb{R}_-$ and $\mathbb{Z}_h^{--} := \mathbb{Z}_h \cap \mathbb{R}^-$ are the apposite elements of $\mathbb{Z}_h$.

3. The following introduces notation for the relevant half-planes of $\mathbb{C}$; the arrow notation is meant to be suggestive of which half-plane is being considered: $\mathbb{C}^\rightarrow := \{z \in \mathbb{C} : \Re z > 0\}$, $\mathbb{C}^\leftarrow := \{z \in \mathbb{C} : \Re z < 0\}$, $\mathbb{C}^\downarrow := \{z \in \mathbb{C} : \Im z < 0\}$ and $\mathbb{C}^\uparrow := \{z \in \mathbb{C} : \Im z > 0\}$. $\overline{\mathbb{C}^\rightarrow}$, $\overline{\mathbb{C}^\leftarrow}$, $\overline{\mathbb{C}^\downarrow}$ and $\overline{\mathbb{C}^\uparrow}$ are then the respective closures of these sets.

4. $\mathbb{N} = \{1, 2, \ldots\}$ and $\mathbb{N}_0 = \mathbb{N} \cup \{0\}$ are the positive and nonnegative integers, respectively. $\lceil x \rceil := \inf\{k \in \mathbb{Z} : k \geq x\}$ ($x \in \mathbb{R}$) is the ceiling function. For $\{a, b\} \subset [-\infty, +\infty]$: $a \wedge b := \min\{a, b\}$ and $a \vee b := \max\{a, b\}$.

5. The Laplace transform of a measure $\mu$ on $\mathbb{R}$, concentrated on $[0, \infty)$, is denoted by $\hat{\mu}$: $\hat{\mu}(\beta) = \int_{[0,\infty)} e^{-\beta x} \mu(dx)$ (for all $\beta \geq 0$ such that this integral is finite). To a nondecreasing right-continuous function $F : \mathbb{R} \to \mathbb{R}$, a measure $dF$ may be associated in the Lebesgue-Stieltjes sense.

The geometric law geom($p$) with success parameter $p \in (0, 1]$ has geom($p$)$(\{k\}) = p(1-p)^k$ ($k \in \mathbb{N}_0$), $1-p$ is then the failure parameter. The exponential law Exp($\beta$) with parameter $\beta > 0$ is specified by the density Exp($\beta$)$(dt) = \beta e^{-\beta t} \mathbb{1}_{(0,\infty)}(t)dt$. A function $f : [0, \infty) \to [0, \infty)$ is said to be of exponential order, if there are $\{\alpha, A\} \subset \mathbb{R}_+$, such that $f(x) \leq Ae^{\alpha x}$ ($x \geq 0$); $f(+\infty) := \lim_{x \to +\infty} f(x)$, when this limit exists. DCT (respectively MCT) stands for the dominated (respectively monotone) convergence theorem. Finally, increasing (respectively decreasing) will mean strictly increasing (respectively strictly decreasing), nondecreasing (respectively nonincreasing) being used for the weaker alternative; we will understand $a/0 = \pm \infty$ for $a \in \pm(0, \infty)$.

## 3. Fluctuation Theory

In the following section, to fully appreciate the similarity (and eventual differences) with the spectrally negative case, the reader is invited to directly compare the exposition of this subsection with that of (Bertoin 1996, sct. VII.1) and (Kyprianou 2006, sct. 8.1).

### 3.1. Laplace Exponent, the Reflected Process, Local Times and Excursions from the Supremum, Supremum Process and Long-Term Behaviour, Exponential Change of Measure

Since the Poisson process admits exponential moments of all orders, it follows that $\mathsf{E}[e^{\beta X_t}] < \infty$ and, in particular, $\mathsf{E}[e^{\beta X_t}] < \infty$ for all $\{\beta, t\} \subset [0, \infty)$. Indeed, it may be seen by a direct computation that for $\beta \in \overline{\mathbb{C}^\rightarrow}$, $t \geq 0$, $\mathsf{E}[e^{\beta X_t}] = \exp\{t\psi(\beta)\}$, where $\psi(\beta) := \int_{\mathbb{R}}(e^{\beta x} - 1)\lambda(dx)$ is the Laplace exponent of $X$. Moreover, $\psi$ is continuous (by the DCT) on $\overline{\mathbb{C}^\rightarrow}$ and analytic in $\mathbb{C}^\rightarrow$ (use the theorems of Cauchy (Rudin 1970, p. 206, 10.13 Cauchy's theorem for triangle), Morera (Rudin 1970, p. 209, 10.17 Morera's theorem) and Fubini).

Next, note that $\psi(\beta)$ tends to $+\infty$ as $\beta \to \infty$ over the reals, due to the presence of the atom of $\lambda$ at $h$. Upon restriction to $[0, \infty)$, $\psi$ is strictly convex, as follows first on $(0, \infty)$ by using differentiation under the integral sign and noting that the second derivative is strictly positive, and then extends to $[0, \infty)$ by continuity.

Denote then by $\Phi(0)$ the largest root of $\psi|_{[0,\infty)}$. Indeed, 0 is always a root, and due to strict convexity, if $\Phi(0) > 0$, then 0 and $\Phi(0)$ are the only two roots. The two cases occur, according as to whether $\psi'(0+) \geq 0$ or $\psi'(0+) < 0$, which is clear. It is less obvious, but nevertheless true, that this right derivative at 0 actually exists, indeed $\psi'(0+) = \int_{\mathbb{R}} x\lambda(dx) \in [-\infty, \infty)$. This follows from the fact that $(e^{\beta x} - 1)/\beta$ is nonincreasing as $\beta \downarrow 0$ for $x \in \mathbb{R}_-$ and hence the monotone convergence applies. Continuing from this, and with a similar justification, one also gets the equality $\psi''(0+) = \int x^2 \lambda(dx) \in (0, +\infty]$ (where we agree $\psi''(0+) = +\infty$ if $\psi'(0+) = -\infty$). In any case, $\psi : [\Phi(0), \infty) \to [0, \infty)$ is continuous and increasing, it is a bijection and we let $\Phi : [0, \infty) \to [\Phi(0), \infty)$ be the inverse bijection, so that $\psi \circ \Phi = \mathrm{id}_{\mathbb{R}_+}$.

With these preliminaries having been established, our first theorem identifies characteristics of the reflected process, the local time of $X$ at the maximum (for a definition of which see e.g., (Kyprianou 2006, p. 140, Definition 6.1)), its inverse, as well as the expected length of excursions and the probability of an infinite excursion therefrom (for definitions of these terms see e.g., (Kyprianou 2006, pp. 140–47); we agree that an excursion (from the maximum) starts immediately after $X$ leaves its running maximum and ends immediately after it returns to it; by its length we mean the amount of time between these two time points).

**Theorem 1** (Reflected process; (inverse) local time; excursions). *Let $q_n := \lambda(\{-nh\})/\lambda(\mathbb{R})$ for $n \in \mathbb{N}$ and $p := \lambda(\{h\})/\lambda(\mathbb{R})$.*

1. *The generator matrix $\tilde{Q}$ of the Markov process $Y := \overline{X} - X$ on $\mathbb{Z}_h^+$ is given by (with $\{s, s'\} \subset \mathbb{Z}_h^+$): $\tilde{Q}_{ss'} = \lambda(\{s-s'\}) - \delta_{ss'}\lambda(\mathbb{R})$, unless $s = s' = 0$, in which case we have $\tilde{Q}_{ss'} = -\lambda((-\infty, 0))$.*
2. *For the reflected process $Y$, $0$ is a holding point. The actual time spent at $0$ by $Y$, which we shall denote $L$, is a local time at the maximum. Its right-continuous inverse $L^{-1}$, given by $L_t^{-1} := \inf\{s \geq 0 : L_s > t\}$ (for $0 \leq t < L_\infty$; $L_t^{-1} := \infty$ otherwise), is then a (possibly killed) compound Poisson subordinator with unit positive drift.*
3. *Assuming that $\lambda((-\infty, 0)) > 0$ to avoid the trivial case, the expected length of an excursion away from the supremum is equal to $\frac{\lambda(\{h\})h - \psi'(0+)}{(\psi'(0+) \vee 0)\lambda((-\infty,0))}$; whereas the probability of such an excursion being infinite is $\frac{\lambda(\{h\})}{\lambda((-\infty,0))}(e^{\Phi(0)h} - 1) =: p^*$.*
4. *Assume again $\lambda((-\infty, 0)) > 0$ to avoid the trivial case. Let $N$, taking values in $\mathbb{N} \cup \{+\infty\}$, be the number of jumps the chain makes before returning to its running maximum, after it has first left it (it does so with probability 1). Then the law of $L^{-1}$ is given by (for $\theta \in [0, +\infty)$):*

$$-\log \mathbb{E}\left[\exp(-\theta L_1^{-1})\mathbb{1}_{\{L_1^{-1} < +\infty\}}\right] = \theta + \lambda((-\infty, 0))\left(1 - \sum_{k=1}^{\infty} P(N = k)\left(\frac{\lambda(\mathbb{R})}{\lambda(\mathbb{R}) + \theta}\right)^k\right).$$

*In particular, $L^{-1}$ has a killing rate of $\lambda((-\infty, 0))p^*$, Lévy mass $\lambda((-\infty, 0))(1 - p^*)$ and its jumps have the probability law on $(0, +\infty)$ given by the length of a generic excursion from the supremum, conditional on it being finite, i.e., that of an independent $N$-fold sum of independent $\mathrm{Exp}(\lambda(\mathbb{R}))$-distributed random variables, conditional on $N$ being finite. Moreover, one has, for $k \in \mathbb{N}$, $P(N = k) = \sum_{l=1}^{k} q_l p_{l,k}$, where the coefficients $(p_{l,k})_{l,k=1}^{\infty}$ satisfy the initial conditions:*

$$p_{l,1} = p\delta_{l1}, \quad l \in \mathbb{N};$$

*the recursions:*

$$p_{l,k+1} = \begin{cases} 0 & \text{if } l = k \text{ or } l > k+1 \\ \sum_{m=1}^{k-1} q_m p_{m+1,k} & \text{if } l = 1 \\ p^{k+1} & \text{if } l = k+1 \\ pp_{l-1,k} + \sum_{m=1}^{k-l} q_m p_{m+l,k} & \text{if } 1 < l < k \end{cases}, \quad \{l, k\} \subset \mathbb{N};$$

*and $p_{l,k}$ may be interpreted as the probability of $X$ reaching level $0$ starting from level $-lh$ for the first time on precisely the $k$-th jump ($\{l, k\} \subset \mathbb{N}$).*

**Proof.** Theorem 1-1 is clear, since, e.g., $Y$ transitions away from $0$ at the rate at which $X$ makes a negative jump; and from $s \in \mathbb{Z}_h^+ \backslash \{0\}$ to $0$ at the rate at which $X$ jumps up by $s$ or more etc.

Theorem 1-2 is standard (Kyprianou 2006, p. 141, Example 6.3 & p. 149, Theorem 6.10).

We next establish Theorem 1-3. Denote, provisionally, by $\beta$ the expected excursion length. Furthermore, let the discrete-time Markov chain $W$ (on the state space $\mathbb{N}_0$) be endowed with the initial distribution $w_j := \frac{q_j}{1-p}$ for $j \in \mathbb{N}$, $w_0 := 0$; and transition matrix $P$, given by $P_{0i} = \delta_{0i}$, whereas for $i \geq 1$: $P_{ij} = p$, if $j = i - 1$; $P_{ij} = q_{j-i}$, if $j > i$; and $P_{ij} = 0$ otherwise ($W$ jumps down with probability $p$, up $i$ steps with probability $q_i$, $i \geq 1$, until it reaches 0, where it gets stuck). Further let $N$ be the first hitting time for $W$ of $\{0\}$, so that a typical excursion length of $X$ is equal in distribution to an independent sum of $N$ (possibly infinite) $\text{Exp}(\lambda(\mathbb{R}))$-random variables. It is Wald's identity that $\beta = (1/\lambda(\mathbb{R}))E[N]$. Then (in the obvious notation, where $\overline{\infty}$ indicates the sum is inclusive of $\infty$), by Fubini: $E[N] = \sum_{n=1}^{\overline{\infty}} n \sum_{l=1}^{\infty} w_l P_l(N = n) = \sum_{l=1}^{\infty} w_l k_l$, where $k_l$ is the mean hitting time of $\{0\}$ for $W$, if it starts from $l \in \mathbb{N}_0$, as in (Norris 1997, p. 12). From the skip-free property of the chain $W$ it is moreover transparent that $k_i = \alpha i$, $i \in \mathbb{N}_0$, for some $0 < \alpha \leq \infty$ (with the usual convention $0 \cdot \infty = 0$). Moreover we know (Norris 1997, p. 17, Theorem 1.3.5) that $(k_i : i \in \mathbb{N}_0)$ is the minimal solution to $k_0 = 0$ and $k_i = 1 + \sum_{j=1}^{\infty} P_{ij} k_j$ ($i \in \mathbb{N}$). Plugging in $k_i = \alpha i$, the last system of linear equations is equivalent to (provided $\alpha < \infty$) $0 = 1 - p\alpha + \alpha \zeta$, where $\zeta := \sum_{j=1}^{\infty} j q_j$. Thus, if $\zeta < p$, the minimal solution to the system is $k_i = i/(p - \zeta)$, $i \in \mathbb{N}_0$, from which $\beta = \zeta/(\lambda((-\infty, 0))(p - \zeta))$ follows at once. If $\zeta \geq p$, clearly we must have $\alpha = +\infty$, hence $E[N] = +\infty$ and thus $\beta = +\infty$.

To establish the probability of an excursion being infinite, i.e., $\sum_{i=1}^{\infty} q_i(1 - \alpha_i) / \sum_{i=1}^{\infty} q_i$, where $\alpha_i := P_i(N < \infty) > 0$, we see that (by the skip-free property) $\alpha_i = \alpha_1^i$, $i \in \mathbb{N}_0$, and by the strong Markov property, for $i \in \mathbb{N}$, $\alpha_i = p\alpha_{i-1} + \sum_{j=1}^{\infty} q_j \alpha_{i+j}$. It follows that $1 = p\alpha_1^{-1} + \sum_{j=1}^{\infty} q_j \alpha_1^j$, i.e., $0 = \psi(\log(\alpha_1^{-1})/h)$. Hence, by Theorem 2-2, whose proof will be independent of this one, $\alpha_1 = e^{-\Phi(0)h}$ (since $\alpha_1 < 1$, if and only if $X$ drifts to $-\infty$).

Finally, Theorem 1-4 is straightforward. □

We turn our attention now to the supremum process $\overline{X}$. First, using the lack of memory property of the exponential law and the skip-free nature of $X$, we deduce from the strong Markov property applied at the time $T_a$, that for every $a, b \in \mathbb{Z}_h^+$, $p > 0$: $P(T_{a+b} < e_p) = P(T_a < e_p)P(T_b < e_p)$. In particular, for any $n \in \mathbb{N}_0$: $P(T_{nh} < e_p) = P(T_h < e_p)^n$. And since for $s \in \mathbb{Z}_h^+$, $\{T_s < e_p\} = \{\overline{X}_{e_p} \geq s\}$ (P-a.s.) one has (for $n \in \mathbb{N}_0$): $P(\overline{X}_{e_p} \geq nh) = P(\overline{X}_{e_p} \geq h)^n$. Therefore $\overline{X}_{e_p}/h \sim \text{geom}(1 - P(\overline{X}_{e_p} \geq h))$.

Next, to identify $P(\overline{X}_{e_p} \geq h)$, $p > 0$, observe that (for $\beta \geq 0$, $t \geq 0$): $E[\exp\{\Phi(\beta)X_t\}] = e^{t\beta}$ and hence $(\exp\{\Phi(\beta)X_t - \beta t\})_{t \geq 0}$ is an $(\mathbb{F}, P)$-martingale by stationary independent increments of $X$, for each $\beta \geq 0$. Then apply the optional sampling theorem at the bounded stopping time $T_x \wedge t$ ($t, x \geq 0$) to get:

$$E[\exp\{\Phi(\beta)X(T_x \wedge t) - \beta(T_x \wedge t)\}] = 1.$$

Please note that $X(T_x \wedge t) \leq h\lceil x/h \rceil$ and $\Phi(\beta)X(T_x \wedge t) - \beta(T_x \wedge t)$ converges to $\Phi(\beta)h\lceil x/h \rceil - \beta T_x$ (P-a.s.) as $t \to \infty$ on $\{T_x < \infty\}$. It converges to $-\infty$ on the complement of this event, P a.s., provided $\beta + \Phi(\beta) > 0$. Therefore we deduce by dominated convergence, first for $\beta > 0$ and then also for $\beta = 0$, by taking limits:

$$E[\exp\{-\beta T_x\}\mathbb{1}_{\{T_x < \infty\}}] = \exp\{-\Phi(\beta)h\lceil x/h \rceil\}. \tag{1}$$

Before we formulate our next theorem, recall also that any non-zero Lévy process either drifts to $+\infty$, oscillates or drifts to $-\infty$ (Sato 1999, pp. 255–56, Proposition 37.10 and Definition 37.11).

**Theorem 2** (Supremum process and long-term behaviour).

1. *The failure probability for the geometrically distributed $\overline{X}_{e_p}/h$ is $\exp\{-\Phi(p)h\}$ ($p > 0$).*
2. *$X$ drifts to $+\infty$, oscillates or drifts to $-\infty$ according as to whether $\psi'(0+)$ is positive, zero, or negative. In the latter case $\overline{X}_\infty/h$ has a geometric distribution with failure probability $\exp\{-\Phi(0)h\}$.*
3. *$(T_{nh})_{n \in \mathbb{N}_0}$ is a discrete-time increasing stochastic process, vanishing at 0 and having stationary independent increments up to the explosion time, which is an independent geometric random variable; it is a killed random walk.*

**Remark 2.** *Unlike in the spectrally negative case (Bertoin 1996, p. 189), the supremum process cannot be obtained from the reflected process, since the latter does not discern a point of increase in X when the latter is at its running maximum.*

**Proof.** We have for every $s \in \mathbb{Z}_h^+$:

$$P(\overline{X}_{e_p} \geq s) = P(T_s < e_p) = E[\exp\{-pT_s\}\mathbb{1}_{\{T_s < \infty\}}] = \exp\{-\Phi(p)s\}. \qquad (2)$$

Thus Theorem 2-1 obtains.

For Theorem 2-2 note that letting $p \downarrow 0$ in (2), we obtain $\overline{X}_\infty < \infty$ (P-a.s.), if and only if $\Phi(0) > 0$, which is equivalent to $\psi'(0+) < 0$. If so, $\overline{X}_\infty/h$ is geometrically distributed with failure probability $\exp\{-\Phi(0)h\}$ and then (and only then) does X drift to $-\infty$.

It remains to consider the case of drifting to $+\infty$ (the cases being mutually exclusive and exhaustive). Indeed, X drifts to $+\infty$, if and only if $E[T_s]$ is finite for each $s \in \mathbb{Z}_h^+$ (Bertoin 1996, p. 172, Proposition VI.17). Using again the nondecreasingness of $(e^{-\beta T_s} - 1)/\beta$ in $\beta \in [0, \infty)$, we deduce from (1), by the monotone convergence, that one may differentiate under the integral sign, to get $E[T_s\mathbb{1}_{\{T_s < \infty\}}] = (\beta \mapsto -\exp\{-\Phi(\beta)s\})'(0+)$. So the $E[T_s]$ are finite, if and only if $\Phi(0) = 0$ (so that $T_s < \infty$ P-a.s.) and $\Phi'(0+) < \infty$. Since $\Phi$ is the inverse of $\psi|_{[\Phi(0),\infty)}$, this is equivalent to saying $\psi'(0+) > 0$.

Finally, Theorem 2-3 is clear. □

Table 1 briefly summarizes for the reader's convenience some of our main findings thus far.

**Table 1.** Connections between the quantities $\psi'(0+)$, $\Phi(0)$, $\Phi'(0+)$, the behaviour of X at large times, and the behaviour of its excursions away from the running supremum (the latter when $\lambda((-\infty, 0)) > 0$).

| $\psi'(0+)$ | $\Phi(0)$ | $\Phi'(0+)$ | Long-Term Behaviour | Excursion Length |
|---|---|---|---|---|
| $\in (0, \infty)$ | 0 | $\in (0, \infty)$ | drifts to $+\infty$ | finite expectation |
| 0 | 0 | $+\infty$ | oscillates | a.s. finite with infinite expectation |
| $\in [-\infty, 0)$ | $\in (0, \infty)$ | $\in (0, \infty)$ | drifts to $-\infty$ | infinite with a positive probability |

We conclude this section by offering a way to reduce the general case of an upwards skip-free Lévy chain to one which necessarily drifts to $+\infty$. This will prove useful in the sequel. First, there is a pathwise approximation of an oscillating X, by (what is again) an upwards skip-free Lévy chain, but drifting to infinity.

**Remark 3.** *Suppose X oscillates. Let (possibly by enlarging the probability space to accommodate for it) N be an independent Poisson process with intensity 1 and $N_t^\epsilon := N_{t\epsilon}$ ($t \geq 0$) so that $N^\epsilon$ is a Poisson process with intensity $\epsilon$, independent of X. Define $X^\epsilon := X + hN^\epsilon$. Then, as $\epsilon \downarrow 0$, $X^\epsilon$ converges to X, uniformly on bounded time sets, almost surely, and is clearly an upwards skip-free Lévy chain drifting to $+\infty$.*

The reduction of the case when X drifts to $-\infty$ is somewhat more involved and is done by a change of measure. For this purpose assume until the end of this subsection, that X is already the coordinate process on the canonical space $\Omega = \mathbb{D}_h$, equipped with the $\sigma$-algebra $\mathcal{F}$ and filtration $\mathbb{F}$ of evaluation maps (so that P coincides with the law of X on $\mathbb{D}_h$ and $\mathcal{F} = \sigma(\text{pr}_s : s \in [0, +\infty))$, while for $t \geq 0$, $\mathcal{F}_t = \sigma(\text{pr}_s : s \in [0, t])$, where $\text{pr}_s(\omega) = \omega(s)$, for $(s, \omega) \in [0, +\infty) \times \mathbb{D}_h$). We make this transition in order to be able to apply the Kolmogorov extension theorem in the proposition, which follows. Note, however, that we are no longer able to assume the standard conditions on $(\Omega, \mathcal{F}, \mathbb{F}, P)$. Notwithstanding this, $(T_x)_{x \in \mathbb{R}}$ remain $\mathbb{F}$-stopping times, since by the nature of the space $\mathbb{D}_h$, for $x \in \mathbb{R}$, $t \geq 0$, $\{T_x \leq t\} = \{\overline{X}_t \geq x\} \in \mathcal{F}_t$.

**Proposition 1** (Exponential change of measure). *Let $c \geq 0$. Then, demanding:*

$$P_c(\Lambda) = E[\exp\{cX_t - \psi(c)t\}\mathbb{1}_\Lambda] \quad (\Lambda \in \mathcal{F}_t, t \geq 0) \qquad (3)$$

*this introduces a unique measure $P_c$ on $\mathcal{F}$. Under the new measure, $X$ remains an upwards skip-free Lévy chain with Laplace exponent $\psi_c = \psi(\cdot + c) - \psi(c)$, drifting to $+\infty$, if $c \geq \Phi(0)$, unless $c = \psi'(0+) = 0$. Moreover, if $\lambda_c$ is the new Lévy measure of $X$ under $P_c$, then $\lambda_c \ll \lambda$ and $\frac{d\lambda_c}{d\lambda}(x) = e^{cx}$ $\lambda$-a.e. in $x \in \mathbb{R}$. Finally, for every $\mathbb{F}$-stopping time $T$, $P_c \ll P$ on restriction to $\mathcal{F}'_T := \{A \cap \{T < \infty\} : A \in \mathcal{F}_T\}$, and:*

$$\frac{dP_c|_{\mathcal{F}'_T}}{dP|_{\mathcal{F}'_T}} = \exp\{cX_T - \psi(c)T\}.$$

**Proof.** That $P_c$ is introduced consistently as a probability measure on $\mathcal{F}$ follows from the Kolmogorov extension theorem (Parthasarathy 1967, p. 143, Theorem 4.2). Indeed, $M := (\exp\{cX_t - \psi(c)t\})_{t \geq 0}$ is a nonnegative martingale (use independence and stationarity of increments of $X$ and the definition of the Laplace exponent), equal identically to 1 at time 0.

Furthermore, for all $\beta \in \overline{\mathbb{C}^\to}$, $\{t, s\} \subset \mathbb{R}_+$ and $\Lambda \in \mathcal{F}_t$:

$$\begin{aligned}
E_c[\exp\{\beta(X_{t+s} - X_t)\}\mathbb{1}_\Lambda] &= E[\exp\{cX_{t+s} - \psi(c)(t+s)\}\exp\{\beta(X_{t+s} - X_t)\}\mathbb{1}_\Lambda] \\
&= E[\exp\{(c+\beta)(X_{t+s} - X_t) - \psi(c)s\}]E[\exp\{cX_t - \psi(c)t\}\mathbb{1}_\Lambda] \\
&= \exp\{s(\psi(c+\beta) - \psi(c))\}P_c(\Lambda).
\end{aligned}$$

An application of the Functional Monotone Class Theorem then shows that $X$ is indeed a Lévy process on $(\Omega, \mathcal{F}, \mathbb{F}, P_c)$ and its Laplace exponent under $P_c$ is as stipulated (that $X_0 = 0$ $P_c$-a.s. follows from the absolute continuity of $P_c$ with respect to $P$ on restriction to $\mathcal{F}_0$).

Next, from the expression for $\psi_c$, the claim regarding $\lambda_c$ follows at once. Then clearly $X$ remains an upwards skip-free Lévy chain under $P_c$, drifting to $+\infty$, if $\psi'(c+) > 0$.

Finally, let $A \in \mathcal{F}_T$ and $t \geq 0$. Then $A \cap \{T \leq t\} \in \mathcal{F}_{T \wedge t}$, and by the Optional Sampling Theorem:

$$P_c(A \cap \{T \leq t\}) = E[M_t \mathbb{1}_{A \cap \{T \leq t\}}] = E[E[M_t \mathbb{1}_{A \cap \{T \leq t\}}|\mathcal{F}_{T \wedge t}]] = E[M_{T \wedge t}\mathbb{1}_{A \cap \{T \leq t\}}] = E[M_T \mathbb{1}_{A \cap \{T \leq t\}}].$$

Using the MCT, letting $t \to \infty$, we obtain the equality $P_c(A \cap \{T < \infty\}) = E[M_T \mathbb{1}_{A \cap \{T < \infty\}}]$. □

**Proposition 2** (Conditioning to drift to $+\infty$). *Assume $\Phi(0) > 0$ and denote $P^\natural := P_{\Phi(0)}$ (see (3)). We then have as follows.*

1. *For every $\Lambda \in \mathcal{A} := \cup_{t \geq 0}\mathcal{F}_t$, $\lim_{n \to \infty} P(\Lambda|\overline{X}_\infty \geq nh) = P^\natural(\Lambda)$.*
2. *For every $x \geq 0$, the stopped process $X^{T_x} = (X_{t \wedge T_x})_{t \geq 0}$ is identical in law under the measures $P^\natural$ and $P(\cdot|T_x < \infty)$ on the canonical space $\mathbb{D}_h$.*

**Proof.** With regard to Proposition 2-1, we have as follows. Let $t \geq 0$. By the Markov property of $X$ at time $t$, the process $\overset{\Delta}{X} := (X_{t+s} - X_t)_{s \geq 0}$ is identical in law with $X$ on $\mathbb{D}_h$ and independent of $\mathcal{F}_t$ under $P$. Thus, letting $\overset{\Delta}{T}_y := \inf\{t \geq 0 : \overset{\Delta}{X}_t \geq y\}$ ($y \in \mathbb{R}$), one has for $\Lambda \in \mathcal{F}_t$ and $n \in \mathbb{N}_0$, by conditioning:

$$P(\Lambda \cap \{t < T_{nh} < \infty\}) = E[E[\mathbb{1}_\Lambda \mathbb{1}_{\{t < T_{nh}\}}\mathbb{1}_{\{\overset{\Delta}{T}_{nh - X_t} < \infty\}}|\mathcal{F}_t]] = E[e^{\Phi(0)(X_t - nh)}\mathbb{1}_{\Lambda \cap \{t < T_{nh}\}}],$$

since $\{\Lambda, \{t < T_{nh}\}\} \cup \sigma(X_t) \subset \mathcal{F}_t$. Next, noting that $\{\overline{X}_\infty \geq nh\} = \{T_{nh} < \infty\}$:

$$
\begin{aligned}
P(\Lambda | \overline{X}_\infty > nh) &= e^{\Phi(0)nh} \left( P(\Lambda \cap \{T_{nh} \leq t\}) + P(\Lambda \cap \{t < T_{nh} < \infty\}) \right) \\
&= e^{\Phi(0)nh} \left( P(\Lambda \cap \{T_{nh} \leq t\}) + E[e^{\Phi(0)(X_t - nh)} \mathbb{1}_{\Lambda \cap \{t < T_{nh}\}}] \right) \\
&= e^{\Phi(0)nh} P(\Lambda \cap \{T_{nh} \leq t\}) + P^\natural(\Lambda \cap \{t < T_{nh}\}).
\end{aligned}
$$

The second term clearly converges to $P^\natural(\Lambda)$ as $n \to \infty$. The first converges to 0, because by (2) $P(\overline{X}_{e_1} \geq nh) = e^{-nh\Phi(1)} = o(e^{-nh\Phi(0)})$, as $n \to \infty$, and we have the estimate $P(T_{nh} \leq t) = P(\overline{X}_t \geq nh) = P(\overline{X}_t \geq nh | e_1 \geq t) \leq P(\overline{X}_{e_1} \geq nh | e_1 \geq t) \leq e^t P(\overline{X}_{e_1} \geq nh)$.

We next show Proposition 2-2. Note first that $X$ is $\mathbb{F}$-progressively measurable (in particular, measurable), hence the stopped process $X^{T_x}$ is measurable as a mapping into $\mathbb{D}_h$ (Karatzas and Shreve 1988, p. 5, Problem 1.16).

Furthermore, by the strong Markov property, conditionally on $\{T_x < \infty\}$, $\mathcal{F}_{T_x}$ is independent of the future increments of $X$ after $T_x$, hence also of $\{T_{x'} < \infty\}$ for any $x' > x$. We deduce that the law of $X^{T_x}$ is the same under $P(\cdot | T_x < \infty)$ as it is under $P(\cdot | T_{x'} < \infty)$ for any $x' > x$. Proposition 2-2 then follows from Proposition 2-1 by letting $x'$ tend to $+\infty$, the algebra $\mathcal{A}$ being sufficient to determine equality in law by a $\pi/\lambda$-argument. □

### 3.2. Wiener-Hopf Factorization

**Definition 2.** We define, for $t \geq 0$, $\overline{G}_t^* := \inf\{s \in [0, t] : X_s = \overline{X}_t\}$, i.e., P-a.s., $\overline{G}_t^*$ is the last time in the interval $[0, t]$ that $X$ attains a new maximum. Similarly we let $\underline{G}_t := \sup\{s \in [0, t] : X_s = \underline{X}_s\}$ be, P-a.s., the last time on $[0, t]$ of attaining the running infimum ($t \geq 0$).

While the statements of the next proposition are given for the upwards skip-free Lévy chain $X$, they in fact hold true for the Wiener-Hopf factorization of *any* compound Poisson process. Moreover, they are (essentially) known in Kyprianou (2006). Nevertheless, we begin with these general observations, in order to (a) introduce further relevant notation and (b) provide the reader with the prerequisites needed to understand the remainder of this subsection. Immediately following Proposition 3, however, we particularize to our the skip-free setting.

**Proposition 3.** *Let $p > 0$. Then:*

1. *The pairs $(\overline{G}_{e_p}^*, \overline{X}_{e_p})$ and $(e_p - \overline{G}_{e_p}^*, \overline{X}_{e_p} - X_{e_p})$ are independent and infinitely divisible, yielding the factorisation:*

$$\frac{p}{p - i\eta - \Psi(\theta)} = \Psi_p^+(\eta, \theta) \Psi_p^-(\eta, \theta),$$

*where for $\{\theta, \eta\} \subset \mathbb{R}$,*

$$\Psi_p^+(\eta, \theta) := E[\exp\{i\eta \overline{G}_{e_p}^* + i\theta \overline{X}_{e_p}\}] \text{ and } \Psi_p^-(\eta, \theta) := E[\exp\{i\eta \underline{G}_{e_p} + i\theta \underline{X}_{e_p}\}].$$

*Duality: $(e_p - \overline{G}_{e_p}^*, \overline{X}_{e_p} - X_{e_p})$ is equal in distribution to $(\underline{G}_{e_p}, -\underline{X}_{e_p})$. $\Psi_p^+$ and $\Psi_p^-$ are the Wiener-Hopf factors.*

2. *The Wiener-Hopf factors may be identified as follows:*

$$E[\exp\{-\alpha \overline{G}_{e_p}^* - \beta \overline{X}_{e_p}\}] = \frac{\kappa^*(p, 0)}{\kappa^*(p + \alpha, \beta)}$$

*and*

$$E[\exp\{-\alpha \underline{G}_{e_p} + \beta \underline{X}_{e_p}\}] = \frac{\hat{\kappa}(p, 0)}{\hat{\kappa}(p + \alpha, \beta)}$$

*for $\{\alpha, \beta\} \subset \overline{\mathbb{C}^\to}$.*

3. Here, in terms of the law of $X$,

$$\kappa^*(\alpha, \beta) := k^* \exp\left(\int_0^\infty \int_{(0,\infty)} (e^{-t} - e^{-\alpha t - \beta x}) \frac{1}{t} P(X_t \in dx) dt\right)$$

and

$$\hat{\kappa}(\alpha, \beta) = \hat{k} \exp\left(\int_0^\infty \int_{(-\infty,0]} (e^{-t} - e^{-\alpha t + \beta x}) \frac{1}{t} P(X_t \in dx) dt\right)$$

for $\alpha \in \mathbb{C}^\to$, $\beta \in \overline{\mathbb{C}^\to}$ and some constants $\{k^*, \hat{k}\} \subset \mathbb{R}^+$.

**Proof.** These claims are contained in the remarks regarding compound Poisson processes in (Kyprianou 2006, pp. 167–68) pursuant to the proof of Theorem 6.16 therein. Analytic continuations have been effected in part Proposition 3-3 using properties of zeros of holomorphic functions (Rudin 1970, p. 209, Theorem 10.18), the theorems of Cauchy, Morera and Fubini, and finally the finiteness/integrability properties of $q$-potential measures (Sato 1999, p. 203, Theorem 30.10(ii)). □

**Remark 4.**

1. (Kyprianou 2006, pp. 157, 168) $\hat{\kappa}$ is also the Laplace exponent of the (possibly killed) bivariate descending ladder subordinator $(\hat{L}^{-1}, \hat{H})$, where $\hat{L}$ is a local time at the minimum, and the descending ladder heights process $\hat{H} = X_{\hat{L}^{-1}}$ (on $\{\hat{L}^{-1} < \infty\}$; $+\infty$ otherwise) is $X$ sampled at its right-continuous inverse $\hat{L}^{-1}$:

$$E[e^{-\alpha \hat{L}_1^{-1} - \beta \hat{H}_1} \mathbb{1}_{\{1 < \hat{L}_\infty\}}] = e^{-\hat{\kappa}(\alpha, \beta)}, \quad \{\alpha, \beta\} \subset \overline{\mathbb{C}^\to}.$$

2. As for the strict ascending ladder heights subordinator $H^* := X_{L^{*-1}}$ (on $L^{*-1} < \infty$; $+\infty$ otherwise), $L^{*-1}$ being the right-continuous inverse of $L^*$, and $L^*$ denoting the amount of time $X$ has spent at a new maximum, we have, thanks to the skip-free property of $X$, as follows. Since $P(T_h < \infty) = e^{-\Phi(0)h}$, $X$ stays at a newly achieved maximum each time for an $\text{Exp}(\lambda(\mathbb{R}))$-distributed amount of time, departing it to achieve a new maximum later on with probability $e^{-\Phi(0)h}$, and departing it, never to achieve a new maximum thereafter, with probability $1 - e^{-\Phi(0)h}$. It follows that the Laplace exponent of $H^*$ is given by:

$$-\log E[e^{-\beta H_1} \mathbb{1}(H_1 < +\infty)] = (1 - e^{-\beta h}) \lambda(\mathbb{R}) e^{-\Phi(0)h} + \lambda(\mathbb{R})(1 - e^{-\Phi(0)h}) = \lambda(\mathbb{R})(1 - e^{-(\beta + \Phi(0))h})$$

(where $\beta \in \mathbb{R}_+$). In other words, $H^*/h$ is a killed Poisson process of intensity $\lambda(\mathbb{R}) e^{-\Phi(0)h}$ and with killing rate $\lambda(\mathbb{R})(1 - e^{-\Phi(0)h})$.

Again thanks to the skip-free nature of $X$, we can expand on the contents of Proposition 3, by offering further details of the Wiener-Hopf factorization. Indeed, if we let $N_t := \overline{X}_t/h$ and $T_k := T_{kh}$ ($t \geq 0$, $k \in \mathbb{N}_0$) then clearly $T := (T_k)_{k \geq 0}$ are the arrival times of a renewal process (with a possibly defective inter-arrival time distribution) and $N := (N_t)_{t \geq 0}$ is the 'number of arrivals' process. One also has the relation: $\overline{G}_t^* = T_{N_t}$, $t \geq 0$ (P-a.s.). Thus the random variables entering the Wiener-Hopf factorization are determined in terms of the renewal process $(T, N)$.

Moreover, we can proceed to calculate explicitly the Wiener-Hopf factors as well as $\hat{\kappa}$ and $\kappa^*$. Let $p > 0$. First, since $\overline{X}_{e_p}/h$ is a geometrically distributed random variable, we have, for any $\beta \in \overline{\mathbb{C}^\to}$:

$$E[e^{-\beta \overline{X}_{e_p}}] = \sum_{k=0}^\infty e^{-\beta h k}(1 - e^{-\Phi(p)h}) e^{-\Phi(p)hk} = \frac{1 - e^{-\Phi(p)h}}{1 - e^{-\beta h - \Phi(p)h}}. \quad (4)$$

Note here that $\Phi(p) > 0$ for all $p > 0$. On the other hand, using conditioning (for any $\alpha \geq 0$):

$$
\begin{aligned}
E\left[e^{-\alpha \overline{G}_{e_p}^*}\right] &= E\left[\left((u,t) \mapsto \sum_{k=0}^{\infty} \mathbb{1}_{[0,\infty)}(t_k) e^{-\alpha t_k} \mathbb{1}_{[t_k, t_{k+1})}(u)\right) \circ (e_p, T)\right] \\
&= E\left[\left(t \mapsto \sum_{k=0}^{\infty} \mathbb{1}_{[0,\infty)}(t_k) e^{-\alpha t_k}(e^{-pt_k} - e^{-pt_{k+1}})\right) \circ T\right], \text{ since } e_p \perp T \\
&= E\left[\sum_{k=0}^{\infty} \mathbb{1}_{\{T_k < \infty\}} \left(e^{-(p+\alpha)T_k} - e^{-(p+\alpha)T_k} e^{-p(T_{k+1}-T_k)}\right)\right] \\
&= E\left[\sum_{k=0}^{\infty} e^{-(p+\alpha)T_k} \mathbb{1}_{\{T_k < \infty\}} \left(1 - e^{-p(T_{k+1}-T_k)}\right)\right].
\end{aligned}
$$

Now, conditionally on $T_k < \infty$, $T_{k+1} - T_k$ is independent of $T_k$ and has the same distribution as $T_1$. Therefore, by (1) and the theorem of Fubini:

$$E[e^{-\alpha \overline{G}_{e_p}^*}] = \sum_{k=0}^{\infty} e^{-\Phi(p+\alpha)hk}(1 - e^{-\Phi(p)h}) = \frac{1 - e^{-\Phi(p)h}}{1 - e^{-\Phi(p+\alpha)h}}. \tag{5}$$

We identify from (4) for any $\beta \in \overline{\mathbb{C}^{\rightarrow}}$: $\frac{\kappa^*(p,0)}{\kappa^*(p,\beta)} = \frac{1-e^{-\Phi(p)h}}{1-e^{-\beta h - \Phi(p)h}}$ and therefore for any $\alpha \geq 0$: $\frac{\kappa^*(p+\alpha,0)}{\kappa^*(p+\alpha,\beta)} = \frac{1-e^{-\Phi(p+\alpha)h}}{1-e^{-\beta h - \Phi(p+\alpha)h}}$. We identify from (5) for any $\alpha \geq 0$: $\frac{\kappa^*(p,0)}{\kappa^*(p+\alpha,0)} = \frac{1-e^{-h\Phi(p)}}{1-e^{-\Phi(p+\alpha)h}}$. Therefore, multiplying the last two equalities, for $\alpha \geq 0$ and $\beta \in \overline{\mathbb{C}^{\rightarrow}}$, the equality:

$$\frac{\kappa^*(p,0)}{\kappa^*(p+\alpha,\beta)} = \frac{1 - e^{-\Phi(p)h}}{1 - e^{-\beta h - \Phi(p+\alpha)h}} \tag{6}$$

obtains. In particular, for $\alpha > 0$ and $\beta \in \overline{\mathbb{C}^{\rightarrow}}$, we recognize for some constant $k^* \in (0, \infty)$: $\kappa^*(\alpha, \beta) = k^*(1 - e^{-(\beta + \Phi(\alpha))h})$. Next, observe that by independence and duality (for $\alpha \geq 0$ and $\theta \in \mathbb{R}$):

$$E[\exp\{-\alpha \overline{G}_{e_p}^* + i\theta \overline{X}_{e_p}\}] E[\exp\{-\alpha \underline{G}_{e_p} + i\theta \underline{X}_{e_p}\}] = \int_0^{\infty} dt\, p e^{-pt} E[\exp\{-\alpha t + i\theta X_t\}] =$$
$$\int_0^{\infty} dt\, p e^{-pt - \alpha t + \Psi(\theta) t} = \frac{p}{p + \alpha - \Psi(\theta)}.$$

Therefore:

$$(p + \alpha - \psi(i\theta)) \frac{\hat{k}(p, 0)}{\hat{k}(p + \alpha, i\theta)} = p \frac{1 - e^{i\theta h - \Phi(p+\alpha)h}}{1 - e^{-\Phi(p)h}}.$$

Both sides of this equality are continuous in $\theta \in \overline{\mathbb{C}^{\downarrow}}$ and analytic in $\theta \in \mathbb{C}^{\downarrow}$. They agree on $\mathbb{R}$, hence agree on $\overline{\mathbb{C}^{\downarrow}}$ by analytic continuation. Therefore (for all $\alpha \geq 0$, $\beta \in \overline{\mathbb{C}^{\rightarrow}}$):

$$(p + \alpha - \psi(\beta)) \frac{\hat{k}(p, 0)}{\hat{k}(p + \alpha, \beta)} = p \frac{1 - e^{\beta h - \Phi(p+\alpha)h}}{1 - e^{-\Phi(p)h}}, \tag{7}$$

i.e., for all $\beta \in \overline{\mathbb{C}^{\rightarrow}}$ and $\alpha \geq 0$ for which $p + \alpha \neq \psi(\beta)$ one has:

$$E[\exp\{-\alpha \underline{G}_{e_p} + \beta \underline{X}_{e_p}\}] = \frac{p}{p + \alpha - \psi(\beta)} \cdot \frac{1 - e^{(\beta - \Phi(p+\alpha))h}}{1 - e^{-\Phi(p)h}}.$$

Moreover, for the unique $\beta_0 > 0$, for which $\psi(\beta_0) = p + \alpha$, one can take the limit $\beta \to \beta_0$ in the above to obtain: $E[\exp\{-\alpha \underline{G}_{e_p} + \beta_0 \underline{X}_{e_p}\}] = \frac{ph}{\psi'(\beta_0)(1 - e^{-\Phi(p)h})} = \frac{ph\Phi'(p+\alpha)}{1 - e^{-\Phi(p)h}}$. We also recognize from (7) for $\alpha > 0$ and $\beta \in \overline{\mathbb{C}^{\rightarrow}}$ with $\alpha \neq \psi(\beta)$, and some constant $\hat{k} \in (0, \infty)$: $\hat{k}(\alpha, \beta) = \hat{k}\frac{\alpha - \psi(\beta)}{1 - e^{(\beta - \Phi(\alpha))h}}$. With $\beta_0 = \Phi(\alpha)$ one can take the limit in the latter as $\beta \to \beta_0$ to obtain: $\hat{k}(\alpha, \beta_0) = \hat{k}\psi'(\beta_0)/h = \frac{\hat{k}}{h\Phi'(\alpha)}$.

In summary:

**Theorem 3** (Wiener-Hopf factorization for upwards skip-free Lévy chains). *We have the following identities in terms of $\psi$ and $\Phi$:*

1. *For every $\alpha \geq 0$ and $\beta \in \overline{\mathbb{C}^{\rightarrow}}$:*

$$\mathsf{E}[\exp\{-\alpha \overline{G}^*_{e_p} - \beta \overline{X}_{e_p}\}] = \frac{1 - e^{-\Phi(p)h}}{1 - e^{-(\beta + \Phi(p+\alpha))h}}$$

*and*

$$\mathsf{E}[\exp\{-\alpha \underline{G}_{e_p} + \beta \underline{X}_{e_p}\}] = \frac{p}{p + \alpha - \psi(\beta)} \frac{1 - e^{(\beta - \Phi(p+\alpha))h}}{1 - e^{-\Phi(p)h}}$$

*(the latter whenever $p + \alpha \neq \psi(\beta)$; for the unique $\beta_0 > 0$ such that $\psi(\beta_0) = p + \alpha$, i.e., for $\beta_0 = \Phi(p+\alpha)$, one has the right-hand side given by $\frac{ph}{\psi'(\beta_0)(1-e^{-\Phi(p)h})} = \frac{ph\Phi'(p+\alpha)}{1-e^{-\Phi(p)h}}$).*

2. *For some $\{k^*, \hat{k}\} \subset \mathbb{R}^+$ and then for every $\alpha > 0$ and $\beta \in \overline{\mathbb{C}^{\rightarrow}}$:*

$$\kappa^*(\alpha, \beta) = k^*(1 - e^{-(\beta + \Phi(\alpha))h})$$

*and*

$$\hat{\kappa}(\alpha, \beta) = \hat{k} \frac{\alpha - \psi(\beta)}{1 - e^{(\beta - \Phi(\alpha))h}}$$

*(the latter whenever $\alpha \neq \psi(\beta)$; for the unique $\beta_0 > 0$ such that $\psi(\beta_0) = \alpha$, i.e., for $\beta_0 = \Phi(\alpha)$, one has the right-hand side given by $\hat{k}\psi'(\beta_0)/h = \frac{\hat{k}}{h\Phi'(\alpha)}$).* □

As a consequence of Theorem 3-1, we obtain the formula for the Laplace transform of the running infimum evaluated at an independent exponentially distributed random time:

$$\mathsf{E}[e^{\beta \underline{X}_{e_p}}] = \frac{p}{p - \psi(\beta)} \frac{1 - e^{(\beta - \Phi(p))h}}{1 - e^{-\Phi(p)h}} \quad (\beta \in \mathbb{R}_+ \setminus \{\Phi(p)\}) \tag{8}$$

(and $\mathsf{E}[e^{\Phi(p)\underline{X}_{e_p}}] = \frac{p\Phi'(p)h}{1-e^{-\Phi(p)h}}$). In particular, if $\psi'(0+) > 0$, then letting $p \downarrow 0$ in (8), one obtains by the DCT:

$$\mathsf{E}[e^{\beta \underline{X}_\infty}] = \frac{e^{\beta h} - 1}{\Phi'(0+)h\psi(\beta)} \quad (\beta > 0). \tag{9}$$

We obtain next from Theorem 3-2 (recall also Remark 1-1), by letting $\alpha \downarrow 0$ therein, the Laplace exponent $\phi(\beta) := -\log \mathsf{E}[e^{-\beta \hat{H}_1} \mathbb{1}(\hat{H}_1 < \infty)]$ of the descending ladder heights process $\hat{H}$:

$$\phi(\beta)(e^{\beta h} - e^{\Phi(0)h}) = \psi(\beta), \quad \beta \in \mathbb{R}_+, \tag{10}$$

where we have set for simplicity $\hat{k} = e^{-\Phi(0)h}$, by insisting on a suitable choice of the local time at the minimum. This gives the following characterization of the class of Laplace exponents of the descending ladder heights processes of upwards skip-free Lévy chains (cf. (Hubalek and Kyprianou 2011, Theorem 1)):

**Theorem 4.** *Let $h \in (0, \infty)$, $\{\gamma, q\} \subset \mathbb{R}_+$, and $(\phi_k)_{k \in \mathbb{N}} \subset \mathbb{R}_+$, with $q + \sum_{k \in \mathbb{N}} \phi_k \in (0, \infty)$. Then:*

*There exists (in law) an upwards skip-free Lévy chain $X$ with values in $\mathbb{Z}_h$ and with (i) $\gamma$ being the killing rate of its strict ascending ladder heights process (see Remark 4-2), and (ii) $\phi(\beta) = q + \sum_{k=1}^\infty \phi_k(1 - e^{-\beta kh})$, $\beta \in \mathbb{R}_+$, being the Laplace exponent of its descending ladder heights process.*

*if and only if the following conditions are satisfied:*

1. $\gamma q = 0$.
2. Setting $x$ equal to 1, when $\gamma = 0$, or to the unique solution of the equation:

$$\gamma = (1 - 1/x)\left(\phi_1 + x \sum_{k \in \mathbb{N}} \phi_k\right)$$

on the interval $x \in (1, \infty)$, otherwise[2]; and then defining $\lambda_1 := q + \sum_{k \in \mathbb{N}} \phi_k$, $\lambda_{-k} := x\phi_k - \phi_{k+1}$, $k \in \mathbb{N}$; it holds:

$$\lambda_{-k} \geq 0, \quad k \in \mathbb{N}.$$

Such an $X$ is then unique (in law), is called the parent process, its Lévy measure is given by $\sum_{k \in \mathbb{N}} \lambda_{-k} \delta_{-kh} + \lambda_1 \delta_h$, and $x = e^{\Phi(0)h}$.

**Remark 5.** *Condition Theorem 4-2 is actually quite explicit. When $\gamma = 0$ (equivalently, the parent process does not drift to $-\infty$), it simply says that the sequence $(\phi_k)_{k \in \mathbb{N}}$ should be nonincreasing. In the case when the parent process $X$ drifts to $-\infty$ (equivalently, $\gamma > 0$ (hence $q = 0$)), we might choose $x \in (1, \infty)$ first, then $(\phi_k)_{k \geq 1}$, and finally $\gamma$.*

**Proof.** Please note that with $\phi(\beta) =: q + \sum_{k=1}^{\infty} \phi_k(1 - e^{-\beta k h})$, $x := e^{\Phi(0)h}$, and comparing the respective Fourier components of the left and the right hand-side, (10) is equivalent to:

1. $q + \sum_{k \in \mathbb{N}} \phi_k = \lambda(\{h\})$.
2. $x(q + \sum_{k \in \mathbb{N}} \phi_k) + \phi_1 = \lambda(\mathbb{R})$.
3. $x\phi_k - \phi_{k+1} = \lambda(\{-kh\}), k \in \mathbb{N}$.

Moreover, the killing rate of the strict ascending ladder heights processes expresses as $\lambda(\mathbb{R})(1 - 1/x)$, whereas (1) and (3) alone, together imply $q + x \sum_{k \in \mathbb{N}} \phi_k + \phi_1 = \lambda(\mathbb{R})$.

Necessity of the conditions. Remark that the strict ascending ladder heights and the descending ladder heights processes cannot simultaneously have a strictly positive killing rate. Everything else is trivial from the above (in particular, we obtain that such an $X$, when it exists, is unique, and has the stipulated Lévy measure and $\Phi(0)$).

Sufficiency of the conditions. The compound Poisson process $X$ whose Lévy measure is given by $\lambda = \sum_{k \in \mathbb{N}} \lambda_{-k} \delta_{-kh} + \lambda_1 \delta_h$ (and whose Laplace exponent we shall denote $\psi$, likewise the largest zero of $\psi$ will be denoted $\Phi(0)$) constitutes an upwards skip-free Lévy chain. Moreover, since $x = 1$, unless $q = 0$, we obtain either way that $\phi(\beta)(e^{\beta h} - x) = \psi(\beta)$ with $\phi(\beta) := q + \sum_{k=1}^{\infty} \phi_k(1 - e^{-\beta k h})$, $\beta \geq 0$. Substituting in this relation $\beta := (\log x)/h$, we obtain at once that if $\gamma > 0$ (so $q = 0$), that then $X$ drifts to $-\infty$, $x = e^{\Phi(0)h}$, and hence $\gamma = (1 - e^{-\Phi(0)})\lambda(\mathbb{R})$ is the killing rate of the strict ascending ladder heights process. On the other hand, when $\gamma = 0$, then $x = 1$, and a direct computation reveals $\psi'(0+) = h\lambda_1 - \sum_{k \in \mathbb{N}} kh(\phi_k - \phi_{k+1}) = h(\lambda_1 - \sum_{k \in \mathbb{N}} \phi_k) = hq \geq 0$. So $X$ does not drift to $-\infty$, and $\Phi(0) = 0$, whence (again) $x = e^{\Phi(0)h}$. Also in this case, the killing rate of the strict ascending ladder heights process is $0 = (1 - x)\lambda(\mathbb{R})$. Finally, and regardless of whether $\gamma$ is strictly positive or not, compared with (10), we conclude that $\phi$ is indeed the Laplace exponent of the descending ladder heights process of $X$. □

## 4. Theory of Scale Functions

Again the reader is invited to compare the exposition of the following section with that of (Bertoin 1996, sct. VII.2) and (Kyprianou 2006, sct. 8.2), which deal with the spectrally negative case.

---

[2] It is part of the condition, that such an $x$ should exist (automatically, given the preceding assumptions, there is at most one).

## 4.1. The Scale Function W

It will be convenient to consider in this subsection the times at which $X$ attains a new maximum. We let $D_1, D_2$ and so on, denote the depths (possibly zero, or infinity) of the excursions below these new maxima. For $k \in \mathbb{N}$, it is agreed that $D_k = +\infty$ if the process $X$ never reaches the level $(k-1)h$. Then it is clear that for $y \in \mathbb{Z}_h^+, x \geq 0$ (cf. (Bühlmann 1970, p. 137, para. 6.2.4(a)) (Doney 2007, sct. 9.3)):

$$P(X_{T_y} \geq -x) = P(D_1 \leq x, D_2 \leq x+h, \ldots, D_{y/h} \leq x+y-h) =$$

$$P(D_1 \leq x) \cdot P(D_1 \leq x+h) \cdots P(D_1 \leq x+y-h) = \frac{\prod_{r=1}^{\lfloor(y+x)/h\rfloor} P(D_1 \leq (r-1)h)}{\prod_{r=1}^{\lfloor x/h\rfloor} P(D_1 \leq (r-1)h)} = \frac{W(x)}{W(x+y)},$$

where we have introduced (up to a multiplicative constant) the *scale function*:

$$W(x) := 1 / \prod_{r=1}^{\lfloor x/h \rfloor} P(D_1 \leq (r-1)h) \quad (x \geq 0). \tag{11}$$

(When convenient, we extend $W$ by 0 on $(-\infty, 0)$.)

**Remark 6.** *If needed, we can of course express* $P(D_1 \leq hk), k \in \mathbb{N}_0$, *in terms of the usual excursions away from the maximum. Thus, let* $\tilde{D}_1$ *be the depth of the first excursion away from the current maximum. By the time the process attains a new maximum (that is to say $h$), conditionally on this event, it will make a total of N departures away from the maximum, where (with $J_1$ the first jump time of X, $p := \lambda(\{h\})/\lambda(\mathbb{R})$, $\tilde{p} := P(X_{J_1} = h | T_h < \infty) = p/P(T_h < \infty)) N \sim \text{geom}(\tilde{p})$. So, denoting $\tilde{\theta}_k := P(\tilde{D}_1 \leq hk)$, one has* $P(D_1 \leq hk) = P(T_h < \infty) \sum_{l=0}^{\infty} \tilde{p}(1-\tilde{p})^l \tilde{\theta}_k^l = \frac{p}{1-(1-e^{\Phi(0)h}p)\tilde{\theta}_k}, k \in \mathbb{N}_0$.

The following theorem characterizes the scale function in terms of its Laplace transform.

**Theorem 5** (The scale function). *For every* $y \in \mathbb{Z}_h^+$ *and* $x \geq 0$ *one has:*

$$P(X_{T_y} \geq -x) = \frac{W(x)}{W(x+y)} \tag{12}$$

*and* $W : [0, \infty) \to [0, \infty)$ *is (up to a multiplicative constant) the unique right-continuous and piecewise continuous function of exponential order with Laplace transform:*

$$\widehat{W}(\beta) = \int_0^\infty e^{-\beta x} W(x) dx = \frac{e^{\beta h} - 1}{\beta h \psi(\beta)} \quad (\beta > \Phi(0)). \tag{13}$$

**Proof.** (For uniqueness see e.g., (Engelberg 2005, p. 14, Theorem 10). It is clear that $W$ is of exponential order, simply from the definition (11).)

Suppose first $X$ tends to $+\infty$. Then, letting $y \to \infty$ in (12) above, we obtain $P(-\underline{X}_\infty \leq x) = W(x)/W(+\infty)$. Here, since the left-hand side limit exists by the DCT, is finite and non-zero at least for all large enough $x$, so does the right-hand side, and $W(+\infty) \in (0, \infty)$.

Therefore $W(x) = W(+\infty)P(-\underline{X}_\infty \leq x)$ and hence the Laplace-Stieltjes transform of $W$ is given by (9)—here we consider $W$ as being extended by 0 on $(-\infty, 0)$:

$$\int_{[0,\infty)} e^{-\beta x} dW(x) = W(+\infty) \frac{e^{\beta h} - 1}{\Phi'(0+)h\psi(\beta)} \quad (\beta > 0).$$

Since (integration by parts (Revuz and Yor 1999, chp. 0, Proposition 4.5)) $\int_{[0,\infty)} e^{-\beta x} dW(x) = \beta \int_{(0,\infty)} e^{-\beta x} W(x) dx$,

$$\int_0^\infty e^{-\beta x} W(x) dx = \frac{W(+\infty)}{\Phi'(0+)} \frac{e^{\beta h} - 1}{\beta h \psi(\beta)} \quad (\beta > 0). \tag{14}$$

Suppose now that $X$ oscillates. Via Remark 3, approximate $X$ by the processes $X^\epsilon$, $\epsilon > 0$. In (14), fix $\beta$, carry over everything except for $\frac{W(+\infty)}{\Phi'(0+)}$, divide both sides by $W(0)$, and then apply this equality to $X^\epsilon$. Then on the left-hand side, the quantities pertaining to $X^\epsilon$ will converge to the ones for the process $X$ as $\epsilon \downarrow 0$ by the MCT. Indeed, for $y \in \mathbb{Z}_h^+$, $P(\underline{X}_{T_y} = 0) = W(0)/W(y)$ and (in the obvious notation): $1/P(\underline{X}^\epsilon_{T_y^\epsilon} = 0) \uparrow 1/P(\underline{X}_{T_y} = 0) = W(y)/W(0)$, since $X^\epsilon \downarrow X$, uniformly on bounded time sets, almost surely as $\epsilon \downarrow 0$. (It is enough to have convergence for $y \in \mathbb{Z}_h^+$, as this implies convergence for all $y \geq 0$, $W$ being the right-continuous piecewise constant extension of $W|_{\mathbb{Z}_h^+}$.) Thus we obtain in the oscillating case, for some $\alpha \in (0, \infty)$ which is the limit of the right-hand side as $\epsilon \downarrow 0$:

$$\int_0^\infty e^{-\beta x} W(x) dx = \alpha \frac{e^{\beta h} - 1}{\beta h \psi(\beta)} \quad (\beta > 0). \tag{15}$$

Finally, we are left with the case when $X$ drifts to $-\infty$. We treat this case by a change of measure (see Proposition 1 and the paragraph immediately preceding it). To this end assume, provisionally, that $X$ is already the coordinate process on the canonical filtered space $\mathbb{D}_h$. Then we calculate by Proposition 2-2 (for $y \in \mathbb{Z}_h^+$, $x \geq 0$):

$$P(\underline{X}_{T_y} \geq -x) = P(T_y < \infty) P(\underline{X}_{T_y} \geq -x | T_y < \infty) = e^{-\Phi(0)y} P(\underline{X}^{T_y}_\infty \geq -x | T_y < \infty) =$$
$$e^{-\Phi(0)y} P^\natural(\underline{X}^{T_y}_\infty \geq -x) = e^{-\Phi(0)y} P^\natural(\underline{X}_{T(y)} \geq -x) = e^{-\Phi(0)y} W^\natural(x)/W^\natural(x+y),$$

where the third equality uses the fact that $(\omega \mapsto \inf\{\omega(s) : s \in [0, \infty)\}) : (\mathbb{D}_h, \mathcal{F}) \to ([-\infty, \infty), \mathcal{B}([-\infty, \infty)))$ is a measurable transformation. Here $W^\natural$ is the scale function corresponding to $X$ under the measure $P^\natural$, with Laplace transform:

$$\int_0^\infty e^{-\beta x} W^\natural(x) dx = \frac{e^{\beta h} - 1}{\beta h \psi(\Phi(0) + \beta)} \quad (\beta > 0).$$

Please note that the equality $P(\underline{X}_{T_y} \geq -x) = e^{-\Phi(0)y} W^\natural(x)/W^\natural(x+y)$ remains true if we revert back to our original $X$ (no longer assumed to be in its canonical guise). This is so because we can always go from $X$ to its canonical counter-part by taking an image measure. Then the law of the process, hence the Laplace exponent and the probability $P(\underline{X}_{T_y} \geq -x)$ do not change in this transformation.

Now define $\tilde{W}(x) := e^{\Phi(0) \lfloor 1 + x/h \rfloor h} W^\natural(x)$ ($x \geq 0$). Then $\tilde{W}$ is the right-continuous piecewise-constant extension of $\tilde{W}|_{\mathbb{Z}_h^+}$. Moreover, for all $y \in \mathbb{Z}_h^+$ and $x \geq 0$, (12) obtains with $W$ replaced by $\tilde{W}$. Plugging in $x = 0$ into (12), $\tilde{W}|_{\mathbb{Z}_h}$ and $W|_{\mathbb{Z}_h}$ coincide up to a multiplicative constant, hence $\tilde{W}$ and $W$ do as well. Moreover, for all $\beta > \Phi(0)$, by the MCT:

$$\begin{aligned}
\int_0^\infty e^{-\beta x} \tilde{W}(x) dx &= e^{\Phi(0)h} \sum_{k=0}^\infty \int_{kh}^{(k+1)h} e^{-\beta x} e^{\Phi(0)kh} W^\natural(kh) dx \\
&= e^{\Phi(0)h} \sum_{k=0}^\infty \frac{1}{\beta} e^{-\beta kh} (1 - e^{-\beta h}) e^{\Phi(0)kh} W^\natural(kh) \\
&= e^{\Phi(0)h} \frac{\beta - \Phi(0)}{\beta} \frac{1 - e^{-\beta h}}{1 - e^{-(\beta - \Phi(0))h}} \int_0^\infty e^{-(\beta - \Phi(0))x} W^\natural(x) dx \\
&= e^{\Phi(0)h} \frac{\beta - \Phi(0)}{\beta} \frac{1 - e^{-\beta h}}{1 - e^{-(\beta - \Phi(0))h}} \frac{e^{(\beta - \Phi(0))h} - 1}{(\beta - \Phi(0)) h \psi(\beta)} = \frac{(e^{\beta h} - 1)}{\beta h \psi(\beta)}.
\end{aligned}$$

□

**Remark 7.** Henceforth the normalization of the scale function $W$ will be understood so as to enforce the validity of (13).

**Proposition 4.** $W(0) = 1/(h\lambda(\{h\}))$, and $W(+\infty) = 1/\psi'(0+)$ if $\Phi(0) = 0$. If $\Phi(0) > 0$, then $W(+\infty) = +\infty$.

**Proof.** Integration by parts and the DCT yield $W(0) = \lim_{\beta \to \infty} \beta \widehat{W}(\beta)$. (13) and another application of the DCT then show that $W(0) = 1/(h\lambda(\{h\}))$. Similarly, integration by parts and the MCT give the identity $W(+\infty) = \lim_{\beta \downarrow 0} \beta \widehat{W}(\beta)$. The conclusion $W(+\infty) = 1/\psi'(0+)$ is then immediate from (13) when $\Phi(0) = 0$. If $\Phi(0) > 0$, then the right-hand side of (13) tends to infinity as $\beta \downarrow \Phi(0)$ and thus, by the MCT, necessarily $W(+\infty) = +\infty$. □

### 4.2. The Scale Functions $W^{(q)}$, $q \geq 0$

**Definition 3.** For $q \geq 0$, let $W^{(q)}(x) := e^{\Phi(q)\lfloor 1+x/h\rfloor h} W_{\Phi(q)}(x)$ ($x \geq 0$), where $W_c$ plays the role of $W$ but for the process $(X, P_c)$ ($c \geq 0$; see Proposition 1). Please note that $W^{(0)} = W$. When convenient we extend $W^{(q)}$ by 0 on $(-\infty, 0)$.

**Theorem 6.** For each $q \geq 0$, $W^{(q)} : [0, \infty) \to [0, \infty)$ is the unique right-continuous and piecewise continuous function of exponential order with Laplace transform:

$$\widehat{W^{(q)}}(\beta) = \int_0^\infty e^{-\beta x} W^{(q)}(x) dx = \frac{e^{\beta h} - 1}{\beta h(\psi(\beta) - q)} \quad (\beta > \Phi(q)). \tag{16}$$

Moreover, for all $y \in \mathbb{Z}_h^+$ and $x \geq 0$:

$$E[e^{-qT_y} \mathbb{1}_{\{\underline{X}_{T_y} \geq -x\}}] = \frac{W^{(q)}(x)}{W^{(q)}(x+y)}. \tag{17}$$

**Proof.** The claim regarding the Laplace transform follows from Proposition 1, Theorem 5 and Definition 3 as it did in the case of the scale function $W$ (cf. final paragraph of the proof of Theorem 5). For the second assertion, let us calculate (moving onto the canonical space $\mathbb{D}_h$ as usual, using Proposition 1 and noting that $X_{T_y} = y$ on $\{T_y < \infty\}$):

$$E[e^{-qT_y} \mathbb{1}_{\{\underline{X}_{T_y} \geq -x\}}] = E[e^{\Phi(q)X_{T_y} - qT_y} \mathbb{1}_{\{\underline{X}_{T_y} \geq -x\}}] e^{-\Phi(q)y} =$$

$$e^{-\Phi(q)y} P_{\Phi(q)}(\underline{X}_{T_y} \geq -x) = e^{-\Phi(q)y} \frac{W_{\Phi(q)}(x)}{W_{\Phi(q)}(x+y)} = \frac{W^{(q)}(x)}{W^{(q)}(x+y)}.$$

□

**Proposition 5.** For all $q > 0$: $W^{(q)}(0) = 1/(h\lambda(\{h\}))$ and $W^{(q)}(+\infty) = +\infty$.

**Proof.** As in Proposition 4, $W^{(q)}(0) = \lim_{\beta \to \infty} \beta \widehat{W^{(q)}}(\beta) = 1/(h\lambda(\{h\}))$. Since $\Phi(q) > 0$, $W^{(q)}(+\infty) = +\infty$ also follows at once from the expression for $\widehat{W^{(q)}}$. □

Moreover:

**Proposition 6.** For $q \geq 0$:

1. If $\Phi(q) > 0$ or $\psi'(0+) > 0$, then $\lim_{x \to \infty} W^{(q)}(x) e^{-\Phi(q)\lfloor 1+x/h\rfloor h} = 1/\psi'(\Phi(q))$.
2. If $\Phi(q) = \psi'(0+) = 0$ (hence $q = 0$), then $W^{(q)}(+\infty) = +\infty$, but $\limsup_{x \to \infty} W^{(q)}(x)/x < \infty$. Indeed, $\lim_{x \to \infty} W^{(q)}(x)/x = 2/m_2$, if $m_2 := \int y^2 \lambda(dy) < \infty$ and $\lim_{x \to \infty} W^{(q)}(x)/x = 0$, if $m_2 = \infty$.

**Proof.** The first claim is immediate from Proposition 4, Definition 3 and Proposition 1. To handle the second claim, let us calculate, for the Laplace transform $\widehat{dW}$ of the measure $dW$, the quantity (using integration by parts, Theorem 5 and the fact that (since $\psi'(0+) = 0$) $\int y\lambda(dy) = 0$):

$$\lim_{\beta \downarrow 0} \beta \widehat{dW}(\beta) = \lim_{\beta \downarrow 0} \frac{\beta^2}{\psi(\beta)} = \frac{2}{m_2} \in [0, +\infty).$$

For:

$$\lim_{\beta \downarrow 0} \int (e^{\beta y} - 1)\lambda(dy)/\beta^2 = \lim_{\beta \downarrow 0} \int \frac{e^{\beta y} - \beta y - 1}{\beta^2 y^2} y^2 \lambda(dy) = \frac{m_2}{2},$$

by the MCT, since $(u \mapsto \frac{e^{-u}+u-1}{u^2})$ is nonincreasing on $(0, \infty)$ (the latter can be checked by comparing derivatives). The claim then follows by the Karamata Tauberian Theorem (Bingham et al. 1987, p. 37, Theorem 1.7.1 with $\rho = 1$). □

### 4.3. The Functions $Z^{(q)}$, $q \geq 0$

**Definition 4.** *For each $q \geq 0$, let $Z^{(q)}(x) := 1 + q \int_0^{\lfloor x/h \rfloor h} W^{(q)}(z)dz$ ($x \geq 0$). When convenient we extend these functions by 1 on $(-\infty, 0)$.*

**Definition 5.** *For $x \geq 0$, let $T_x^- := \inf\{t \geq 0 : X_t < -x\}$.*

**Proposition 7.** *In the sense of measures on the real line, for every $q > 0$:*

$$\mathsf{P}_{-\underline{X}_{e_q}} = \frac{qh}{e^{\Phi(q)h} - 1} dW^{(q)} - qW^{(q)}(\cdot - h) \cdot \Delta,$$

*where $\Delta := h \sum_{k=1}^\infty \delta_{kh}$ is the normalized counting measure on $\mathbb{Z}_h^{++} \subset \mathbb{R}$, $\mathsf{P}_{-\underline{X}_{e_q}}$ is the law of $-\underline{X}_{e_q}$ under $\mathsf{P}$, and $(W^{(q)}(\cdot - h) \cdot \Delta)(A) = \int_A W^{(q)}(y - h)\Delta(dy)$ for Borel subsets $A$ of $\mathbb{R}$.*

**Theorem 7.** *For each $x \geq 0$,*

$$\mathsf{E}[e^{-qT_x^-} \mathbb{1}_{\{T_x^- < \infty\}}] = Z^{(q)}(x) - \frac{qh}{e^{\Phi(q)h} - 1} W^{(q)}(x) \tag{18}$$

*when $q > 0$, and $\mathsf{P}(T_x^- < \infty) = 1 - W(x)/W(+\infty)$. The Laplace transform of $Z^{(q)}$, $q \geq 0$, is given by:*

$$\widehat{Z^{(q)}}(\beta) = \int_0^\infty Z^{(q)}(x) e^{-\beta x} dx = \frac{1}{\beta}\left(1 + \frac{q}{\psi(\beta) - q}\right), \quad (\beta > \Phi(q)). \tag{19}$$

**Proofs of Proposition 7 and Theorem 7.** First, with regard to the Laplace transform of $Z^{(q)}$, we have the following derivation (using integration by parts, for every $\beta > \Phi(q)$):

$$\int_0^\infty Z^{(q)}(x) e^{-\beta x} dx = \int_0^\infty \frac{e^{-\beta x}}{\beta} dZ^{(q)}(x) = \frac{1}{\beta}\left(1 + q\sum_{k=1}^\infty e^{-\beta kh} W^{(q)}((k-1)h)h\right)$$

$$= \frac{1}{\beta}\left(1 + \frac{qe^{-\beta h}\beta h}{1 - e^{-\beta h}} \sum_{k=1}^\infty \frac{(1 - e^{-\beta h})}{\beta} e^{-\beta(k-1)h} W^{(q)}((k-1)h)\right)$$

$$= \frac{1}{\beta}\left(1 + q\frac{\beta h}{e^{\beta h} - 1}\widehat{W^{(q)}}(\beta)\right) = \frac{1}{\beta}\left(1 + \frac{q}{\psi(\beta) - q}\right).$$

Next, to prove Proposition 7, note that it will be sufficient to check the equality of the Laplace transforms (Bhattacharya and Waymire 2007, p. 109, Theorem 8.4). By what we have just shown, (8), integration by parts, and Theorem 6, we then only need to establish, for $\beta > \Phi(q)$:

$$\frac{q}{\psi(\beta)-q}\frac{e^{(\beta-\Phi(q))h}-1}{1-e^{-\Phi(q)h}} = \frac{qh}{e^{\Phi(q)h}-1}\frac{\beta(e^{\beta h}-1)}{(\psi(\beta)-q)\beta h} - \frac{q}{\psi(\beta)-q},$$

which is clear.

Finally, let $x \in \mathbb{Z}_h^+$. For $q > 0$, evaluate the measures in Proposition 7 at $[0, x]$, to obtain:

$$\begin{aligned}
\mathsf{E}[e^{-qT_x^-}\mathbb{1}_{\{T_x^- < \infty\}}] &= \mathsf{P}(e_q \geq T_x^-) = \mathsf{P}(\underline{X}_{e_q} < -x) = 1 - \mathsf{P}(\underline{X}_{e_q} \geq -x) \\
&= 1 + q\int_0^x W^{(q)}(z)dz - \frac{qh}{e^{\Phi(q)h}-1}W^{(q)}(x),
\end{aligned}$$

whence the claim follows. On the other hand, when $q = 0$, the following calculation is straightforward: $\mathsf{P}(T_x^- < \infty) = \mathsf{P}(\underline{X}_\infty < -x) = 1 - \mathsf{P}(\underline{X}_\infty \geq -x) = 1 - W(x)/W(+\infty)$ (we have passed to the limit $y \to \infty$ in (12) and used the DCT on the left-hand side of this equality). □

**Proposition 8.** *Let $q \geq 0$, $x \geq 0$, $y \in \mathbb{Z}_h^+$. Then:*

$$\mathsf{E}[e^{-qT_x^-}\mathbb{1}_{\{T_x^- < T_y\}}] = Z^{(q)}(x) - Z^{(q)}(x+y)\frac{W^{(q)}(x)}{W^{(q)}(x+y)}.$$

**Proof.** Observe that $\{T_x^- = T_y\} = \emptyset$, P-a.s. The case when $q = 0$ is immediate and indeed contained in Theorem 5, since, P-a.s., $\Omega\backslash\{T_x^- < T_y\} = \{T_x^- \geq T_y\} = \{\underline{X}_{T_y} \geq -x\}$. For $q > 0$ we observe that by the strong Markov property, Theorem 6 and Theorem 7:

$$\begin{aligned}
\mathsf{E}[e^{-qT_x^-}\mathbb{1}_{\{T_x^- < T_y\}}] &= \mathsf{E}[e^{-qT_x^-}\mathbb{1}_{\{T_x^- < \infty\}}] - \mathsf{E}[e^{-qT_x^-}\mathbb{1}_{\{T_y < T_x^- < \infty\}}] \\
&= Z^{(q)}(x) - \frac{qh}{e^{\Phi(q)h}-1}W^{(q)}(x) - \mathsf{E}[e^{-qT_y}\mathbb{1}_{\{T_y < T_x^-\}}]\mathsf{E}[e^{-qT_{x+y}^-}\mathbb{1}_{\{T_{x+y}^- < \infty\}}] \\
&= Z^{(q)}(x) - \frac{qh}{e^{\Phi(q)h}-1}W^{(q)}(x) - \frac{W^{(q)}(x)}{W^{(q)}(x+y)}\left(Z^{(q)}(x+y) - \frac{qh}{e^{\Phi(q)h}-1}W^{(q)}(x+y)\right) \\
&= Z^{(q)}(x) - Z^{(q)}(x+y)\frac{W^{(q)}(x)}{W^{(q)}(x+y)},
\end{aligned}$$

which completes the proof. □

### 4.4. Calculating Scale Functions

In this subsection it will be assumed for notational convenience, but without loss of generality, that $h = 1$. We define:

$$\gamma := \lambda(\mathbb{R}), \quad p := \lambda(\{1\})/\gamma, \quad q_k := \lambda(\{-k\})/\gamma, \ k \geq 1.$$

Fix $q \geq 0$. Then denote, provisionally, $e_{m,k} := \mathsf{E}[e^{-qT_k}\mathbb{1}_{\{\underline{X}_{T_k} \geq -m\}}]$, and $e_k := e_{0,k}$, where $\{m,k\} \subset \mathbb{N}_0$ and note that, thanks to Theorem 6, $e_{m,k} = \frac{e_{m+k}}{e_m}$ for all $\{m,k\} \subset \mathbb{N}_0$. Now, $e_0 = 1$. Moreover, by the strong Markov property, for each $k \in \mathbb{N}_0$, by conditioning on $\mathcal{F}_{T_k}$ and then on $\mathcal{F}_J$, where $J$ is the time of the first jump after $T_k$ (so that, conditionally on $T_k < \infty$, $J - T_k \sim \text{Exp}(\gamma)$):

$$e_{k+1} = \mathrm{E}\Big[e^{-qT_k}\mathbb{1}_{\{X_{T_k}\geq 0\}}e^{-q(J-T_k)}\big(\mathbb{1}(\text{next jump after } T_k \text{ up}) +$$
$$\mathbb{1}(\text{next jump after } T_k \text{ 1 down, then up 2 before down more than } k-1) + \cdots +$$
$$\mathbb{1}(\text{next jump after } T_k \text{ } k \text{ down \& then up } k+1 \text{ before down more than } 0)\big)e^{-q(T_{k+1}-J)}\Big]$$
$$= e_k\frac{\gamma}{\gamma+q}[p+q_1 e_{k-1,2}+\cdots+q_k e_{0,k+1}] = e_k\frac{\gamma}{\gamma+q}[p+q_1\frac{e_{k+1}}{e_{k-1}}+\cdots+q_k\frac{e_{k+1}}{e_0}].$$

Upon division by $e_k e_{k+1}$, we obtain:
$$W^{(q)}(k) = \frac{\gamma}{\gamma+q}[pW^{(q)}(k+1) + q_1 W^{(q)}(k-1) + \cdots + q_k W^{(q)}(0)].$$

Put another way, for all $k \in \mathbb{Z}_+$:
$$pW^{(q)}(k+1) = \left(1+\frac{q}{\gamma}\right)W^{(q)}(k) - \sum_{l=1}^{k} q_l W^{(q)}(k-l). \tag{20}$$

Coupled with the initial condition $W^{(q)}(0) = 1/(\gamma p)$ (from Proposition 5 and Proposition 4), this is an explicit recursion scheme by which the values of $W^{(q)}$ obtain (cf. (De Vylder and Goovaerts 1988, sct. 4, eq. (6) & (7)) (Dickson and Waters 1991, sct. 7, eq. (7.1) & (7.5)) (Marchal 2001, p. 255, Proposition 3.1)). We can also see the vector $W^{(q)} = (W^{(q)}(k))_{k\in\mathbb{Z}}$ as a suitable eigenvector of the transition matrix $P$ associated with the jump chain of $X$. Namely, we have for all $k \in \mathbb{Z}_+$: $\left(1+\frac{q}{\gamma}\right)W^{(q)}(k) = \sum_{l\in\mathbb{Z}} P_{kl} W^{(q)}(l)$.

Now, with regard to the function $Z^{(q)}$, its values can be computed directly from the values of $W^{(q)}$ by a straightforward summation, $Z^{(q)}(n) = 1 + q\sum_{k=0}^{n-1} W^{(q)}(k)$ ($n \in \mathbb{N}_0$). Alternatively, (20) yields immediately its analogue, valid for each $n \in \mathbb{Z}^+$ (make a summation $\sum_{k=0}^{n-1}$ and multiply by $q$, using Fubini's theorem for the last sum):
$$pZ^{(q)}(n+1) - p - pqW^{(q)}(0) = \left(1+\frac{q}{\gamma}\right)(Z^{(q)}(n) - 1) - \sum_{l=1}^{n-1} q_l(Z^{(q)}(n-l) - 1),$$

i.e., for all $k \in \mathbb{Z}_+$:
$$pZ^{(q)}(k+1) + \left(1-p-\sum_{l=1}^{k-1} q_l\right) = \left(1+\frac{q}{\gamma}\right)Z^{(q)}(k) - \sum_{l=1}^{k-1} q_l Z^{(q)}(k-l). \tag{21}$$

Again this can be seen as an eigenvalue problem. Namely, for all $k \in \mathbb{Z}_+$: $\left(1+\frac{q}{\gamma}\right)Z^{(q)}(k) = \sum_{l\in\mathbb{Z}} P_{kl} Z^{(q)}(l)$. In summary:

**Proposition 9** (Calculation of $W^{(q)}$ and $Z^{(q)}$). *Let $h=1$ and $q \geq 0$. Seen as vectors, $W^{(q)} := (W^{(q)}(k))_{k\in\mathbb{Z}}$ and $Z^{(q)} := (Z^{(q)}(k))_{k\in\mathbb{Z}}$ satisfy, entry-by-entry (P being the transition matrix associated with the jump chain of $X$; $\lambda_q := 1 + q/\lambda(\mathbb{R})$):*
$$(PW^{(q)})|_{\mathbb{Z}_+} = \lambda_q W^{(q)}|_{\mathbb{Z}_+} \text{ and } (PZ^{(q)})|_{\mathbb{Z}_+} = \lambda_q Z^{(q)}|_{\mathbb{Z}_+}, \tag{22}$$

*i.e., (20) and (21) hold true for $k \in \mathbb{Z}_+$. Additionally, $W^{(q)}|_{\mathbb{Z}_-} = 0$ with $W^{(q)}(0) = 1/\lambda(\{1\})$, whereas $Z^{(q)}|_{\mathbb{Z}_-} = 1$.*

An alternative form of recursions (20) and (21) is as follows:

**Corollary 1.** *We have for all $n \in \mathbb{N}_0$:*

$$W^{(q)}(n+1) = W^{(q)}(0) + \sum_{k=1}^{n+1} W^{(q)}(n+1-k)\frac{q+\lambda(-\infty,-k]}{\lambda(\{1\})}, \quad W^{(q)}(0) = 1/\lambda(\{1\}), \quad (23)$$

*and for $\widetilde{Z^{(q)}} := Z^{(q)} - 1$,*

$$\widetilde{Z^{(q)}}(n+1) = (n+1)\frac{q}{\lambda(\{1\})} + \sum_{k=1}^{n} \widetilde{Z^{(q)}}(n+1-k)\frac{q+\lambda(-\infty,-k]}{\lambda(\{1\})}, \quad \widetilde{Z^{(q)}}(0) = 0. \quad (24)$$

**Proof.** Recursion (23) obtains from (20) as follows (cf. also (Asmussen and Albrecher 2010, (proof of) Proposition XVI.1.2)):

$$pW^{(q)}(n+1) + \sum_{k=1}^{n} q_k W^{(q)}(n-k) = v_q W^{(q)}(n), \forall n \in \mathbb{N}_0 \Rightarrow$$

$$pW^{(q)}(k+1) + \sum_{m=0}^{k-1} q_{k-m} W^{(q)}(m) = v_q W^{(q)}(k), \forall k \in \mathbb{N}_0 \Rightarrow \text{(making a summation } \sum_{k=0}^{n})$$

$$p\sum_{k=0}^{n} W^{(q)}(k+1) + \sum_{k=0}^{n}\sum_{m=0}^{k-1} q_{k-m} W^{(q)}(m) = v_q \sum_{k=0}^{n} W^{(q)}(k), \forall n \in \mathbb{N}_0 \Rightarrow \text{(Fubini)}$$

$$pW^{(q)}(n+1) + p\sum_{k=0}^{n} W^{(q)}(k) + \sum_{m=0}^{n-1} W^{(q)}(m)\sum_{k=m+1}^{n} q_{k-m} = pW^{(q)}(0) + v_q\sum_{k=0}^{n} W^{(q)}(k), \forall n \in \mathbb{N}_0 \Rightarrow \text{(relabeling)}$$

$$pW^{(q)}(n+1) + p\sum_{k=0}^{n} W^{(q)}(k) + \sum_{k=0}^{n-1} W^{(q)}(k)\sum_{l=1}^{n-k} q_l = pW^{(q)}(0) + (1+q/\gamma)\sum_{k=0}^{n} W^{(q)}(k), \forall n \in \mathbb{N}_0 \Rightarrow \text{(rearranging)}$$

$$W^{(q)}(n+1) = W^{(q)}(0) + \sum_{k=0}^{n} W^{(q)}(k)\frac{q+\gamma\sum_{l=n-k+1}^{\infty} q_l}{p\gamma}, \forall n \in \mathbb{N}_0 \Rightarrow \text{(relabeling)}$$

$$W^{(q)}(n+1) = W^{(q)}(0) + \sum_{k=1}^{n+1} W^{(q)}(n+1-k)\frac{q+\gamma\sum_{l=k}^{\infty} q_l}{p\gamma}, \forall n \in \mathbb{N}_0.$$

Then (24) follows from (23) by another summation from $n = 0$ to $n = w - 1$, $w \in \mathbb{N}_0$, say, and an interchange in the order of summation for the final sum. □

Now, given these explicit recursions for the calculation of the scale functions, searching for those Laplace exponents of upwards skip-free Lévy chains (equivalently, their descending ladder heights processes, cf. Theorem 4), that allow for an inversion of (16) in terms of some or another (more or less exotic) *special function*, appears less important. This is in contrast to the spectrally negative case, see e.g., Hubalek and Kyprianou (2011).

That said, when the scale function(s) can be expressed in terms of *elementary functions*, this is certainly note-worthy. In particular, whenever the support of $\lambda$ is bounded from below, then (20) becomes a homogeneous linear difference equation with constant coefficients of some (finite) order, which can always be solved for explicitly in terms of elementary functions (as long as one has control over the zeros of the characteristic polynomial). The minimal example of this situation is of course when $X$ is skip-free to the left also. For simplicity let us only consider the case $q = 0$.

- **Skip-free chain.** Let $\lambda = p\delta_1 + (1-p)\delta_{-1}$. Then $W(k) = \frac{1}{1-2p}\left[\left(\frac{1-p}{p}\right)^{k+1} - 1\right]$, unless $p = 1/2$, in which case $W(k) = 2(1+k)$, $k \in \mathbb{N}_0$.

Indeed one can in general *reverse-engineer* the Lévy measure, so that the zeros of the characteristic polynomial of (20) (with $q = 0$) are known *a priori*, as follows. Choose $l \in \mathbb{N}$ as being $-\inf \text{supp}(\lambda)$; $p \in (0,1)$ as representing the probability of an up-jump; and then the numbers $\lambda_1, \ldots, \lambda_{l+1}$ (real, or not), in such a way that the polynomial (in $x$) $p(x-\lambda_1)\cdots(x-\lambda_{l+1})$ coincides with the characteristic polynomial of (20) (for $q = 0$):

$$px^{l+1} - x^l + q_1 x^{l-1} + \cdots + q_l$$

of *some* upwards skip-free Lévy chain, which can jump down by at most (and does jump down by) $l$ units (this imposes some set of algebraic restrictions on the elements of $\{\lambda_1, \ldots, \lambda_{l+1}\}$). A priori one then has access to the zeros of the characteristic polynomial, and it remains to use the linear recursion in order to determine the first $l+1$ values of $W$, thereby finding (via solving a set of linear equations of dimension $l+1$) the sought-after particular solution of (20) (with $q = 0$), that is $W$. A particular parameter set for the zeros is depicted in Figure 1 and the following is a concrete example of this procedure.

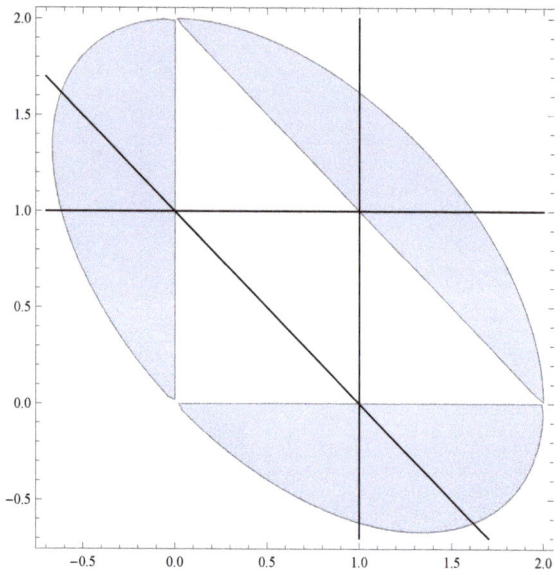

**Figure 1.** Consider the possible zeros $\lambda_1$, $\lambda_2$ and $\lambda_3$ of the characteristic polynomial of (20) (with $q = 0$), when $l := -\inf \mathrm{supp}(\lambda) = 2$ and $p = 1/2$. Straightforward computation shows they are precisely those that satisfy (o) $\lambda_3 = 2 - \lambda_1 - \lambda_2$; (i) $(\lambda_1 - 1)(\lambda_2 - 1)(\lambda_1 + \lambda_2 - 1) = 0$ and (ii) $\lambda_1\lambda_2 + (\lambda_1 + \lambda_2)(2 - \lambda_1 - \lambda_2) \geq 0$ & $\lambda_1\lambda_2(2 - \lambda_1 - \lambda_2) < 0$. In the plot one has $\lambda_1$ as the abscissa, $\lambda_2$ as the ordinate. The shaded area (an ellipse missing the closed inner triangle) satisfies (ii), the black lines verify (i). Then $q_1 = (\lambda_1\lambda_2 + (\lambda_1 + \lambda_2)(2 - \lambda_1 - \lambda_2))/2$ and $q_2 = (-\lambda_1\lambda_2(2 - \lambda_1 - \lambda_2))/2$.

- **"Reverse-engineered" chain.** Let $l = 2$, $p = \frac{1}{2}$ and, with reference to (the caption of) Figure 1, $\lambda_1 = 1$, $\lambda_2 = -\frac{1}{2}$, $\lambda_3 = \frac{3}{2}$. Then this corresponds (in the sense that has been made precise above) to an upwards skip-free Lévy chain with $\lambda/\lambda(\mathbb{R}) = \frac{1}{2}\delta_1 + \frac{1}{8}\delta_{-1} + \frac{3}{8}\delta_{-2}$ and with $W(n) = A + B(-\frac{1}{2})^n + C(\frac{3}{2})^n$, for all $n \in \mathbb{Z}_+$, for some $\{A, B, C\} \subset \mathbb{R}$. Choosing (say) $\lambda(\mathbb{R}) = 2$, we have from Proposition 4, $W(0) = 1$; and then from (20), $W(1) = 2$, $W(2) = \frac{15}{4}$. This renders $A = -\frac{4}{3}$, $B = \frac{1}{12}$, $C = \frac{9}{4}$.

An example in which the support of $\lambda$ is not bounded, but one can still obtain closed form expressions in terms of elementary functions, is the following.

- **"Geometric" chain.** Assume $p \in (0,1)$, take an $a \in (0,1)$, and let $q_l = (1-p)(1-a)a^{l-1}$ for $l \in \mathbb{N}$. Then (20) implies for $z(k) := W(k)/a^k$ that $paz(k+1) = z(k) - \sum_{l=1}^{k}(1-p)(1-a)z(k-l)/a$, i.e., for $\gamma(k) := \sum_{l=0}^{k} z(l)$ the relation $pa^2\gamma(k+1) - (a+pa^2)\gamma(k) + (1-p+pa)\gamma(k-1) = 0$, a homogeneous second order linear difference equation with constant coefficients. Specialize now to $p = a = \frac{1}{2}$ and take $\gamma = \lambda(\mathbb{R}) = 2$. Solving the difference equation with the initial conditions that are got from the known values of $W(0)$ and $W(1)$ leads to $W(k) = 2(\frac{3}{2})^k - 1$, $k \in \mathbb{Z}_+$.

This example is further developed in Section 5, in the context of the modeling of the capital surplus process of an insurance company.

Beyond this "geometric" case it seems difficult to come up with other Lévy measures for $X$ that have unbounded support and for which $W$ could be rendered explicit in terms of elementary functions.

We close this section with the following remark and corollary (cf. (Biffis and Kyprianou 2010, eq. (12)) and (Avram et al. 2004, Remark 5), respectively, for their spectrally negative analogues): for them we no longer assume that $h = 1$.

**Remark 8.** *Let $L$ be the infinitesimal generator (Sato 1999, p. 208, Theorem 31.5) of $X$. It is seen from (22), that for each $q \geq 0$, $((L-q)W^{(q)})|_{\mathbb{R}_+} = ((L-q)Z^{(q)})|_{\mathbb{R}_+} = 0$.*

**Corollary 2.** *For each $q \geq 0$, the stopped processes $Y$ and $Z$, defined by $Y_t := e^{-q(t \wedge T_0^-)} W^{(q)} \circ X_{t \wedge T_0^-}$ and $Z_t := e^{-q(t \wedge T_0^-)} W^{(q)} \circ X_{t \wedge T_0^-}$, $t \geq 0$, are nonnegative P-martingales with respect to the natural filtration $\mathbb{F}^X = (\mathcal{F}_s^X)_{s \geq 0}$ of $X$.*

**Proof.** We argue for the case of the process $Y$, the justification for $Z$ being similar. Let $(H_k)_{k \geq 1}$, $H_0 := 0$, be the sequence of jump times of $X$ (where, possibly by discarding a P-negligible set, we may insist on all of the $T_k$, $k \in \mathbb{N}_0$, being finite and increasing to $+\infty$ as $k \to \infty$). Let $0 \leq s < t$, $A \in \mathcal{F}_s^X$. By the MCT it will be sufficient to establish for $\{l, k\} \subset \mathbb{N}_0$, $l \leq k$, that:

$$\mathsf{E}[\mathbb{1}(H_l \leq s < H_{l+1}) \mathbb{1}_A Y_t \mathbb{1}(H_k \leq t < H_{k+1})] = \mathsf{E}[\mathbb{1}(H_l \leq s < H_{l+1}) \mathbb{1}_A Y_s \mathbb{1}(H_k \leq t < H_{k+1})]. \quad (25)$$

On the left-hand (respectively right-hand) side of (25) we may now replace $Y_t$ (respectively $Y_s$) by $Y_{H_k}$ (respectively $Y_{H_l}$) and then harmlessly insist on $l < k$. Moreover, up to a completion, $\mathcal{F}_s^X \subset \sigma((H_m \wedge s, X(H_m \wedge s))_{m \geq 0})$. Therefore, by a $\pi/\lambda$-argument, we need only verify (25) for sets $A$ of the form: $A = \bigcap_{m=1}^M \{H_m \wedge s \in A_m\} \cap \{X(H_m \wedge s) \in B_m\}$, $A_m$, $B_m$ Borel subsets of $\mathbb{R}$, $1 \leq m \leq M$, $M \in \mathbb{N}$. Due to the presence of the indicator $\mathbb{1}(H_l \leq s < H_{l+1})$, we may also take, without loss of generality, $M = l$ and hence $A \in \mathcal{F}_{H_l}^X$. Furthermore, $\mathcal{H} := \sigma(H_{l+1} - H_l, H_k - H_l, H_{k+1} - H_l)$ is independent of $\mathcal{F}_{H_l}^X \vee \sigma(Y_{H_k})$ and then $\mathsf{E}[Y_{H_k}|\mathcal{F}_{H_l}^X \vee \mathcal{H}] = \mathsf{E}[Y_{H_k}|\mathcal{F}_{H_l}^X] = Y_{H_l}$, P-a.s. (as follows at once from (22) of Proposition 9), whence (25) obtains. □

## 5. Application to the Modeling of an Insurance Company's Risk Process

Consider an insurance company receiving a steady but temporally somewhat uncertain stream of premia—the uncertainty stemming from fluctuations in the number of insurees and/or simply from the randomness of the times at which the premia are paid in—and which, independently, incurs random claims. For simplicity assume all the collected premia are of the same size $h > 0$ and that the claims incurred and the initial capital $x \geq 0$ are all multiples of $h$. A possible, if somewhat simplistic, model for the aggregate capital process of such a company, *net of initial capital*, is then precisely the upwards skip-free Lévy chain $X$ of Definition 1.

Fix now the $X$. We retain the notation of the previous sections, and in particular of Section 4.4, assuming still that $h = 1$ (of course this just means that we are expressing all monetary sums in the unit of the sizes of the received premia).

As an illustration we may then consider the computation of the Laplace transform (and hence, by inversion, of the density) of the time until ruin of the insurance company, which is to say of the time $T_x^-$.

To make it concrete let us take the parameters as follows. The masses of the Lévy measure on the down jumps: $\lambda(\{-k\}) = (\frac{1}{2})^k$, $k \in \mathbb{N}$; mass of Lévy measure on the up jump: $\lambda(\{1\}) = \frac{1}{2} + \sum_{n=1}^\infty n \cdot (\frac{1}{2})^n = \frac{5}{2}$ /positive "safety loading" $s := \frac{1}{2}$/; initial capital: $x = 10$. This is a special case of the "geometric" chain from Section 4.4 with $\gamma = \frac{7}{2}$, $p = \frac{5}{7}$ and $a = \frac{1}{2}$ (see p. 20 for $a$). Setting, for $k \in \mathbb{N}_0$, $\gamma^{(q)}(k) := \sum_{l=0}^k W^{(q)}(l) 2^l$ produces the following difference equation: $5\gamma^{(q)}(k+1) - (19 +$

$4q)\gamma^{(q)}(k) + (18 + 4q)\gamma^{(q)}(k-1) = 0$, $k \in \mathbb{N}$. The initial conditions are $\gamma^{(q)}(0) = W^{(q)}(0) = \frac{2}{5}$ and $\gamma^{(q)}(1) = \gamma^{(q)}(0) + 2W^{(q)}(1) = \frac{2}{5} + (\frac{2}{5})^2(7 + 2q)$. Finishing the tedious computation with the help of *Mathematica* produces the results reported in Figure 2.

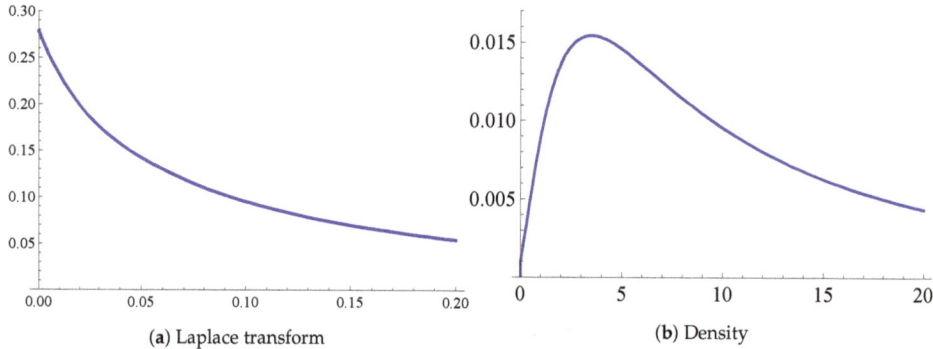

(a) Laplace transform

(b) Density

**Figure 2.** (**a**): The Laplace transform $l := ([0, \infty) \ni q \mapsto E[e^{-qT_x^-}; T_x^- < \infty])$ for the parameter set described in the body of the text, on the interval $[0, 0.2]$. The probability of ruin is $P(T_x^- < \infty) = l(0) = 1 - W(10)/W(\infty) = 1 - \psi'(0+)W(x) = 1 - sW(x) \doteq 0.28$ and the mean ruin time conditionally on ruin is $E[T_x^-|T_x^- < \infty] = -l'(0+)/l(0) \doteq 21.8$ (graphically this is one over where the tangent to $l$ at zero meets the abscissa); (**b**): Density of $T_x^-$ on $\{T_x^- < \infty\}$, plotted on the interval $[0, 20]$, and obtained by means of numerically inverting the Laplace transform $l$ (the Lebesgue integral of this density on $[0, \infty)$ is equal to $P(T_x^- < \infty)$).

On a final note, we should point out that the assumptions made above concerning the risk process are, strictly speaking, unrealistic. Indeed (i) the collected premia will typically not all be of the same size, and, moreover, (ii) the initial capital, and incurred claims will not be a multiple thereof. Besides, there is no reason to believe (iii) that the times that elapse between the accrual of premia are (approximately) i.i.d. exponentially distributed. Nevertheless, these objections can be reasonably addressed to some extent. For (ii) one just need to choose $h$ small enough so that the error committed in "rounding off" the initial capital and the claims is negligible (of course even a priori the monetary units are not infinitely divisible, but e.g., $h = 0.01$ €, may not be the most computationally efficient unit to consider in this context). Concerning (i) and (iii) we would typically prefer to see a premium drift (with slight stochastic deviations). This can be achieved by taking $\lambda(\{h\})$ sufficiently large: we will then be witnessing the arrival of premia with very high-intensity, which by the law of large numbers on a large enough time scale will look essentially like premium drift (but slightly stochastic), interdispersed with the arrivals of claims. This is basically an approximation of the Cramér-Lundberg model in the spirit of Mijatović et al. (2015), which however (because we are not ultimately effecting the limits $h \downarrow 0$, $\lambda(\{h\}) \to \infty$) retains some stochasticity in the premia. Keeping this in mind, it would be interesting to see how the upwards skip-free model behaves when fitted against real data, but this investigation lies beyond the intended scope of the present text.

**Funding:** The support of the Slovene Human Resources Development and Scholarship Fund under contract number 11010-543/2011 is acknowledged.

**Acknowledgments:** I thank Andreas Kyprianou for suggesting to me some of the investigations in this paper. I am also grateful to three anonymous Referees whose comments and suggestions have helped to improve the presentation as well as the content of this paper. Finally my thanks goes to Florin Avram for inviting me to contribute to this special issue of Risks.

**Conflicts of Interest:** The author declares no conflict of interest.

## References

Asmussen, Søren, and Hansjörg Albrecher. 2010. *Ruin Probabilities*. Advanced series on statistical science and applied probability. Singapore: World Scientific.

Avram, Florin, Andreas E. Kyprianou, and Martijn R. Pistorius. 2004. Exit Problems for Spectrally Negative Lévy Processes and Applications to (Canadized) Russian Options. *The Annals of Applied Probability* 14: 215–38.

Avram, Florin, Zbigniew Palmowski, and Martijn R. Pistorius. 2007. On the optimal dividend problem for a spectrally negative Lévy process. *The Annals of Applied Probability* 17: 156–80. [CrossRef]

Avram, Florin, and Matija Vidmar. 2017. First passage problems for upwards skip-free random walks via the $\Phi, W, Z$ paradigm. *arXiv* arXiv:1708.0608.

Bao, Zhenhua, and He Liu. 2012. The compound binomial risk model with delayed claims and random income. *Mathematical and Computer Modelling* 55: 1315–23. [CrossRef]

Bertoin, Jean. 1996. *Lévy Processes*. Cambridge Tracts in Mathematics. Cambridge: Cambridge University Press.

Bhattacharya, Rabindra Nath, and Edward C. Waymire. 2007. *A Basic Course in Probability Theory*. New York: Springer.

Biffis, Enrico, and Andreas E. Kyprianou. 2010. A note on scale functions and the time value of ruin for Lévy insurance risk processes. *Insurance: Mathematics and Economics* 46: 85–91.

Bingham, Nicholas Hugh, Charles M. Goldie, and Jozef L. Teugels. 1987. *Regular Variation*. Encyclopedia of Mathematics and its Applications. Cambridge: Cambridge University Press.

Brown, Mark, Erol A. Peköz, and Sheldon M. Ross. 2010. Some results for skip-free random walk. *Probability in the Engineering and Informational Sciences* 24: 491–507. [CrossRef]

Bühlmann, Hans. 1970. *Mathematical Methods in Risk Theory*. Grundlehren der mathematischen Wissenschaft: A series of comprehensive studies in mathematics. Berlin/ Heidelberg: Springer.

Chiu, Sung Nok, and Chuancun Yin. 2005. Passage times for a spectrally negative Lévy process with applications to risk theory. *Bernoulli* 11: 511–22. [CrossRef]

De Vylder, Florian, and Marc J. Goovaerts. 1988. Recursive calculation of finite-time ruin probabilities. *Insurance: Mathematics and Economics* 7: 1–7. [CrossRef]

Dickson, David C. M., and Howard R. Waters. 1991. Recursive calculation of survival probabilities. *ASTIN Bulletin* 21: 199–221. [CrossRef]

Doney, Ronald A. 2007. *Fluctuation Theory for Lévy Processes: Ecole d'Eté de Probabilités de Saint-Flour XXXV-2005*. Edited by Jean Picard. Number 1897 in Ecole d'Eté de Probabilités de Saint-Flour. Berlin/Heidelberg: Springer.

Engelberg, Shlomo. 2005. *A Mathematical Introduction to Control Theory*. Series in Electrical and Computer Engineering. London: Imperial College Press, vol. 2.

Hubalek, Friedrich, and Andreas E. Kyprianou. 2011. Old and New Examples of Scale Functions for Spectrally Negative Lévy Processes. In *Seminar on Stochastic Analysis, Random Fields and Applications VI*. Edited by Robert Dalang, Marco Dozzi and Francesco Russo. Basel: Springer, pp. 119–45.

Kallenberg, Olav. 1997. *Foundations of Modern Probability*. Probability and Its Applications. New York and Berlin/Heidelberg: Springer.

Karatzas, Ioannis, and Steven E. Shreve. 1988. *Brownian Motion and Stochastic Calculus*. Graduate Texts in Mathematics. New York: Springer.

Kyprianou, Andreas E. 2006. *Introductory Lectures on Fluctuations of Lévy Processes with Applications*. Berlin/ Heidelberg: Springer.

Marchal, Philippe. 2001. A Combinatorial Approach to the Two-Sided Exit Problem for Left-Continuous Random Walks. *Combinatorics, Probability and Computing* 10: 251–66. [CrossRef]

Mijatović, Aleksandar, Matija Vidmar, and Saul Jacka. 2014. Markov chain approximations for transition densities of Lévy processes. *Electronic Journal of Probability* 19: 1–37. [CrossRef]

Mijatović, Aleksandar, Matija Vidmar, and Saul Jacka. 2015. Markov chain approximations to scale functions of Lévy processes. *Stochastic Processes and their Applications* 125: 3932–57. [CrossRef]

Norris, James R. 1997. *Markov Chains*. Cambridge series in statistical and probabilistic mathematics. Cambridge: Cambridge University Press.

Parthasarathy, Kalyanapuram Rangachari. 1967. *Probability Measures on Metric Spaces*. New York and London: Academic Press.

Quine, Malcolm P. 2004. On the escape probability for a left or right continuous random walk. *Annals of Combinatorics* 8: 221–23. [CrossRef]

Revuz, Daniel, and Marc Yor. 1999. *Continuous Martingales and Brownian Motion*. Berlin/Heidelberg: Springer.

Rudin, Walter. 1970. *Real and Complex Analysis*. International student edition. Maidenhead: McGraw-Hill.

Sato, Ken-iti. 1999. *Lévy Processes and Infinitely Divisible Distributions*. Cambridge studies in advanced mathematics. Cambridge: Cambridge University Press.

Spitzer, Frank. 2001. *Principles of Random Walk*. Graduate texts in mathematics. New York: Springer.

Vidmar, Matija. 2015. Non-random overshoots of Lévy processes. *Markov Processes and Related Fields* 21: 39–56.

Wat, Kam Pui, Kam Chuen Yuen, Wai Keung Li, and Xueyuan Wu. 2018. On the compound binomial risk model with delayed claims and randomized dividends. *Risks* 6: 6. [CrossRef]

Xiao, Yuntao, and Junyi Guo. 2007. The compound binomial risk model with time-correlated claims. *Insurance: Mathematics and Economics* 41: 124–33. [CrossRef]

Yang, Hailiang, and Lianzeng Zhang. 2001. Spectrally negative Lévy processes with applications in risk theory. *Advances in Applied Probability* 33: 281–91. [CrossRef]

© 2018 by the author. Licensee MDPI, Basel, Switzerland. This article is an open access article distributed under the terms and conditions of the Creative Commons Attribution (CC BY) license (http://creativecommons.org/licenses/by/4.0/).

Article

# The $W, Z$ Paradigm for the First Passage of Strong Markov Processes without Positive Jumps

Florin Avram [1,*], Danijel Grahovac [2] and Ceren Vardar-Acar [3]

[1] Laboratoire de Mathématiques Appliquées, Université de Pau, 64012 Pau, France
[2] Department of Mathematics, University of Osijek, 31000 Osijek, Croatia; dgrahova@mathos.hr
[3] Department of Statistics, Middle East Technical University, Ankara 06800, Turkey; cvardar@metu.edu.tr
[*] Correspondence: florin.avram@univ-Pau.fr

Received: 21 November 2018; Accepted: 13 February 2019; Published: 19 February 2019

**Abstract:** As is well-known, the benefit of restricting Lévy processes without positive jumps is the "$W, Z$ scale functions paradigm", by which the knowledge of the scale functions $W, Z$ extends immediately to other risk control problems. The same is true largely for strong Markov processes $X_t$, with the notable distinctions that (a) it is more convenient to use as "basis" differential exit functions $\nu, \delta$, and that (b) it is not yet known how to compute $\nu, \delta$ or $W, Z$ beyond the Lévy, diffusion, and a few other cases. The unifying framework outlined in this paper suggests, however, via an example that the spectrally negative Markov and Lévy cases are very similar (except for the level of work involved in computing the basic functions $\nu, \delta$). We illustrate the potential of the unified framework by introducing a new objective (33) for the optimization of dividends, inspired by the de Finetti problem of maximizing expected discounted cumulative dividends until ruin, where we replace ruin with an optimally chosen Azema-Yor/generalized draw-down/regret/trailing stopping time. This is defined as a hitting time of the "draw-down" process $Y_t = \sup_{0 \leq s \leq t} X_s - X_t$ obtained by reflecting $X_t$ at its maximum. This new variational problem has been solved in a parallel paper.

**Keywords:** first passage; drawdown process; spectrally negative process; scale functions; dividends; de Finetti valuation objective; variational problem

---

## 1. A Brief Review of First Passage Theory for Strong Markov Processes without Positive Jumps and Their Draw-Downs

**Motivation.** First passage times intervene in the control of reserves/risk processes. The rough idea is that when below low levels $a$, the reserves should be replenished at some cost, and when above high levels $b$, the reserves should be invested to yield dividends—see for example Albrecher and Asmussen (2010). There is a wide variety of first passage control problems (involving absorption, reflection and other boundary mechanisms), and it has been known for a long while that these problems are simpler in the "completely asymmetric" case when all jumps go in the same direction. In recent years it has become clearer that most first passage problems can be reduced to the two basic problems of going up before down, or vice versa, and that their answers may usually be ergonomically expressed in terms of two basic "scale functions" $W, Z$ (Albrecher et al. (2016); Avram et al. (2004, 2007, 2015, 2017a, 2017b, 2018a, 2018b); Avram and Zhou (2017); Bertoin (1997); Ivanovs and Palmowski (2012); Kyprianou (2014); Landriault et al. (2017b); Li et al. (2017); Li and Zhou (2018); Suprun (1976)). The proofs require typically not much more than the strong Markov property; it is natural, therefore, to develop extensions to strong Markov processes. This has been achieved already in particular spectrally negative cases such as random walks Avram and Vidmar (2017), Markov additive processes Ivanovs and Palmowski (2012), Lévy processes with $\Omega$ state-dependent killing Ivanovs and Palmowski (2012), certain Lévy processes with state-dependent drift Czarna et al. (2017), and is in fact possible in general.

However, characterizing the functions $W, Z$ is still an open problem, even for simple classic processes such as the Ornstein-Uhlenbeck and the Feller branching diffusion with jumps.

Let $X_t$ denote a one-dimensional strong Markov process without positive jumps, defined on a filtered probability space $(\Omega, \{\mathcal{F}_t\}_{t\geq 0}, \mathbb{P})$. Denote its first passage times above and below by

$$T_{b,+} = T_{b,+}(X) = \inf\{t \geq 0 : X_t > b\}, \quad T_{a,-} = T_{a,-}(X) = \inf\{t \geq 0 : X_t < a\},$$

with $\inf \emptyset = +\infty$.

Recall that first passage theory for diffusions and spectrally negative or spectrally positive Lévy processes is considerably simpler than that for processes which may jump both ways. For these two families, a large variety of first passage problems may be reduced to the computation of two monotone "scale functions" $W, Z$ (by simple arguments such as the strong Markov property). See Albrecher et al. (2016); Avram et al. (2004, 2007, 2015, 2017a, 2018a); Avram and Zhou (2017); Bertoin (1997); Ivanovs and Palmowski (2012); Li and Zhou (2018); Suprun (1976) for the introduction and applications of $W, Z$ in the Lévy case. For diffusions, the most convenient basic functions are the monotone solutions $\varphi_+, \varphi_-$ of the Sturm-Liouville equation—see Borovkov (2012). Finally, for spectrally negative or spectrally positive Lévy processes and diffusions, off-shelf computer programs could easily produce the answer to a large variety of problems, once approximations for the basic functions associated with the process have been produced. This continues to be true in principle for non-homogeneous Markov processes with one-sided jumps (by a simple application of the strong Markov property at the smooth crossing exit from an interval). However, there are very few papers proposing methods to compute $W, Z$ for non-Lévy processes (see though Czarna et al. (2017), and Jacobsen and Jensen (2007), where the case of Ornstein-Uhlenbeck processes with phase-type jumps is studied).

The two sided exit functions. The most important first passage functions are the solutions of the two-sided upward and downward exit problems from a bounded interval $[a,b]$:

$$\begin{cases} \overline{\Psi}_{q,\theta}^b(x,a) := \mathbb{E}_x\left[e^{-qT_{b,+} - \theta(X_{T_{b,+}} - b)} \mathbf{1}_{\{T_{b,+} < T_{a,-}\}}\right] \\ \Psi_{q,\theta}^b(x,a) := \mathbb{E}_x\left[e^{-qT_{a,-} + \theta(X_{T_{a,-}} - a)} \mathbf{1}_{\{T_{a,-} < T_{b,+}\}}\right] \end{cases} \quad q, \theta \geq 0, \ a \leq x \leq b. \quad (1)$$

We will also call them killed survival and ruin first passage probabilities, respectively. Note that these are functions of five variables, very hard to compute in general. For processes with one-sided jumps, one of the exits must be smooth (without overshoot); in this case, the parameter $\theta$ is unnecessary and will be omitted. Also, when $a = 0$, it will be omitted, to simplify the notation.

For diffusions and Lévy processes with one-sided jumps, the two sided exit functions have well-known explicit formulas.

For spectrally negative Lévy processes, the simplest is the smooth survival probability, whose factors are:

$$\overline{\Psi}_q^b(x,a) = \frac{W_q(x-a)}{W_q(b-a)} = e^{-\int_x^b v_q(s-a)ds}. \quad (2)$$

$W_q(x)$ is called the scale function Bertoin (1998); Suprun (1976)[1]. We will assume throughout that $W_q$ is differentiable (see Chan et al. (2011) for information on the smoothness of scale functions). Then, $v_q(s) = \frac{W_q'(s)}{W_q(s)}$ is the logarithmic derivative of $W_q$, and may be interpreted as the "survival function of excursions lengths" Bertoin (1998). The non-smooth ruin probability has a more complicated explicit formula involving a second scale function $Z_q$ Avram et al. (2004)—see Remark 1 below.

---

[1] The fact that the survival probability has the multiplicative structure (2) is equivalent to the absence of positive jumps, by the strong Markov property.

The draw-down/regret/loss/process. Motivated by applications in statistics, mathematical finance and risk theory, there has been increased interest recently in the study of the running maximum and of the draw-down/regret/loss/process reflected at the maximum, defined by

$$Y_t = \overline{X}_t - X_t, \quad \overline{X}_t := \sup_{0 \leq t' \leq t} X_s.$$

Of equal interest is the infimum, and the draw-up/gain/process reflected at the infimum, defined by

$$\underline{Y}_t = X_t - \underline{X}_t, \quad \underline{X}_t = \inf_{0 \leq t' \leq t} X_s.$$

See Landriault et al. (2015, 2017a); Mijatovic and Pistorius (2012) for references to the numerous applications of draw-downs and draw-ups.

Draw-down and draw-up times are first passage times for the reflected processes:

$$\begin{aligned}\tau_d &:= \inf\{t \geq 0 : \overline{X}_t - X_t > d\}, \\ \underline{\tau}_d &:= \inf\{t \geq 0 : X_t - \underline{X}_t > d\}, \, d > 0.\end{aligned} \quad (3)$$

Such times turn out to be optimal in several stopping problems, in statistics Page (1954) in mathematical finance/risk theory—see for example Avram et al. (2004); Carr (2014); Lehoczky (1977); Shepp and Shiryaev (1993); Taylor (1975)—and in queueing. More specifically, they figure in risk theory problems involving capital injections or dividends at a fixed boundary, and idle times until a buffer reaches capacity in queueing theory.

**Remark 1.** *The second scale function Z Avram et al. (2004); Ivanovs and Palmowski (2012); Pistorius (2004) useful for solving the spectrally negative non-smooth ruin probability (and many other problems) is best defined via the solution of the non-smooth total discounted "regulation" problem.*

*Let $X_t^{[0} = X_t + L_t$ denote the process $X_t$ modified by Skorohod reflection at 0, with regulator $L_t = -\underline{X}_t$, let $\mathbb{E}_x^{[0}$ denote expectation for this process and let*

$$T_b^{[0} = T_{b,+} \mathbb{1}_{\{T_{b,+} < T_{0,-}\}} + \underline{T}_b \mathbb{1}_{\{T_{0,-} < T_{b,+}\}} \quad (4)$$

*denote the first passage to b of $X_t^{[0}$.*

*(a) The Laplace transform of the total regulation ("capital injections/bailouts") into the process reflected non-smoothly at 0, until the first smooth up-crossing of a level b, may be factored as (Ivanovs and Palmowski 2012, Thm. 2):*

$$\mathbb{E}_x^{[0} \left[ e^{-qT_b^{[0} - \theta L_{T_b^{[0}}} \right] = \begin{cases} \dfrac{Z_{q,\theta}(x)}{Z_{q,\theta}(b)}, & \theta < \infty \\ \mathbb{E}_x \left[ e^{-qT_b^{[0}}; T_{b,+} < T_{0,-} \right] = \dfrac{W_q(x)}{W_q(b)}, & \theta = \infty \end{cases} \quad (5)$$

*with $Z_{q,\theta}(x)$ determined up to a multiplying constant.*

*(b) Decomposing (5) at $\min(T_b^+, T_{0,-})$ yields a formula (1) for the ruin probability Ivanovs and Palmowski (2012). Indeed:*

$$\mathbb{E}_x^{[0} \left[ e^{-qT_b^{[0} - \theta L_{T_b^{[0}}} \right] = \frac{Z_{q,\theta}(x)}{Z_{q,\theta}(b)} = \frac{W_q(x)}{W_q(b)} + \mathbb{E}_x \left[ e^{-qT_{0,-} + \theta X_{T_{0,-}}}; T_{0,-} < T_{b,+} \right] \frac{Z_{q,\theta}(0)}{Z_{q,\theta}(b)} \Longrightarrow \quad (6)$$

$$\Psi_{q,\theta}^b(x) Z_{q,\theta}(0) = \mathbb{E}_x \left[ e^{-qT_{0,-} + \theta X_{T_{0,-}}}; T_{0,-} < T_{b,+} \right] Z_{q,\theta}(0) = Z_{q,\theta}(x) - W_q(x) W_q(b)^{-1} Z_{q,\theta}(b). \quad (7)$$

*To simplify this formula, it is customary to choose $Z_{q,\theta}(0) = 1$.*

For non-homogeneous spectrally negative Markov processes, it is possible Avram et al. (2017a) to extend the equalities (2), (7) to analogue expressions involving scale functions of two variables

$$\overline{\Psi}_q^b(x,a) = \frac{W_q(x,a)}{W_q(b,a)}, \quad \Psi_{q,\theta}^b(x,a) = Z_{q,\theta}(x,a) - W_q(x,a)W_q(b,a)^{-1}Z_{q,\theta}(b,a). \tag{8}$$

However, it is simpler to start, following Landriault et al. (2017b), with differential versions, whose existence will be assumed throughout this paper.

**Assumption 1.** *For all $q, \theta \geq 0$ and $y \leq x$ fixed, assume that $\overline{\Psi}_q^b(x,y)$ and $\Psi_{q,\theta}^b(x,y)$ are differentiable in $b$ at $b = x$, and in particular that the following limits exist:*

$$\nu_q(x,y) := \lim_{\varepsilon \downarrow 0} \frac{1 - \overline{\Psi}_q^{x+\varepsilon}(x,y)}{\varepsilon} \tag{9}$$

and

$$\delta_{q,\theta}(x,y) := \lim_{\varepsilon \downarrow 0} \frac{\Psi_{q,\theta}^{x+\varepsilon}(x,y)}{\varepsilon} \tag{10}$$

**Remark 2.** *A necessary condition for Assumption 1 to hold is that X is upward regular and creeping upward at every x in the state space—see (Landriault et al. 2017b, Rem. 3.1). Within this class, it seems difficult to provide examples where Assumption 1 is not satisfied.*

It turns out that the differentiability of the two-sided ruin and survival probabilities as functions of the upper limit provides a method for computing other first passage quantities; for example, (12) and (23) below may be computed by solving the first order ODE's in Theorem 2. Informally, we may say that the pillar of first passage theory for spectrally negative Markov processes is proving the existence of $\nu, \delta$.

In the Lévy case note that by (2) $\nu_q(x,y) = \frac{W_q'(x-y)}{W_q(x-y)} = \nu_q(x-y)$, and $\delta_{q,\theta}(x,y) = \delta_{q,\theta}(x-y)$ where Avram et al. (2017a)

$$\delta_{q,\theta}(x) := Z_{q,\theta}(x) - W_q(x)\frac{Z_{q,\theta}'(x)}{W_q'(x)}. \tag{11}$$

**Remark 3.** *For diffusions, $W_q(x,a)$ is a certain Wronskian—see for example Borovkov (2012). Also, for Langevin type processes with decreasing state-dependent drifts, $W_q(x,a)$ solves a certain renewal equation Czarna et al. (2017). The case of Ornstein-Uhlenbeck/Segerdahl-Tichy processes with exponential jumps is currently under study in Avram and Garmendia (2019). Some information about the generalization to Ornstein-Uhlenbeck processes with phase-type jumps can be found in Jacobsen and Jensen (2007). Beyond that, computing $W_q(x,a)$ or $\nu_q(x,a)$ is an open problem. This is an important problem, and we conjecture that the method of Jacobsen and Jensen (2007) may be extended, at least to affine diffusions with phase-type jumps, and possibly to all diffusions with phase-type jumps.*

The drawdown exit functions. Recently, control results with drawdown times $\tau_d$ replacing classic first passage times started being investigated—see for example Landriault et al. (2017a); Mijatovic and Pistorius (2012). Two natural objects of interest for studying $\tau_d$ are the two sided exit times

$$T_{b+,d} = \min(\tau_d, T_{b,+}), \quad T_{a-,d} = \min(\tau_d, T_{a,-}).$$

In terms of the two-dimensional process $t \mapsto (X_t, Y_t)$, these are the first exit times from the regions $(-\infty, b] \times [0, d]$ and $[a, \infty) \times [0, d]$.

Fundamental in the study of say $T_{b+,d}$ are the following two Laplace transforms $UbD/DbU$ (up-crossing before draw-down/draw-down before up-crossing), which are analogues of the killed survival and ruin probabilities:

$$UbD_{q,\theta,d}^b(x) = \mathbb{E}_x\left[e^{-qT_{b,+}-\theta(X_{T_{b,+}}-b)}; T_{b,+} < \tau_d\right] = \mathbb{E}_x\left[e^{-qT_{b,+}-\theta(X_{T_{b,+}}-b)}; \overline{X}_{\tau_d} > b\right]$$
$$DbU_{q,\theta,d}^b(x) = \mathbb{E}_x\left[e^{-q\tau_d-\theta(Y_{\tau_d}-d)}; \tau_d < T_{b,+}\right] = \mathbb{E}_x\left[e^{-q\tau_d-\theta(Y_{\tau_d}-d)}; \overline{X}_{\tau_d} < b\right]. \quad (12)$$

For spectrally negative Lévy processes, these have again simple formulas:

1.
$$UbD_{q,d}^b(x) := \mathbb{E}_x\left[e^{-qT_{b,+}}; T_{b,+} \leq \tau_d\right] = e^{-(b-x)\frac{W_q'(d)}{W_q(d)}}, \quad (13)$$

2. The function $DbU$ may be obtained by integrating the fundamental law (Mijatovic and Pistorius 2012, Thm 1), (Landriault et al. 2017a, Thm 3.1)[2]

$$\delta_{q,\theta}(d,x,s) := \mathbb{E}_x\left[e^{-q\tau_d-\theta(Y_{\tau_d}-d)}; \overline{X}_{\tau_d} \in ds\right] = \left(v_q(d) e^{-v_q(d)(s-x)_+} ds\right)\delta_{q,\theta}(d)$$
$$\Leftrightarrow \mathbb{E}_x\left[e^{-q\tau_d-\theta(Y_{\tau_d}-d)-\vartheta(\overline{X}_{\tau_d}-x)}\right] = \frac{v_q(d)}{\vartheta + v_q(d)}\delta_{q,\theta}(d) \quad (14)$$

where $\delta_{q,\theta}(d)$ is given by (11). Integrating yields

$$DbU_{q,\theta,d}^b(x) = \left(1 - e^{-(b-x)\frac{W_q'(d)}{W_q(d)}}\right)\delta_{q,\theta}(d). \quad (15)$$

**Remark 4.** *The probabilistic interpretation of $v_q$, the logarithmic derivative of $W_q$. Taking $a = 0$ for simplicity, the last formula in (2) has the interesting interpretation as the probability that no arrival has occurred between times $x$ and $b$, for a non-homogeneous Poisson process of rate $v_q(s), s \in [x,b]$. Alternatively, differentiating (2) yields*

$$\frac{d}{ds}\overline{\Psi}_q^b(s) - v_q(s)\overline{\Psi}_q^b(s) = 0, \qquad \overline{\Psi}_q^b(b) = 1. \quad (16)$$

*This equation coincides the Kolmogorov equation for the probability that a deterministic process $\tilde{Y}_s = s$, killed at rate $v_q(s)$, reaches $b$ before killing, when starting at $s$. It turns out, by excursion theory, that such a process $\tilde{Y}_s$ may be constructed by excising the negative excursions from $X_t$, and by taking the running maximum $s$ as time parameter.*

*The logarithmic derivative $v_q(s)$ will be needed below in the de Finetti problem (17), where we will use the fact that the expected dividends $v_q(b)$ paid at a fixed barrier $b$, starting from $b$, equal the expected discounted time until killing, which is exponential with parameter $v_q(b)$, being therefore simply the reciprocal of the killing parameter $v_q(b)$:*

$$v_q(b) := \mathbb{E}_b\left[\int_0^{T_{0,-}^{b|}} e^{-qt}d(\overline{X}_t - b)\right] = v_q(b)^{-1}. \quad (17)$$

---

[2] Please note that (Mijatovic and Pistorius 2012, Thm. 1) give a more complicated "sextuple law" with two cases, and that (Landriault et al. 2017a, Thm 3.1) use an alternative to the function $Z_q(x,\theta)$, so that some computing is required to get (11) and (14).

We see in the equation above and others that $v_q$ may serve as a convenient alternative characteristic of a spectrally negative Markov process, replacing $W_q$. Just as $W_q$, it may be extended to the case of generalized drawdown killing introduced in Avram et al. (2017b); Li et al. (2017).

**Contents.** We start in Section 2 by presenting a pedagogic first passage example illustrating the $W, Z$ paradigm: the first time

$$T_R = T_{a,b,d} = T_{a,-} \wedge T_{b,+} \wedge \tau_d. \tag{18}$$

when $(X, Y)$ with $X$ Lévy leaves a rectangular region $R = [a, b] \times [0, d]$.

**Remark 5.** *Please note that letting $a \to -\infty, b \to \infty$ reduces $T_{a,b,d}$ to $\tau_d$, and letting $d \to \infty, b \to \infty$ reduces $T_{a,b,d}$ to $T_{a,-}$. Hence both classic first passage and drawdown times appear as special cases of $T_{a,b,d}$. For finite $a, b, d$, our region has two classic and one drawdown exit boundary.[3]*

In Section 3 we provide geometric considerations which reduce computations of the Laplace transforms of the "three-sided" exit times of $(X, Y)$ to that of Laplace transforms of two-sided exit problems involving $T_{a,-}$, $T_{b,+}$ and $\tau_d$ (like (1) and (12))—see Figure 1.

Only the strong Markov property is used; however, for the sake of simple notations we restricted the exposition to the family of Lévy processes (which have also the convenient feature that the scale functions $W, Z$ may be computed by inverting Laplace transforms Avram et al. (2004, 2015); Bertoin (1998); Ivanovs and Palmowski (2012); Kyprianou (2014)).

In Section 4 we enlarge the framework to that of generalized drawdown times Avram et al. (2017b); Li et al. (2017). This immediately entails that $v, \delta$ become functions of two variables defined in (9) and (10), and the extension to the spectrally negative Markov case becomes natural. We turn therefore to exits from certain trapezoidal-type regions in Section 5, under the spectrally negative Markov model.

In Section 6 we consider processes reflected at an upper barrier and formulate a Finetti's optimal dividends type objective with combined ruin and generalized drawdown stopping; this involves adding one reflecting vertex to our trapezoidal region. Included here is a new variational problem for de Finetti's dividends with generalized drawdown stopping (33); since the solution is not immediate even in the Lévy case, this has been provided in the parallel paper Avram and Goreac (2018).

## 2. Geometric Considerations Concerning the Joint Evolution of a Lévy Process and Its Draw-Down in a Rectangle

To study the process $(X_t, Y_t)$, it is useful to start with its evolution in a rectangular region $R := [a, b] \times [0, d] \subset \mathbb{R} \times \mathbb{R}_+$, where $a < b$ and $d > 0$. Define

$$T_R = T_{a,b,d} := \inf\{t : (X_t, Y_t) \notin R\} = \tau_d \wedge T_{a,-} \wedge T_{b,+}.$$

A sample path of $(X, Y)$, where $X$ is chosen to be a spectrally negative Lévy process, and the region $R$ is depicted in Figure 1.

---

[3] Choosing $a, b, d$ optimally in various control problems involving optimal dividends and capital injections should be of interest, and will be pursued in further work.

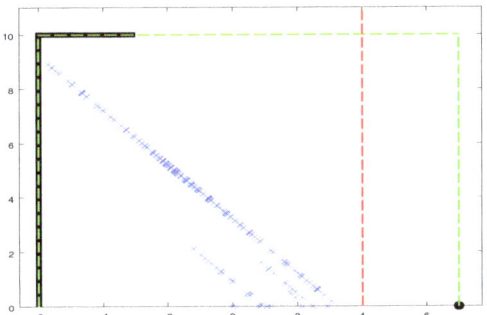

**Figure 1.** A sample path of $(X,Y)$ with $X$ a spectrally negative Lévy process. The region $R$ has $d = 10$, $a = -6$ and $b = 7$; the dark boundary shows the possible exit points of $(X,Y)$ from $R$. The base of the red line separates $R$ in two parts with different behavior.

As is clear from the figure and from its definition, the process $(X,Y)$ has very particular dynamics on $R$: away from the boundary $\partial_1 := \{(x_1, x_2) \in \mathbb{R} \times \mathbb{R}_+ : x_2 = 0\}$ it oscillates during negative excursions from the maximum on line segments $l_{\overline{X}_t}$ where, for $c \in \mathbb{R}$, $l_c := \{(x_1, x_2) \in \mathbb{R} \times \mathbb{R}_+ : x_1 + x_2 = c\}$.

As $\overline{X}_t$ increases, the line segment $l_{\overline{X}_t}$ on which $(X,Y)$ oscillates advances to the right—continuously, in the spectrally negative case, and in general possibly with jumps.

On $\partial_1$, we observe the Markovian upward ladder process, i.e., the maximum $\overline{X}$ with downward excursions excised, with extra spatial killing upon exiting $R$. If only time killing was present, with $d = \infty$, this would be a killed drift subordinator, with Laplace exponent $\kappa(s) = s + \Phi_q$ (as a consequence of the Wiener-Hopf decomposition Kyprianou (2014)). In the rectangle, in the spectrally negative case, the ladder process becomes a killed drift with generator $\mathcal{G}\varphi(s) := \varphi'(s) - \nu_q(d)\varphi(s)$ Albrecher et al. (2014); Avram et al. (2017b). Finally, with generalized drawdown (when the upper boundary is replaced by one determined by certain parametrizations $(\widehat{d}(s), d(s))$—see below), the generator will have state-dependent killing:

$$\mathcal{G}\varphi(s) := \varphi'(s) - \nu_q(d(s))\varphi(s). \tag{19}$$

Several functionals (ruin, dividends, tax, etc.) of the original process may be expressed as functionals of the killed ladder process. This explains the prevalence of first order ODE's—see (25) for one example—when working with spectrally negative processes. Several implications for $T_R$ are immediately clear from these dynamics: for example, the process $(X,Y)$ can leave $R$ only through $\partial R \cap \{(x_1, x_2) \in \mathbb{R} \times \mathbb{R}_+ : x_1 \leq b - d\}$ or through the point $(b, 0)$ (see the shaded region in Figure 1). Also,

1. If $b \leq a + d$, it is impossible for the process to leave $R$ through the upper boundary of $\partial R$ and for these parameter values $T_R$ reduces to $T_{a,-} \wedge T_{b,+}$. Here it suffices to know the functions (1) to obtain the Laplace transform of $T_R$.
2. If $a + d \leq x$, it is impossible for the process to leave $R$ through the left boundary of $\partial R$, and $T_R$ reduces to $T_{b,+} \wedge \tau_d$. Here it suffices to apply the spectrally negative drawdown formulas provided in Landriault et al. (2017a); Mijatovic and Pistorius (2012).
3. In the remaining case $x \leq a + d \leq b$, both drawdown and classic exits are possible. For the latter case, see Figure 1. The key observation here is that drawdown [classic] exits occur iff $X_t$ does [does not] cross the line $x_1 = d + a$. The final answers will combine these two cases.

## 3. The Three Laplace Transforms of the Exit Time out of a Rectangle for Lévy Processes without Positive Jumps

In this section we provide Laplace transforms of $T_R$ and of the eventual overshoot at $T_R$. One can break down the analysis of $T_R$ to nine cases, depending on which of the three exit boundaries $T_{a,-}$, $T_{b,+}$ or $\tau_d$ occurred, and on the three relations between $x$, $a$, $b$ and $d$ described above.

The following results are immediate applications of the strong Markov property and of known first passage and draw-down results.

**Theorem 1.** *Consider a spectrally negative Lévy process $X$ with differentiable scale function $W_q$. Then, for fixed $d \geq 0$ and $a \leq x \leq b$, letting $UbD$, $DbU$ denote the functions defined in (13), (15), we have:*

|  | $a+d \leq x \leq b$ | $x \leq a+d \leq b$ | $b \leq a+d$ |
|---|---|---|---|
| $\mathbb{E}_x\left[e^{-qT_{b,+}}; T_{b,+} \leq \min(\tau_d, T_{a,-})\right] =$ | $UbD^b_{q,d}(x)$ | $\overline{\Psi}^{(a+d)}_q(x,a) UbD^b_{q,d}(a+d)$ | $\overline{\Psi}^b_q(x,a)$ |
| $\mathbb{E}_x\left[e^{-qT_{a,-}+\theta(X_{T_{a,-}}-a)}; T_{a,-} \leq \min(\tau_d, T_{b,+})\right] =$ | $0$ | $\Psi^{(a+d)}_{q,\theta}(x,a)$ | $\Psi^b_{q,\theta}(x,a)$ |
| $\mathbb{E}_x\left[e^{-q\tau_d-\theta(Y_{\tau_d}-d)}; \tau_d \leq \min(T_{b,+}, T_{a,-})\right] =$ | $DbU^b_{q,\theta,d}(x)$ | $\overline{\Psi}^{(a+d)}_q(x,a) DbU^b_{q,\theta,d}(a+d)$ | $0$ |

(20)

**Proof.** Please note that in the third column the $d$ boundary is invisible and does not appear in the results, and in the first column the $a$ boundary is invisible and does not appear in the results. These two cases follow therefore by applying already known results.

The middle column holds by breaking the path at the first crossing of $a + d$. The main points here are that

1. the middle case may happen only if $X_t$ visits $a$ before $a + d$;
2. the first case (exit through $b$) and the third case (drawdown exit) may happen only if $X_t$ visits first $a + d$, with the drawdown barrier being invisible, and that subsequently the lower first passage barrier $a$ becomes invisible.

The results follow then due to the smooth crossing upward and the strong Markov property. □

**Proof.** Let us check the first and third row of the second column. Applying the strong Markov property at $T_{a+d,+}$ yields

$$\mathbb{E}_x\left[e^{-qT_{b,+}}; T_{b,+} \leq \min(\tau_d, T_{a,-})\right] = \mathbb{E}_x\left[e^{-qT_{b,+}}; T_{a+d,+} \leq T_{a,-}\right] \mathbb{E}_{a+d}\left[e^{-qT_{b,+}}; T_{b,+} \leq \tau_d\right]$$
$$= \frac{W_q(x-a)}{W_q(d)} e^{-(b-a-d)\frac{W'_q(d)}{W_q(d)}}$$

and

$$\mathbb{E}_x\left[e^{-q\tau_d-\theta(Y_{\tau_d}-d)}; \tau_d \leq \min(T_{b,+}, T_{a,-})\right] = \mathbb{E}_x\left[e^{-q\tau_d-\theta(Y_{\tau_d}-d)}; T_{a+d,+} \leq T_{a,-}\right] \mathbb{E}_{a+d}\left[e^{-q\tau_d-\theta(Y_{\tau_d}-d)}; \tau_d \leq T_{b,+}\right]$$
$$= \frac{W_q(x-a)}{W_q(d)} \delta_{q,\theta}(d) \left(1 - e^{-(b-a-d))\frac{W'_q(d)}{W_q(d)}}\right).$$

□

## 4. Generalized Draw-Down Stopping for Processes without Positive Jumps

Generalized drawdown times appear naturally in the Azema Yor solution of the Skorokhod embedding problem Azéma and Yor (1979), and in the Dubbins-Shepp-Shiryaev, and

Peskir-Hobson-Egami optimal stopping problems Dubins et al. (1994); Egami and Oryu (2015); Hobson (2007); Peskir (1998). Importantly, they allow a unified treatment of classic first passage and drawdown times (see also Avram et al. (2018b) for a further generalization to taxed processes)—see Avram et al. (2017b); Li et al. (2017). The idea is to replace the upper side of the rectangle $R$ by a parametrized curve

$$(x_1, x_2) = (\widehat{d}(s), d(s)), \quad \widehat{d}(s) = s - d(s),$$

where $s = x_1 + x_2$ represents the value of $\overline{X}_t$ during the excursion which intersects the upper boundary at $(x_1, x_2)$ (see Figure 2). Alternatively, parametrizing by $x$ yields

$$y = h(x), \quad h(x) = \widehat{d}^{-1}(x) - x.$$

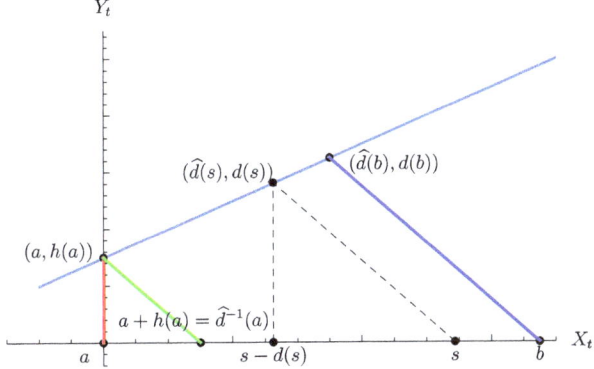

**Figure 2.** Affine drawdown exit of $(X, Y)$ $d(s) = \frac{1}{3}s + 1$.

**Definition 1.** *Li et al. (2017) For any function $d(s) > 0$ such that $\widehat{d}(s) = s - d(s)$ is nondecreasing, a generalized drawdown time is defined by*

$$T_{\widehat{d}(\cdot)} := \inf\{t \geq 0 : Y_t > d(\overline{X}_t)\} = \inf\left\{t \geq 0 : X_t < \widehat{d}(\overline{X}_t)\right\}. \tag{21}$$

Such times provide a natural unification of classic and drawdown times.
Introduce

$$\widetilde{Y}_t := Y_t - d(\overline{X}_t), \quad t \geq 0$$

to be called draw-down type process. Please note that we have $\widetilde{Y}_0 = -\widehat{d}(X_0) < 0$, and that the process $\widetilde{Y}_t$ is in general non-Markovian. However, it is Markovian during each negative excursion of $X_t$, along one of the oblique lines in the geometric decomposition sketched in Figure 1.

**Example 1.** *With affine functions*

$$d(s) = (1 - \xi)s + d \Leftrightarrow \widehat{d}(s) = \xi s - d, \quad \xi \in [0, 1], d > 0, \tag{22}$$

*we obtain the affine draw-down/regret times studied in Avram et al. (2017b).*

*Affine drawdown times reduce to a classic drawdown time (3) when $\xi = 1$, $d(s) = d$, and to a ruin time when $\xi = 0$, $\widehat{d}(s) = -d$, $d(s) = s + d$. When $\xi$ varies, we are dealing with the pencil of lines passing through $(x_1, x_2) = (-d, d)$. In particular, for $\xi = 1$ we obtain the rectangle case from section 3, and for $\xi = 0$ we have an infinite strip with a vertical boundary at $x_1 = -d$.*

*One of the merits of affine drawdown times is that they allow unifying the classic first passage theory with the drawdown theory Avram et al. (2017b); in particular, the generalized drawdown functions (23) below unify* the classic and drawdown survival and ruin probabilities (and have relatively simple formulas as well—see Avram et al. (2017a)).

Introduce now generalized drawdown analogues of the drawdown survival and ruin probabilities (12) for which we will use the same notation:

$$\begin{aligned} UbD^b_{q,\widehat{a}(\cdot)}(x) &= \mathbb{E}_x \left[ e^{-qT_{b,+}} ; T_{b,+} \leq \tau_{\widehat{a}(\cdot)} \right] \\ DbU^b_{q,\theta,\widehat{a}(\cdot)}(x)) &= \mathbb{E}_x \left[ e^{-q\tau_{\widehat{a}(\cdot)} - \theta \widetilde{Y}_{\tau_{\widehat{a}(\cdot)}}} ; \tau_{\widehat{a}(\cdot)} < T_b^+ \right]. \end{aligned} \qquad (23)$$

**Remark 6.** *In the spectrally negative case, these functions may be represented as integrals:*

$$\begin{aligned} UbD^b_{q,\widehat{a}(\cdot)}(x) &= e^{-\int_x^b v_q(s,\widehat{a}(s))ds}, \\ DbU^b_{q,\theta,\widehat{a}(\cdot)}(x) &= \int_x^b e^{-\int_x^y v_q(s,\widehat{a}(s))ds} v_q(y,\widehat{a}(y)) \delta_{q,\theta}(y,\widehat{a}(y)) dy, \end{aligned} \qquad (24)$$

where $v_q(y,\widehat{a}(y))$, $\delta_{q,\theta}(y,\widehat{a}(y))$ are defined in (9), (10).

This is already apparent in (Landriault et al. 2017b, Cor 3.1), and may be easily understood probabilistically from Figure 2: the first equation is the probability of no occurrence in a non-homogeneous Poisson process, and the second decomposes the transform of the deficit, by conditioning on the point $y \in [x,b]$ where the maximum occurred.

We provide now a heuristic proof valid for the Lévy case when $v_q(y,\widehat{a}(y)) = v_q(y - \widehat{a}(y)) = v_q(d(y))$ and $\delta_{q,\theta}(y,\widehat{a}(y)) = \delta_{q,\theta}(y - \widehat{a}(y)) = \delta_{q,\theta}(d(y))$.

1. Due to creeping, UbD is a product of infinitesimal events

$$\overline{\Psi}_q^{y+\epsilon}(y, y-d(y)) = \frac{W_q(d(y))}{W_q(d(y)+\epsilon)} \sim 1 - \epsilon v_q(d(y)) \sim e^{-\epsilon v_q(d(y))}.$$

Taking product, with $\epsilon = dy$, yields (24).

2. Informally, we condition on the density $\overline{X}_t \in dy$. The integrand of DbU is obtained multiplying survival infinitesimal events up to level $y$ by an infinitesimal termination event in $[y, y+dy]$. The probability of this event, conditioned on survival up to $y$, is given by the deficit formula

$$\begin{aligned} \Psi_{q,\theta}^{y+\epsilon}(y, y-d(y)) &= Z_{q,\theta}(d(y)) - W_q(d(y)) \frac{Z_{q,\theta}(d(y)+\epsilon)}{W_q(d(y)+\epsilon)} \\ &\sim \epsilon(-Z'_{q,\theta}(d(y)) + v_q(d(y))Z_{q,\theta}(d(y))) = \epsilon v_q(d(y)) \delta_{q,\theta}(d(y)) \end{aligned}$$

For a rigorous (rather intricate) proof, see Avram et al. (2018b).

The end result for generalized drawdown times is (Avram et al. 2018b, Thm1):

**Theorem 2.** *Consider a process X for which the functions $\Psi, \overline{\Psi}$ are differentiable in the upper variable b. Assume $d(x) > 0$ and $\widehat{a}(x) = x - d(x)$ nondecreasing. Then, $\forall q, \theta \geq 0, b \in \mathbb{R}$, the functions $UbD(x) = UbD^b_q(x,\widehat{a}(\cdot))$, $DbU(x) = DbU^b_{q,\theta}(x,\widehat{a}(\cdot))$ satisfy (24). Alternatively, they satisfy the ODE's*

$$UbD'(y) - v_q(y,\widehat{a}(y))UbD(y) = 0, \quad UbD(b) = 1, \qquad (25)$$

$$DbU'(y) - v_q(y,\widehat{a}(y))DbU(y) + \delta_{q,\theta}(y,\widehat{a}(y)) = 0, \quad DbU(b) = 0. \qquad (26)$$

**Remark 7.** *The operator involved in the ODE's above is the generator of the upward ladder process, under time and spatial killing, and with the downward excursions excised. Once this known, variations involving different boundary conditions are easily obtained as well.*

## 5. The Three Laplace Transforms of the Exit Time out of a Curved Trapezoid, for Processes without Positive Jumps

We will replace now the classic drawdown time in Section 3 by a generalized one. Similar geometric considerations, with $d(\cdot), a + h(a)$ replacing $d, a + d$ in Theorem 1, yield:

**Theorem 3.** *Consider a spectrally negative Lévy process $X$ with differentiable scale function $W_q$. Then, for $a \leq x \leq b$ and $d(\cdot)$ satisfying the conditions of Definition 1, we have:*

| | $a + h(a) \leq x$ | $x \leq a + h(a) \leq b$ | $b \leq a + h(a)$ |
|---|---|---|---|
| $\mathbb{E}_x\left[e^{-qT_{b,+}}; T_{b,+} \leq \min(\tau_{\widehat{d}(\cdot)}, T_{a,-})\right] =$ | $UbD^b_{q,\widehat{d}(\cdot)}(x)$ | $\overline{\Psi}^{a+h(a)}_q(x,a) UbD^b_{q,\widehat{d}(\cdot)}(a + h(a))$ | $\overline{\Psi}^b_q(x,a)$ |
| $\mathbb{E}_x\left[e^{-qT_{a,-} + \theta(X_{T_{a,-}} - a)}; T_{a,-} \leq \min(\tau_{\widehat{d}(\cdot)}, T_{b,+})\right] =$ | $0$ | $\Psi^{a+h(a)}_{q,\theta}(x,a)$ | $\Psi^b_{q,\theta}(x,a)$ |
| $\mathbb{E}_x\left[e^{-q\tau_{\widehat{d}(\cdot)} - \theta(Y_{\tau_{\widehat{d}(\cdot)}} - d)}; \tau_{\widehat{d}(\cdot)} \leq \min(T_{b,+}, T_{a,-})\right] =$ | $DbU^b_{q,\theta,\widehat{d}(\cdot)}(x)$ | $\overline{\Psi}^{a+h(a)}_q(x,a) DbU^b_{q,\theta,\widehat{d}(\cdot)}(a+h(a))$ | $0$ |

**Proof.** Note that if $b \leq a + h(a)$ (narrow band), it is again impossible for the process to leave $R$ through the upper boundary of $\partial R$, and $T_R$ reduces to $T_{a,-} \wedge T_{b,+}$, and nothing changes. Similarly, if $a + h(a) \leq x$ (flat band), it is impossible for the process to leave $R$ through the left boundary of $\partial R$, and $T_R$ reduces to $T_{b,+} \wedge \tau_d$. Finally, the two zones in the intermediate case are separated by $a + h(a)$ (instead of $a + d$). □

## 6. de Finetti's Optimal Dividends for Spectrally Negative Markov Processes with Generalized Draw-Down Stopping

In this section, we revisit the de Finetti's optimal dividend problem for spectrally negative Markov processes with the point $b$ becoming a reflecting boundary, instead of absorbing, as it was in Section 3.

Define the Skorokhod reflected/constrained process at first passage times below or above by:

$$X^{[a}_t = X_t + L_t, \quad X^{b]}_t = X_t - U_t. \tag{27}$$

Here

$$L_t = L^{[a}_t = -(\underline{X}_t - a)_-, \quad U_t = U^{b]}_t = (\overline{X}_t - b)_+ \tag{28}$$

are the minimal "Skorohod regulators" constraining $X_t$ to be bigger than $a$, and smaller than $b$, respectively.

Let now

$$V^{b]}(x) = V^{b]}_{q,\widehat{d}(\cdot)}(x) := \mathbb{E}_x\left[\int_0^{\tau_{\widehat{d}(\cdot)} \wedge T_{a,-}} e^{-qt} dU^{b]}_t\right] \tag{29}$$

denote the present value of all dividend payments at $b$, until the first passage time either below $a$, or below the drawdown boundary for the process $X^{b]}_t$ reflected at $b$, starting from $x \leq b$ (a generalization of the famous de Finetti objective). By the strong Markov property, it holds that

$$V^{b]}(x) = \mathbb{E}_x\left[e^{-qT_{b,+}}; T_{b,+} \leq \min(\tau_{\widehat{d}(\cdot)}, T_{a,-})\right] v(b), \quad v(b) = v_q(b, \widehat{d}(b)) := \mathbb{E}_b\left[\int_0^{\tau_{\widehat{d}(\cdot)}} e^{-qt} dU^{b]}_t\right]. \tag{30}$$

**Remark 8.** *The function $v(b)$, the expected discounted time until killing for the reflected process, when starting from $b$, equals the time the process reflected at $b$ spends at point $(b,0)$ in Figure 2, before a downward excursion beyond $\hat{d}(b)$ kills the process. In the Lévy case, it is well-known Kyprianou (2014) that this time is exponential with parameter $v_q(b,\hat{d}(b))$, and thus its expectation is the reciprocal of the killing parameter $v_q(b,\hat{d}(b))$, i.e.,*

$$v(b) = v_q(b,\hat{d}(b))^{-1} \tag{31}$$

Excursion theoretic arguments show that (31) continues to hold in the spectrally negative Markov case (for a proof under a similar setup, see (Czarna et al. 2018, sct. 4)).

Furthermore, by (Avram et al. 2018b, Thm. 1) included above as (24), it holds that

$$\mathbb{E}_x\left[e^{-qT_{b,+}} \mathbf{1}_{\{T_{b,+} < \tau_{d(\cdot)}\}}\right] = e^{-\int_x^b v_q(z,\hat{d}(z))dz}. \tag{32}$$

When $a = -\infty$, we arrive finally to an explicit formula

$$V^{b]}(x) = \frac{e^{-\int_x^b v_q(y,\hat{d}(y))ds}}{v_q(b,\hat{d}(b))} \tag{33}$$

expressing the expected dividends in terms of $v_q(y,\hat{d}(y))$. Please note that in the Lévy case Equation (33) simplifies to:

$$V^{b]}(x) = \frac{W_q(d(x))}{W_q(d(b))} v_q(d(b))^{-1}$$

(using $x - l(x) = d(x)$), which checks with (Wang and Zhou 2018, Lem. 3.1–3.2).

The problem of choosing a drawdown boundary to optimize dividends in (33) is solved in Avram and Goreac (2018) via Pontryaghin's maximum principle.

## 7. Example: Affine Draw-Down Stopping for Brownian Motion

Consider optimizing expected dividends $V^{b]}(x)$ given in Equation (29) with respect to the optimal dividend barrier $b$ for Brownian motion with drift $X(t) = \sigma B_t + \mu t$ and with affine drawdown stopping $d(x) = (1 - \xi)x + d$, where $\xi \in [0,1]$, $d \geq 0$, $a \leq x \leq b$.

Please note that if $a + h(a) > b$, where $h(x) = d(x)/\xi$, then the drawdown constraint is invisible, and the problem reduces to the classical de Finetti objective. Hence, we consider $a + h(a) \leq b$.

The scale function of Brownian motion is

$$W_q(x) = \frac{2e^{-\mu x/\sigma^2}}{\Delta}\sinh(x\Delta/\sigma^2) = \frac{1}{\Delta}[e^{(-\mu+\Delta)x/\sigma^2} - e^{-(\mu+\Delta)x/\sigma^2}],$$

where $\Delta = \sqrt{\mu^2 + 2q\sigma^2}$. Assume that $x \geq a + h(a) = a + \frac{d(a)}{\xi} = \frac{a+d}{\xi}$, then as a special case of spectrally negative Levy process, the expected dividends for Brownian motion equals

$$V^{b]}(x) = \mathbb{E}_x\left[e^{-qT_{b,+}}; T_{b,+} \leq \min(\tau_{\hat{d}(\cdot)}, T_{a,-})\right]v(b) = \left(\frac{W_q(d(x))}{W_q(d(b))}\right)^{\frac{1}{1-\xi}} \frac{W_q(d(b))}{W'_q(d(b))}, \tag{34}$$

see (Avram et al. 2017b, Thm. 1.1), with tax parameter $\gamma = 0$, and (Avram et al. 2017b, Rem. 7), with tax parameter $\gamma = 1$. The barrier influence function which must be optimized in $b$ becomes

$$BI(b,d,\xi) = \frac{W_q((1-\xi)x+d)^{1-\frac{1}{1-\xi}}}{W'_q((1-\xi)x+d)} = \frac{\sigma^2}{2} \cdot \frac{e^{x\mu/\sigma^2}\operatorname{csch}\left(x\sqrt{\mu^2+2q\sigma^2}/\sigma^2\right)}{\coth\left((d+x-x\xi)\sqrt{\mu^2+2q\sigma^2}/\sigma^2\right) - \mu/\sqrt{\mu^2+2q\sigma^2}}. \tag{35}$$

The critical point $b^*$ satisfies

$$\frac{W_q'' W_q}{(W_q')^2}((1-\xi)b^* + d) = -\frac{\xi}{1-\xi}, \tag{36}$$

that is $b^*$ satisfies

$$-\frac{q\sigma^2 + \mu^2 + \mu\sqrt{2q\sigma^2+\mu^2}\sinh\left(\frac{2b^*\sqrt{2q\sigma^2+\mu^2}}{\sigma^2}\right) - (q\sigma^2+\mu^2)\cosh\left(\frac{2b^*\sqrt{2q\sigma^2+\mu^2}}{\sigma^2}\right)}{\left(\sqrt{2q\sigma^2+\mu^2}\cosh\left(\frac{b^*\sqrt{2q\sigma^2+\mu^2}}{\sigma^2}\right) - \mu\sinh\left(\frac{b^*\sqrt{2q\sigma^2+\mu^2}}{\sigma^2}\right)\right)^2} = -\frac{\xi}{1-\xi}.$$

In Figure 3 given below, we have an illustration of plot of barrier influence function and its derivative for Brownian motion with drift $\mu = 1/2$ and $\sigma = 1$.

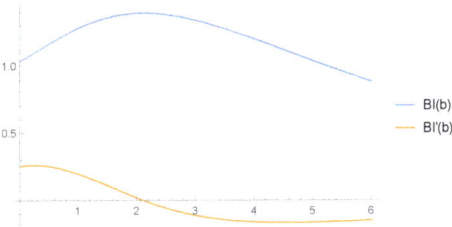

**Figure 3.** Optimizing dividends with affine drawdown stopping where $\mu = 1/2$, $q = 1/10$, $\sigma = 1$, $\xi = 1/3$, $b = 20$, $d = 1$. The critical point $b^* = 2.12445$.

**Remark 9.** *Please note that once $\xi$ is fixed, we get nontrivial results for the optimal barrier. However, if we maximize over $\xi$ as well, the optimum is achieved by the classical de Finetti solution $\xi = 0 \implies W_q''(b^* + d) = 0$, corresponding to forced stopping below $-d$ ($d$ is just a shift of the origin, with respect to the classical solution $W_q''(b^*) = 0$) Avram and Goreac (2018). In the diffusion case, it is not yet known whether examples in which the generalized de Finetti problem improves on the classic de Finetti solution are possible.*

**Remark 10.** *Let us note now that Equation (36) holds in fact for any spectrally negative Lévy process. Similar computations may be therefore performed for any spectrally negative Levy process, by plugging exact or approximate formulas for the scale function into the function*

$$\frac{W_q'' W_q}{(W_q')^2} \tag{37}$$

*which is required to solve (36).*

The easiest case is the Cramér-Lundberg process with phase-type claims, since in this case the scale function is a sum of exponentials. For example, for a Cramér-Lundberg process with premium rate $c > 0$, Poisson arrivals of intensity $\lambda$ and exponential claims with mean $1/\mu$, the scale function is $W_q(x) = c^{-1}(\frac{\mu + \Delta_+}{\Delta_+ - \Delta_-}e^{\Delta_+ x} - \frac{\mu + \Delta_-}{\Delta_+ - \Delta_-}e^{\Delta_- x})$, $x \geq 0$, where $\Delta_\pm = \frac{q + \lambda - \mu c \pm \sqrt{(q + \lambda - \mu c)^2 + 4cq\mu}}{2c}$, and similar computations may be performed (see also (Wang and Zhou 2018, Example 5.2)).

**Author Contributions:** Conceptualization, F.A.; Formal analysis, F.A. and C.V.-A.; Project administration, F.A.; Supervision, F.A.; Validation, D.G., C.V.-A.; Writing—original draft, F.A. and C.V.-A.; Writing—review, editing, D.G., C.V.-A.

**Funding:** This work is supported by the Scientific and Technological Research Council of Turkey, TUBITAK Project No. 117F273.

**Acknowledgments:** We thank Ali Devin Sezer for useful discussions.

**Conflicts of Interest:** The authors declare no conflict of interest.

## References

Albrecher, Hansjörg, and Sören Asmussen. 2010. *Ruin Probabilities*. Singapore: World Scientific, vol. 14.

Albrecher, Hansjörg, Florin Avram, Corina Constantinescu, and Jevgenijs Ivanovs. 2014. The tax identity for Markov additive risk processes. *Methodology and Computing in Applied Probability* 16: 245–58. [CrossRef]

Albrecher, Hansjörg, Jevgenijs Ivanovs, and Xiaowen Zhou. 2016. Exit identities for Lévy processes observed at Poisson arrival times. *Bernoulli* 22: 1364–82. [CrossRef]

Avram, Florin, and J. P. Garmendia. 2019. Some first passage theory for the Segerdahl-Tichy risk process, and open problems. Forthcoming.

Avram, Florin, and Dan Goreac. 2018. A pontryaghin maximum principle approach for the optimization of dividends/consumption of spectrally negative Markov processes, until a generalized draw-down time. *arXiv* arXiv:1812.08438.

Avram, Florin, Andreas E. Kyprianou, and Martijn R. Pistorius. 2004. Exit problems for spectrally negative Lévy processes and applications to (Canadized) Russian options. *The Annals of Applied Probability* 14: 215–38.

Avram, Florin, Zbigniew Palmowski, and Martijn R. Pistorius. 2007. On the optimal dividend problem for a spectrally negative Lévy process. *The Annals of Applied Probability* 17: 156–80. [CrossRef]

Avram, Florin, Zbigniew Palmowski, and Martijn R. Pistorius. 2015. On Gerber–Shiu functions and optimal dividend distribution for a Lévy risk process in the presence of a penalty function. *The Annals of Applied Probability* 25: 1868–935. [CrossRef]

Avram, Florin, Danijel Grahovac, and Ceren Vardar-Acar. 2017a. The $W, Z$ scale functions kit for first passage problems of spectrally negative Lévy processes, and applications to the optimization of dividends. *arXiv* arXiv:1706.06841.

Avram, Florin, Nhat Linh Vu, and Xiaowen Zhou. 2017b. On taxed spectrally negative Lévy processes with draw-down stopping. *Insurance: Mathematics and Economics* 76: 69–74. [CrossRef]

Avram, Florin, José-Luis Pérez, and Kazutoshi Yamazaki. 2018a. Spectrally negative Lévy processes with Parisian reflection below and classical reflection above. *Stochastic Processes and Their Applications* 128: 255–90. [CrossRef]

Avram, Florin, Bin Li, and Shu Li. 2018b. A unified analysis of taxed draw-down spectrally negative Markov processes. Forthcoming.

Avram, Florin, and Matija Vidmar. 2017. First passage problems for upwards skip-free random walks via the $\Phi, W, Z$ paradigm. *arXiv* arXiv:1708.06080.

Avram, Florin, and Xiaowen Zhou. 2017. On fluctuation theory for spectrally negative Lévy processes with parisian reflection below, and applications. *Theory of Probability and Mathematical Statistics* 95: 17–40. [CrossRef]

Azéma, Jacques, and Marc Yor. 1979. Une solution simple au probleme de Skorokhod. In *Séminaire de Probabilités XIII*. Berlin: Springer, pp. 90–115.

Bertoin, Jean. 1997. Exponential decay and ergodicity of completely asymmetric Lévy processes in a finite interval. *The Annals of Applied Probability* 7: 156–69. [CrossRef]

Bertoin, Jean. 1998. *Lévy Processes*. Cambridge: Cambridge University Press, vol. 121.

Borovkov, Alexandr. 2012. *Stochastic Processes in Queueing Theory*. Berlin: Springer Science & Business Media, vol. 4.

Carr, Peter. 2014. First-order calculus and option pricing. *Journal of Financial Engineering* 1: 1450009. [CrossRef]

Chan, Terence, Andreas E. Kyprianou, and Mladen Savov. 2011. Smoothness of scale functions for spectrally negative Lévy processes. *Probability Theory and Related Fields* 150: 691–708. [CrossRef]

Czarna, Irmina, José-Luis Pérez, Tomasz Rolski, and Kazutoshi Yamazaki. 2017. Fluctuation theory for level-dependent Lévy risk processes. *arXiv* arXiv:1712.00050.

Czarna, Irmina, Adam Kaszubowski, Shu Li, and Zbigniew Palmowski. 2018. Fluctuation identities for omega-killed Markov additive processes and dividend problem. *arXiv* arXiv:1806.08102.

Dubins, Lester E., Larry A. Shepp, and Albert Nikolaevich Shiryaev. 1994. Optimal stopping rules and maximal inequalities for Bessel processes. *Theory of Probability & Its Applications* 38: 226–61.

Egami, Masahiko, and Tadao Oryu. 2015. An excursion-theoretic approach to regulator's bank reorganization problem. *Operations Research* 63: 527–39. [CrossRef]

Hobson, David. 2007. Optimal stopping of the maximum process: A converse to the results of Peskir. *Stochastics An International Journal of Probability and Stochastic Processes* 79: 85–102. [CrossRef]

Ivanovs, Jevgenijs, and Zbigniew Palmowski. 2012. Occupation densities in solving exit problems for Markov additive processes and their reflections. *Stochastic Processes and Their Applications* 122: 3342–60. [CrossRef]

Jacobsen, Martin, and Anders Tolver Jensen. 2007. Exit times for a class of piecewise exponential Markov processes with two-sided jumps. *Stochastic Processes and Their Applications* 117: 1330–56. [CrossRef]

Kyprianou, Andreas. 2014. *Fluctuations of Lévy Processes with Applications: Introductory Lectures*. Berlin: Springer Science & Business Media.

Landriault, David, Bin Li, and Shu Li. 2015. Analysis of a drawdown-based regime-switching Lévy insurance model. *Insurance: Mathematics and Economics* 60: 98–107. [CrossRef]

Landriault, David, Bin Li, and Hongzhong Zhang. 2017a. On magnitude, asymptotics and duration of drawdowns for Lévy models. *Bernoulli* 23: 432–58. [CrossRef]

Landriault, David, Bin Li, and Hongzhong Zhang. 2017b. A unified approach for drawdown (drawup) of time-homogeneous Markov processes. *Journal of Applied Probability* 54: 603–26. [CrossRef]

Lehoczky, John P. 1977. Formulas for stopped diffusion processes with stopping times based on the maximum. *The Annals of Probability* 5: 601–7. [CrossRef]

Li, Bo, Linh Vu, and Xiaowen Zhou. 2017. Exit problems for general draw-down times of spectrally negative Lévy processes. *arXiv* arXiv:1702.07259.

Li, Bo, and Xiaowen Zhou. 2018. On weighted occupation times for refracted spectrally negative Lévy processes. *Journal of Mathematical Analysis and Applications* 466: 215–37. [CrossRef]

Mijatovic, Aleksandar, and Martijn R Pistorius. 2012. On the drawdown of completely asymmetric Lévy processes. *Stochastic Processes and Their Applications* 122: 3812–36. [CrossRef]

Page, Ewan S. 1954. Continuous inspection schemes. *Biometrika* 41: 100–15. [CrossRef]

Peskir, Goran. 1998. Optimal stopping of the maximum process: The maximality principle. *Annals of Probability* 26: 1614–40. [CrossRef]

Pistorius, Martijn R. 2004. On exit and ergodicity of the spectrally one-sided Lévy process reflected at its infimum. *Journal of Theoretical Probability* 17: 183–220. [CrossRef]

Shepp, Larry, and Albert N. Shiryaev. 1993. The Russian option: Reduced regret. *The Annals of Applied Probability* 3: 631–40. [CrossRef]

Suprun, V. N. 1976. Problem of destruction and resolvent of a terminating process with independent increments. *Ukrainian Mathematical Journal* 28: 39–51. [CrossRef]

Taylor, Howard M. 1975. A stopped Brownian motion formula. *The Annals of Probability* 3: 234–46. [CrossRef]

Wang, Wenyuan, and Xiaowen Zhou. 2018. General drawdown-based de Finetti optimization for spectrally negative Lévy risk processes. *Journal of Applied Probability* 55: 513–42. [CrossRef]

© 2019 by the authors. Licensee MDPI, Basel, Switzerland. This article is an open access article distributed under the terms and conditions of the Creative Commons Attribution (CC BY) license (http://creativecommons.org/licenses/by/4.0/).

Article

# On the Laplace Transforms of the First Hitting Times for Drawdowns and Drawups of Diffusion-Type Processes

Pavel V. Gapeev [1,*], Neofytos Rodosthenous [2] and V. L. Raju Chinthalapati [3]

[1] Department of Mathematics, London School of Economics, Houghton Street, London WC2A 2AE, UK
[2] School of Mathematical Sciences, Queen Mary University of London, Mile End Road, London E1 4NS, UK
[3] Southampton Business School, University of Southampton, Southampton SO17 1BJ, UK
* Correspondence: p.v.gapeev@lse.ac.uk

Received: 27 May 2019; Accepted: 30 July 2019; Published: 5 August 2019

**Abstract:** We obtain closed-form expressions for the value of the joint Laplace transform of the running maximum and minimum of a diffusion-type process stopped at the first time at which the associated drawdown or drawup process hits a constant level before an independent exponential random time. It is assumed that the coefficients of the diffusion-type process are regular functions of the current values of its running maximum and minimum. The proof is based on the solution to the equivalent inhomogeneous ordinary differential boundary-value problem and the application of the normal-reflection conditions for the value function at the edges of the state space of the resulting three-dimensional Markov process. The result is related to the computation of probability characteristics of the take-profit and stop-loss values of a market trader during a given time period.

**Keywords:** Laplace transform; first hitting time; diffusion-type process; running maximum and minimum processes; boundary-value problem; normal reflection.

---

## 1. Introduction

The aim of this paper is to derive closed-form expressions for the joint Laplace transform (4) of the first time to a fixed drawdown occurring before a fixed drawup of the diffusion-type process $X$ and its running maximum and minimum $S$ and $Q$ defined in (1)–(2) considered up to a random exponentially distributed time $\eta$, which is independent of the driving standard Brownian motion. We consider a model for the diffusion-type process $X$ with the coefficients being regular functions of the current values of the process $X$ itself as well as of its running maximum and minimum $S$ and $Q$. The value function in (4) provides the Laplace transform of the value function in (6) which is the joint Laplace transform of the same random variables representing functionals of the diffusion-type process $X$ stopped before a fixed time. We derive a closed-form solution to the equivalent inhomogeneous ordinary differential boundary-value problem for the value of the joint Laplace transform as a stopping problem for the resulting three-dimensional continuous Markov process $(X, S, Q)$. This result can therefore be interpreted as the computation of the probability characteristics of the random variables associated with the take-profit and stop-loss values of a market trader on a fixed-time interval. The problem of computation of the Laplace transform of the same random times and variables in a model in which the coefficients of the original diffusion-type process depend on the current values of the running maximum and minimum as well as on the maximum drawdown and maximum drawup was explicitly solved in Gapeev and Rodosthenous (2015) on the infinite time interval. Other functionals of diffusion processes evaluated at independent exponential times were computed in Borodin and Salminen (Borodin and Salminen 2002, Part II) among others.

The joint Laplace transform of the first time at which a Brownian motion with linear drift hits a given drawdown value and the running maximum stopped at the same time was computed by Taylor (1975). The joint distribution of the same random variables was obtained by Lehoczky (1977). The mean value and the density of the maximum drawdown of a Brownian motion with linear drift were explicitly derived by (Douady et al. 2000; Magdon-Ismail et al. 2004), respectively. More recently, Pospisil et al. (2009) computed the probability of the event that the drawdown of a one-dimensional diffusion reaches a fixed value occurs before the drawup of the same process reaches another fixed value. Mijatović and Pistorius (2012) obtained the distribution laws of the first-passage times of spectrally positive and negative Lévy processes over constant levels and derived explicit expressions for several related characteristics for the drawdowns and drawups in those models. An extensive overview of various probabilistic and practically applied aspects of drawdowns such as the speed of market crashes and others was recently provided in the monograph of Zhang (2018).

The diffusion-type processes can be considered as immediate generalisations of the diffusion processes particularly arising in the so-called local volatility models introduced by Dupire (1997), where the local drift and diffusion coefficients depend only on the running value of the original process. Other generalisations of the original processes with diffusion coefficients depending on the running values of the initial processes and their running minima were constructed by Forde (2011) for given joint laws of the terminal level and supremum at an independent exponential time (see also Forde et al. 2013; Zhang 2014) for other important probability characteristics of processes of such type). The valuation functional equations for general functional path-dependent volatility models were derived in (Cont and Fournié 2013; Fournié 2010), who also considered the sensitivity analysis of path-dependent financial derivative securities. Henry-Labordère (2009) and Ren et al. (2007), among others, considered the option pricing and calibration problems in models of stochastic interest rates and volatility based on diffusion-type processes with tractable path-dependent coefficients.

Optimal stopping problems for running maxima of some diffusion processes were studied by (Jacka 1991; Dubins et al. 1993; Peskir 1998; Peskir and Shiryaev 2006, chp. V) among others. Discounted optimal stopping problems for certain payoff functions depending on the current values of the running maxima of geometric Brownian motions were initiated by (Shepp and Shiryaev 1993, 1994) and then taken further by (Pedersen 2000; Guo and Shepp 2001; Guo and Zervos 2010, Glover et al. 2013; Rodosthenous and Zervos 2017) among others. Moreover, Peskir (2012, 2014) studied optimal stopping problems for three-dimensional Markov processes having the initial diffusion process as well as its maximum and minimum as state space components. Other three-dimensional optimal stopping problems for continuous Markov processes of such type were studied in (Gapeev and Rodosthenous 2014, 2016) among others. The main feature of the resulting optimal stopping problems and their equivalent free-boundary problems was the application of the normal-reflection conditions for the value functions at the edges of the multi-dimensional state spaces to derive systems of first-order nonlinear ordinary differential equations for the optimal stopping boundaries depending on the current values of the running extremal processes. Optimal stopping problems for diffusion and spectrally negative Lévy processes on random time intervals were considered in (Carr 1998; Avram et al. 2004; Agarwal et al. 2016) among others. It turned out that the resulting value functions and optimal stopping boundaries in models with exponentially distributed time horizons independent of the underlying processes are analytically more tractable than those obtained in models with fixed time horizons. Other optimal stopping problems for exponentially distributed time horizons which are dependent of the underlying Lévy process were recently considered in Rodosthenous and Zhang (2018).

Glattfelder et al. (2011) suggested a new paradigm, the directional changes, that summarises the price dynamics in the financial market. Unlike interval based summary along the physical time, the new paradigm summarizes the price movements along the intrinsic time scale of the market that is driven by the events in the market. The events in the market are identified by the a priori defined significant percentage of price moves known as thresholds. For a given threshold, the price movements are summarised by identifying the local price extremes from where there has been a percentage drop (or rise) in price that accedes the threshold. The process of price drop (or rise) from a local price extreme to the point where the price is dropped (risen) by the threshold is defined as directional change event. The price movement that continues after directional change event in the same direction beyond the threshold is considered as overshoot. Roughly speaking, directional changes and overshoots summarise the upward or downward trends in the market according to the prescribed thresholds. It is obvious that the summary of the directional changes is depending on the selected threshold. Using the high frequency foreign exchange data, in Glattfelder et al. (2011) scaling laws were demonstrated in intrinsic times for the variables like average times that are taken for directional changes, event thresholds, average overshoots, etc. The authors of Glattfelder et al. (2011) have identified 12 scaling laws across 13 currency pairs that are consistent over varying time intervals. The scaling laws throw light on market physics of moving prices. Each scaling law encapsulates certain stylised facts of the market. The scaling law that describes the relationship between the directional change and overshoot sections of the total price move has drawn quite a lot of attention. Even though the empirical evidence of the scaling laws is demonstrated in the literature (see, e.g., Bakhach et al. 2018; Bakhach et al. 2018; Tsang et al. 2017), the required theoretical framework is not developed yet. We believe that the present work on first hitting times for drawdowns and drawups on diffusion-type processes on random time horizons throws light on the underlying theoretical aspects of the scaling laws that are presented in financial data.

The paper is organised as follows. In Section 2, we introduce the setting and notation of the model with a three-dimensional continuous Markov process, whose state space components are the original process and its running maximum and minimum processes. We define the value function of the joint Laplace transform of the first time to a fixed drawdown occurring before the first time of a fixed drawup and an independent exponential time together with the running maximum and minimum processes stopped at the earliest of those times. In Section 3, we obtain a closed-form solution to the associated inhomogeneous ordinary differential boundary-value problem and show that the value function represents a linear combination of the solutions to the systems of first-order partial differential equations which arise from the application of the normal-reflection conditions for this function at the edges of the three-dimensional state space. We also illustrate the results on several examples of the original processes representing locally a Brownian motion with drift, or a mean-reverting Ornstein-Uhlenbeck process, or the logarithm of a Feller square root process. In Section 4, we formulate the result of the paper and prove that the solution to the boundary-value problem provides the required joint Laplace transform.

## 2. Preliminaries

In this section, we give a precise formulation of the model and the three-dimensional stopping problem as well as its equivalent boundary-value problem.

### 2.1. Formulation of the Problem

Let us consider a probability space $(\Omega, \mathcal{F}, P)$ with a standard Brownian motion $B = (B_t)_{t \geq 0}$ and a positive random time $\eta$ such that $P(\eta > t) = e^{-\alpha t}$, for all $t \geq 0$ and some $\alpha > 0$ fixed ($B$ and $\eta$ are supposed to be independent). Assume that there exists a process $X = (X_t)_{t \geq 0}$ solving the stochastic differential equation

$$dX_t = \mu(X_t, S_t, Q_t)\, dt + \sigma(X_t, S_t, Q_t)\, dB_t \quad (X_0 = x) \tag{1}$$

where $x \in \mathbb{R}$ is fixed, and $\mu(x, s, q)$ and $\sigma(x, s, q) > 0$ are continuously differentiable functions on $[-\infty, \infty]^3$ which are of at most linear growth in $x$ and uniformly bounded in $s$ and $q$.

Here, the associated with $X$ running maximum process $S = (S_t)_{t \geq 0}$ and the *running minimum* process $Q = (Q_t)_{t \geq 0}$ are defined by

$$S_t = s \vee \max_{0 \leq u \leq t} X_u \quad \text{and} \quad Q_t = q \wedge \min_{0 \leq u \leq t} X_u \tag{2}$$

for arbitrary $q \leq x \leq s$. It follows from the result of (Liptser and Shiryaev [1977] 2001, chp. IV, Theorem 4.8) that the equation in (1) admits a pathwise unique (strong) solution. We also define the associated first hitting (stopping) times

$$\tau_a = \inf\{t \geq 0 \,|\, S_t - X_t \geq a\} \quad \text{and} \quad \zeta_b = \inf\{t \geq 0 \,|\, X_t - Q_t \geq b\} \tag{3}$$

for some $a, b > 0$ fixed.

The purpose of the present paper is to derive closed-form expressions for the joint Laplace transform of the random time $\tau_a \wedge \zeta_b \wedge \eta$ and the random variables $S_{\tau_a \wedge \zeta_b \wedge \eta}$ and $Q_{\tau_a \wedge \zeta_b \wedge \eta}$. We therefore need to compute the value function of the following stopping problem for the (time-homogeneous strong) Markov process $(X, S, Q) = (X_t, S_t, Q_t)_{t \geq 0}$ given by

$$V_*(x, s, q) = E_{x,s,q}\left[e^{-\lambda(\tau_a \wedge \eta) - \theta S_{\tau_a \wedge \eta} - \kappa Q_{\tau_a \wedge \eta}} I(\tau_a < \zeta_b)\right] \tag{4}$$

for any $(x, s, q) \in E^3$ and some $\lambda, \theta, \kappa > 0$ fixed, where $I(\cdot)$ denotes the indicator function. Here, $E_{x,s,q}$ denotes the expectation under the assumption that the (three-dimensional) Markov process $(X, S, Q)$ defined in (1)–(2) starts at $(x, s, q) \in E^3$, where we assume that the state space of $(X, S, Q)$ is essentially $E^3 = \{(x, s, q) \in \mathbb{R}^3 \,|\, q \leq x \leq s\}$ with its border planes $d_1^3 = \{(x, s, q) \in \mathbb{R}^3 \,|\, x = s\}$ and $d_2^3 = \{(x, s, q) \in \mathbb{R}^3 \,|\, x = q\}$.

It follows from the independence of the process $X$ and the random time $\eta$ that the value function in (4) admits the representation

$$V_*(x, s, q) = \int_0^\infty W_*(T; x, s, q)\, \alpha e^{-\alpha T}\, dT \tag{5}$$

where we set

$$W_*(T; x, s, q) = E_{x,s,q}\left[e^{-\lambda(\tau_a \wedge T) - \theta S_{\tau_a \wedge T} - \kappa Q_{\tau_a \wedge T}} I(\tau_a < \zeta_b)\right] \tag{6}$$

for any $(x, s, q) \in E^3$, and each $T > 0$ fixed.

## 2.2. The Boundary-Value Problems

By means of standard arguments based on the application of Itô's formula (see, e.g., Karatzas and Shreve 1991, chp. V, sct. 5.1), it is shown that the infinitesimal operator $\mathbb{L}$ of the process $(X, S, Q)$ acts on a function $F(x, s, q)$ from the class $C^{2,1,1}$ on the interior of $E^3$ according to the rule

$$(\mathbb{L}F)(x, s, q) = \mu(x, s, q)\, \partial_x F(x, s, q) + \frac{\sigma^2(x, s, q)}{2}\, \partial_{xx} F(x, s, q) \tag{7}$$

for all $q < x < s$. It follows from the results of general theory of Markov processes (see, e.g., Dynkin 1965, chp. V) that the value function $W_*(T; x, s, q)$ in (6) solves the equivalent parabolic-type boundary-value problem

$$(\mathbb{L}W - \lambda W - \partial_T W)(T; x, s, q) = 0 \quad \text{for} \quad (s-a) \vee q < x < s \wedge (q+b) \tag{8}$$

$$W(T; x, s, q)|_{x=(s-a)+} = e^{-\theta s - \kappa q} \quad \text{for} \quad s - q \geq a \tag{9}$$

$$W(T; x, s, q)|_{x=(q+b)-} = 0 \quad \text{for} \quad s - q \geq b \tag{10}$$

$$\partial_q W(T; x, s, q)|_{x=q+} = 0 \quad \text{for} \quad 0 < s - q < a \tag{11}$$

$$\partial_s W(T; x, s, q)|_{x=s-} = 0 \quad \text{for} \quad 0 < s - q < b \tag{12}$$

for all $T > 0$. In this case, using the integration-by-parts formula, and taking into account the assumption that the value function in (6) is bounded, we have

$$\int_0^\infty \partial_T W(T; x, s, q) \, \alpha \, e^{-\alpha T} \, dT \tag{13}$$

$$= \left[ W(T; x, s, q) \, \alpha \, e^{-\alpha T} \right]_0^\infty + \int_0^\infty W(T; x, s, q) \, \alpha^2 \, e^{-\alpha T} \, dT$$

$$= -\alpha \, e^{-\theta s - \kappa q} + \int_0^\infty W(T; x, s, q) \, \alpha^2 \, e^{-\alpha T} \, dT = -\alpha \, e^{-\theta s - \kappa q} + \alpha \, V(x, s, q)$$

$$\int_0^\infty \partial_x W(T; x, s, q) \, \alpha \, e^{-\alpha T} \, dT = \partial_x V(x, s, q) \tag{14}$$

$$\int_0^\infty \partial_{xx} W(T; x, s, q) \, \alpha \, e^{-\alpha T} \, dT = \partial_{xx} V(x, s, q) \tag{15}$$

$$\int_0^\infty \partial_q W(T; x, s, q) \, \alpha \, e^{-\alpha T} \, dT = \partial_q V(x, s, q) \tag{16}$$

and

$$\int_0^\infty \partial_s W(T; x, s, q) \, \alpha \, e^{-\alpha T} \, dT = \partial_s V(x, s, q) \tag{17}$$

for all $(x, s, q) \in E^3$. Hence, it follows from the boundary-value problem in (8)–(12), that the value function $V_*(x, s, q)$ in (6) solves the equivalent inhomogeneous ordinary boundary-value problem

$$(\mathbb{L}V - (\alpha + \lambda) V)(x, s, q) = -\alpha \, e^{-\theta s - \kappa q} \quad \text{for} \quad (s-a) \vee q < x < s \wedge (q+b) \tag{18}$$

$$V(x, s, q)|_{x=(s-a)+} = e^{-\theta s - \kappa q} \quad \text{for} \quad s - q \geq a \tag{19}$$

$$V(x, s, q)|_{x=(q+b)-} = 0 \quad \text{for} \quad s - q \geq b \tag{20}$$

$$\partial_q V(x, s, q)|_{x=q+} = 0 \quad \text{for} \quad 0 < s - q < a \tag{21}$$

$$\partial_s V(x, s, q)|_{x=s-} = 0 \quad \text{for} \quad 0 < s - q < b \tag{22}$$

for $a, b > 0$ fixed. Note that the homogeneous version of the ordinary differential boundary-value problem in (18)–(22) in a model with more general diffusion-type processes $X$ was explicitly solved in (Gapeev and Rodosthenous 2015, sct. 3).

## 3. Solutions to the Boundary-Value Problem

In this section, we obtain closed-form solutions to the boundary-value problem in (18)–(22) under various relations on the parameters of the model.

## 3.1. The General Solution of the Ordinary Differential Equation

We first observe that the general solution of the equation in (18) has the form

$$V(x,s,q) = C_1(s,q)\Psi_1(x,s,q) + C_2(s,q)\Psi_2(x,s,q) + \frac{\alpha}{\alpha+\lambda}e^{-\theta s - \kappa q} \quad (23)$$

where $C_i(s,q)$, $i=1,2$, are some arbitrary continuously differentiable functions, and $\Psi_i(x,s,q)$, $i=1,2$, are the two fundamental positive solutions (i.e., nontrivial linearly independent particular solutions) of the homogeneous version of the second-order ordinary differential equation in (18). Without loss of generality, we may assume that $\Psi_1(x,s,q)$ and $\Psi_2(x,s,q)$ are the (strictly) increasing and decreasing (convex) functions, respectively. Note that these solutions should satisfy the properties $\Psi_1(r,r,r) \uparrow \infty$ and $\Psi_2(r,r,r) \downarrow 0$ as $r \uparrow \infty$ and $\Psi_1(r,r,r) \downarrow 0$ and $\Psi_2(r,r,r) \uparrow \infty$ as $r \downarrow -\infty$ on the state space $E^3$ of the process $(X,S,Q)$. These functions can be represented as the functionals

$$\Psi_1(x,s,q) = \begin{cases} E_{x,s,q}[e^{-\lambda \zeta'}I(\zeta' < \infty)], & \text{if } x \leq x' \\ 1/E_{x',s,q}[e^{-\lambda \zeta}I(\zeta < \infty)], & \text{if } x \geq x' \end{cases} \quad (24)$$

and

$$\Psi_2(x,s,q) = \begin{cases} 1/E_{x',s,q}[e^{-\lambda \zeta}I(\zeta < \infty)], & \text{if } x \leq x' \\ E_{x,s,q}[e^{-\lambda \zeta'}I(\zeta' < \infty)], & \text{if } x \geq x' \end{cases} \quad (25)$$

of the first hitting times $\zeta = \inf\{t \geq 0 \mid X_t = x\}$ and $\zeta' = \inf\{t \geq 0 \mid X_t = x'\}$ of the process $X$ solving the stochastic differential equation in (1) and started at $x$ and $x'$ such that $(x,s,q), (x',s,q) \in E^3$, respectively (see, e.g., Rogers and Williams 1987, chp. V, sct. 50 for further details).

Hence, by applying the conditions of (19)–(22) to the function in (23), we obtain the equalities

$$C_1(s,q)\Psi_1(s-a,s,q) + C_2(s,q)\Psi_2(s-a,s,q) = \frac{\lambda}{\alpha+\lambda}e^{-\theta s - \kappa q} \quad (26)$$

for $s - q \geq a$,

$$C_1(s,q)\Psi_1(q+b,s,q) + C_2(s,q)\Psi_2(q+b,s,q) = -\frac{\alpha}{\alpha+\lambda}e^{-\theta s - \kappa q} \quad (27)$$

for $s - q \geq b$,

$$\sum_{i=1}^{2}\left(\partial_q C_i(s,q)\Psi_i(q,s,q) + C_i(s,q)\partial_q \Psi_i(x,s,q)\big|_{x=q}\right) = \frac{\alpha \kappa}{\alpha+\lambda}e^{-\theta s - \kappa q} \quad (28)$$

for $0 < s - q < a$,

$$\sum_{i=1}^{2}\left(\partial_s C_i(s,q)\Psi_i(s,s,q) + C_i(s,q)\partial_s\partial_s \Psi_i(x,s,q)\big|_{x=s}\right) = \frac{\alpha \theta}{\alpha+\lambda}e^{-\theta s - \kappa q} \quad (29)$$

for $0 < s - q < b$.

## 3.2. The Solution to the Boundary-Value Problem

We now derive the solution of the boundary-value problem in (18)–(22). For this purpose, we recall that the second and third components of the process $(X,S,Q)$ can increase and decrease only at the planes $d_1^3$ and $d_2^3$, that is, when $X_t = S_t$ and $X_t = Q_t$ for $t \geq 0$, respectively.

(i) Let us first consider the domain $a \vee b \leq s - q \leq a + b$. In this case, solving the system of equations in (26) and (27), we conclude that the candidate value function admits the representation

$$V(x,s,q;\infty) = C_1(s,q;\infty)\,\Psi_1(x,s,q) + C_2(s,q;\infty)\,\Psi_2(x,s,q) + \frac{\alpha}{\alpha+\lambda}e^{-\theta s - \kappa q} \qquad (30)$$

in the region $R^3(\infty) = \{(x,s,q) \in E^3 \mid q \leq s - a \leq x \leq q + b \leq s\}$, with

$$C_1(s,q;\infty) = \frac{e^{-\theta s - \kappa q}(\lambda \Psi_2(q+b,s,q) + \alpha \Psi_2(s-a,s,q))/(\alpha+\lambda)}{\Psi_1(s-a,s,q)\Psi_2(q+b,s,q) - \Psi_1(q+b,s,q)\Psi_2(s-a,s,q)} \qquad (31)$$

and

$$C_2(s,q;\infty) = \frac{e^{-\theta s - \kappa q}(\lambda \Psi_1(q+b,s,q) + \alpha \Psi_1(s-a,s,q))/(\alpha+\lambda)}{\Psi_1(q+b,s,q)\Psi_2(s-a,s,q) - \Psi_1(s-a,s,q)\Psi_2(q+b,s,q)} \qquad (32)$$

for all $q + a \vee b \leq s \leq q + a + b$ (see Figures 1 and 2 below).

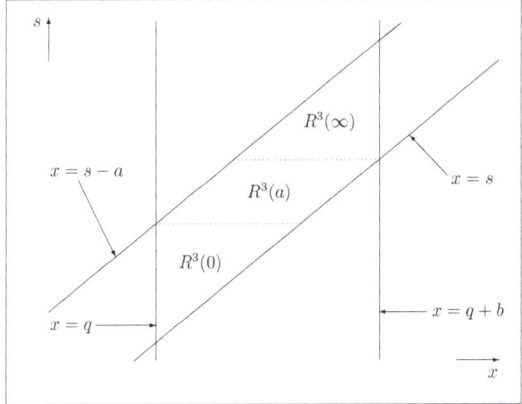

**Figure 1.** A computer drawing of the state space of the process $(X,S,Q)$, for some $q \in \mathbb{R}$ fixed and $a < b$.

(ii) Let us now consider the domain $a \leq s - q < b$. In this case, it follows from the equations in (26) and (29) that the candidate value function admits the representation

$$V(x,s,q;a) = C_1(s,q;a)\,\Psi_1(x,s,q) + C_2(s,q;a)\,\Psi_2(x,s,q) + \frac{\alpha}{\alpha+\lambda}e^{-\theta s - \kappa q} \qquad (33)$$

in the region $R^3(a) = \{(x,s,q) \in E^3 \mid q \leq s - a \leq x \leq s < q + b\}$, with

$$C_2(s,q;a) = \frac{\lambda}{\alpha+\lambda}\frac{e^{-\theta s - \kappa q}}{\Psi_2(s-a,s,q)} - C_1(s,q;a)\frac{\Psi_1(s-a,s,q)}{\Psi_2(s-a,s,q)} \qquad (34)$$

for $q + a \leq s < q + b$, where $C_1(s,q;a)$ solves the first-order linear ordinary differential equation

$$\partial_s C_1(s,q;a)\,H_{1,2}(s,q;a) + C_1(s,q;a)\,H_{1,1}(s,q;a) = H_{1,0}(s,q;a) \qquad (35)$$

with

$$H_{1,2}(s,q;a) = \Psi_1(s,s,q) - \Psi_2(s,s,q) \frac{\Psi_1(s-a,s,q)}{\Psi_2(s-a,s,q)} \tag{36}$$

$$H_{1,1}(s,q;a) = \partial_s \Psi_1(x,s,q)|_{x=s} \tag{37}$$

$$- \partial_s \left( \frac{\Psi_1(s-a,s,q)}{\Psi_2(s-a,s,q)} \right) \Psi_2(s,s,q) - \frac{\Psi_1(s-a,s,q)}{\Psi_2(s-a,s,q)} \partial_s \Psi_1(x,s,q)|_{x=s}$$

$$H_{1,0}(s,q;a) = \frac{\lambda}{\alpha+\lambda} \left( \theta e^{-\theta s - \kappa q} \right. \tag{38}$$

$$\left. - \partial_s \left( \frac{e^{-\theta s - \kappa q}}{\Psi_2(s-a,s,q)} \right) \Psi_2(s,s,q) - \frac{e^{-\theta s - \kappa q}}{\Psi_2(s-a,s,q)} \partial_s \Psi_2(x,s,q)|_{x=s} \right)$$

for all $q + a \leq s < q + b$. Observe that the process $(X, S, Q)$ can exit the region $R^3(a)$ by passing to the region $R^3(\infty)$ in part (i) of this subsection only through the point $x = s = q + b$, by hitting the plane $d_1^3$ so that increasing its second component $S$. Thus, the candidate function $V(x, s, q)$ should be continuous at the point $(q + b, q + b, q)$, that is expressed by the equality

$$C_1(q+b,q;a)\Psi_1(q+b,q+b,q) + C_2(q+b,q;a)\Psi_2(q+b,q+b,q) = -\frac{\alpha}{\alpha+\lambda} e^{-\theta(q+b)-\kappa q} \tag{39}$$

for all $q \in \mathbb{R}$ (see Figure 1 above). Hence, solving the differential equation in (35) together with the system of equations in (34) with $s = q + b$ and (39), we obtain

$$C_1(s,q;a) = C_1(q+b,q;a) \exp\left( \int_s^{q+b} \frac{H_{1,1}(u,q;a)}{H_{1,2}(u,q;a)} du \right) \tag{40}$$

$$- \int_s^{q+b} \frac{H_{1,0}(u,q;a)}{H_{1,2}(u,q;a)} \exp\left( \int_s^u \frac{H_{1,1}(v,q;a)}{H_{1,2}(v,q;a)} dv \right) du$$

for all $q + a \leq s < q + b$, where $C_1(q + b, q; a)$ is given by

$$C_1(q+b,q;a) \tag{41}$$

$$= \frac{e^{-\theta(q+b)-\kappa q}(\lambda \Psi_2(q+b,q+b,q) + \alpha \Psi_2(q+b-a,q+b,q))/(\alpha+\lambda)}{\Psi_1(q+b-a,q+b,q)\Psi_2(q+b,q+b,q) - \Psi_1(q+b,q+b,q)\Psi_2(q+b-a,q+b,q)}$$

for all $q \in \mathbb{R}$.

Note that in the case in which $\mu(s,q) = \mu(s)$ and $\sigma(s,q) = \sigma(s)$ in (1) as well as $\kappa = 0$ and $b = \infty$ in (6), the candidate value function admits the representation of (33) with $V(x,s,q;a) = U(x,s;a)$ and $C_i(s,q;a) = D_i(s;a)$ as well as $\Psi_i(x,s,q) = \Phi_i(x,s)$, $i = 1,2$. Moreover, we observe that $D_1(\infty;a) = 0$ should hold in (33), since otherwise $U(x,s;a) \to \pm \infty$ as $x = s \uparrow \infty$, which must be excluded, by virtue of the obvious fact that the value function $V_*(x,s,q) = U_*(x,s)$ in (6) is bounded. Therefore, using arguments similar to the ones above, we conclude that the function $C_2(s,q;a) = D_2(s;a)$ has the form of (34) with $C_1(s,q;a) = D_1(s;a)$ given by

$$D_1(s;a) = -\int_s^\infty \frac{G_{1,0}(u;\infty)}{G_{1,2}(u;\infty)} \exp\left( \int_s^u \frac{G_{1,1}(v;\infty)}{G_{1,2}(v;\infty)} dv \right) du \tag{42}$$

and $H_{1,j}(s,q;a) = G_{1,j}(s;a)$, $j = 0, 1, 2$, from (36)–(38), for all $s \in \mathbb{R}$.

(iii) Let us now consider the domain $b \leq s - q < a$. In this case, it follows from the equations in (27) and (28) that the candidate value function admits the representation

$$V(x,s,q;b) = C_1(s,q;b)\Psi_1(x,s,q) + C_2(s,q;b)\Psi_2(x,s,q) + \frac{\alpha}{\alpha+\lambda} e^{-\theta s - \kappa q} \tag{43}$$

in the region $R^3(b) = \{(x,s,q) \in E^3 \mid s - a < q \leq x \leq q + b \leq s\}$, with

$$C_2(s,q;b) = -\frac{\alpha}{\alpha+\lambda}\frac{e^{-\theta s - \kappa q}}{\Psi_2(q+b,s,q)} - C_1(s,q;b)\frac{\Psi_1(q+b,s,q)}{\Psi_2(q+b,s,q)} \tag{44}$$

for $q + b \leq s < q + a$, where $C_1(s,q;b)$ solves the first-order linear ordinary differential equation

$$\partial_q C_1(s,q;b) H_{2,2}(s,q;b) + C_1(s,q;b) H_{2,1}(s,q;b) = H_{2,0}(s,q;b) \tag{45}$$

with

$$H_{2,2}(s,q;b) = \Psi_1(q,s,q) - \Psi_2(q,s,q)\frac{\Psi_1(q+b,s,q)}{\Psi_2(q+b,s,q)} \tag{46}$$

$$H_{2,1}(s,q;b) = \partial_q \Psi_1(x,s,q)|_{x=q} \tag{47}$$

$$- \partial_q\left(\frac{\Psi_1(q+b,s,q)}{\Psi_2(q+b,s,q)}\right)\Psi_2(q,s,q) - \frac{\Psi_1(q+b,s,q)}{\Psi_2(q+b,s,q)}\partial_q\Psi_2(x,s,q)|_{x=q}$$

$$H_{2,0}(s,q;b) = \frac{\alpha}{\alpha+\lambda}\left(\kappa e^{-\theta s - \kappa q}\right. \tag{48}$$

$$\left. + \partial_q\left(\frac{e^{-\theta s - \kappa q}}{\Psi_2(q+b,s,q)}\right)\Psi_2(q,s,q) + \frac{e^{-\theta s - \kappa q}}{\Psi_2(q+b,s,q)}\partial_q\Psi_2(x,s,q)|_{x=q}\right)$$

for all $q + b \leq s < q + a$. Observe that the process $(X, S, Q)$ can exit $R^3(b)$ by passing to the region $R^3(\infty)$ in part (i) of this subsection only through the point $x = q = s - a$, by hitting the plane $d_2^3$ so that decreasing its third component $Q$. Then, the candidate value function should be continuous at the point $(s - a, s, s - a)$, that is expressed by the equality

$$C_1(s, s-a; b)\Psi_1(s-a, s, s-a) + C_2(s, s-a; b)\Psi_2(s-a, s, s-a) = -\frac{\alpha}{\alpha+\lambda}e^{-\theta s - \kappa(s-a)} \tag{49}$$

for all $s \in \mathbb{R}$ (see Figure 2 below). Hence, solving the differential equation in (45) together with the system of equations in (44) with $q = s - a$ and (49), we obtain

$$C_1(s,q;b) = C_1(s,s-a;b)\exp\left(-\int_{s-a}^q \frac{H_{2,1}(s,u;b)}{H_{2,2}(s,u;b)}du\right) \tag{50}$$

$$+ \int_{s-a}^q \frac{H_{2,0}(s,u;b)}{H_{2,2}(s,u;b)}\exp\left(-\int_u^q \frac{H_{2,1}(s,v;b)}{H_{2,2}(s,v;b)}dv\right)du$$

for all $q + b \leq s < q + a$, where $C_1(s, s - a; b)$ is given by

$$C_1(s, s-a; b) \tag{51}$$

$$= \frac{e^{-\theta s - \kappa(s-a)}(\lambda \Psi_2(s-a+b,s,s-a) + \alpha \Psi_2(s-a,s,s-a))/(\alpha+\lambda)}{\Psi_1(s-a,s,s-a)\Psi_2(s-a+b,s,s-a) - \Psi_1(s-a+b,s,s-a)\Psi_2(s-a,s,s-a)}$$

for $s \in \mathbb{R}$.

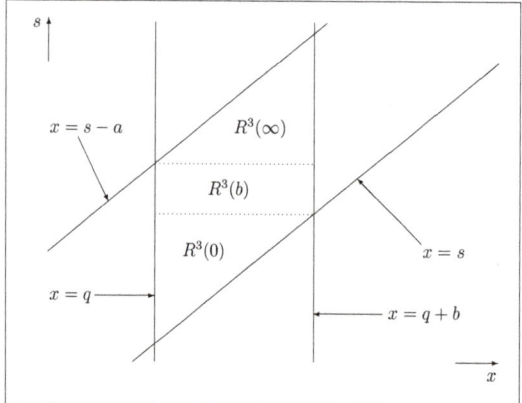

**Figure 2.** A computer drawing of the state space of the process $(X, S, Q)$, for some $q \in \mathbb{R}$ fixed and $b \leq a$.

(iv) Let us now consider the domain $0 \leq s - q < a \wedge b$. In this case, it follows that the candidate value function admits the representation

$$V(x, s, q; 0) = C_1(s, q; 0) \Psi_1(x, s, q) + C_2(s, q; 0) \Psi_2(x, s, q) + \frac{\alpha}{\alpha + \lambda} e^{-\theta s - \kappa q} \tag{52}$$

in the region $R^3(0) = \{(x, s, q) \in E^3 \mid s - a < q \leq x \leq s < q + b\}$, where $C_i(s, q; 0)$, $i = 1, 2$, solve the first-order linear partial differential equations in (28) and (29), for all $0 < s - q < a \wedge b$. Observe that, the process $(X, S, Q)$ can exit $R^3(0)$ by passing to the region $R^3(a \wedge b)$ in part (ii) or (iii) of this subsection only through the points $x = s = q + a \wedge b$ and $x = q = s - a \wedge b$, by hitting the plane $d_1^3$ or $d_2^3$, so that increasing its second or third components, $S$ or $Q$, respectively. Then, the candidate value function should be continuous at the points $(q + a \wedge b, q + a \wedge b, q)$ and $(s - a \wedge b, s, s - a \wedge b)$, that is expressed by the equalities

$$\begin{aligned}
& C_1(q + a \wedge b, q; 0) \Psi_1(q + a \wedge b, q + a \wedge b, q) \\
& + C_2(q + a \wedge b, q; 0) \Psi_2(q + a \wedge b, q + a \wedge b, q) \\
& = C_1(q + a \wedge b, q; a \wedge b) \Psi_1(q + a \wedge b, q + a \wedge b, q) \\
& + C_2(q + a \wedge b, q; a \wedge b) \Psi_2(q + a \wedge b, q + a \wedge b, q)
\end{aligned} \tag{53}$$

for all $q \in \mathbb{R}$, and

$$\begin{aligned}
& C_1(s, s - a \wedge b; 0) \Psi_1(s - a \wedge b, s, s - a \wedge b) \\
& + C_2(s, s - a \wedge b; 0) \Psi_2(s - a \wedge b, s, s - a \wedge b) \\
& = C_1(s, s - a \wedge b; a \wedge b) \Psi_1(s - a \wedge b, s, s - a \wedge b) \\
& + C_2(s, s - a \wedge b; a \wedge b) \Psi_2(s - a \wedge b, s, s - a \wedge b)
\end{aligned} \tag{54}$$

for all $s \in \mathbb{R}$, where $C_i(q + a \wedge b, q; a \wedge b)$ and $C_i(s, s - a \wedge b; a \wedge b)$, $i = 1, 2$, are found in (34) + (40) or (44) + (50). Moreover, we have the property $C_2(r, r; 0) \to 0$ as $r \downarrow -\infty$, since otherwise $V(r, r, r; 0) \to \pm \infty$, that must be excluded by virtue of the obvious fact that the value function in (6) is bounded (see Figures 1 and 2 above). We may therefore conclude that the candidate value function admits the representation of (52) in the region $R^3(0)$ above, where $C_i(s, q; 0)$, $i = 1, 2$, provide a unique solution of the two-dimensional system of first-order linear partial differential equations in (21) and (22) with the boundary conditions of (53)–(54) and $C_2(r, r; 0) \to 0$ as $r \downarrow -\infty$. Here, the existence and uniqueness of solutions to such special kinds of systems of equations follow from the classical existence and uniqueness results of solutions to appropriate boundary-value problems for first-order linear partial differential equations.

### 3.3. Some Examples

Let us finally consider some examples of processes $X$ from (1) and present explicit expressions for the fundamental solutions $\Psi_i(x,s,q)$, $i = 1,2$, of the homogeneous version of the second-order ordinary differential equation in (18).

**Example 1.** Let $\mu(x,s,q) = \beta(s,q)$ and $\sigma(x,s,q) = v(s,q)$, for all $(x,s,q) \in E^3$ and some continuously differentiable functions $\beta(s,q)$ and $v(s,q) > 0$ on $[-\infty,\infty]^2$, so that the process $X$ from (1) represents locally a Brownian motion with linear drift. In this case, we have $\Psi_i(x,s,q) = e^{\gamma_i(s,q)x}$ with

$$\gamma_i(s,q) = -\frac{\beta(s,q)}{v^2(s,q)} - (-1)^i \sqrt{\frac{\beta^2(s,q)}{v^4(s,q)} + \frac{2(\alpha+\lambda)}{v^2(s,q)}} \tag{55}$$

for every $i = 1,2$, so that $\gamma_2(s,q) < 0 < \gamma_1(s,q)$, for all $q \leq s$.

**Example 2.** Let $\mu(x,s,q) = \beta(s,q) - \delta(s,q)x$ and $\sigma(x,s,q) = v(s,q)$, for all $(x,s,q) \in E^3$ and some continuously differentiable functions $\beta(s,q)$, $\delta(s,q) \neq 0$, and $v(s,q) > 0$ on $[-\infty,\infty]^2$, so that the process $X$ from (1) represents locally a mean-reverting Ornstein-Uhlenbeck process. In this case, we have

$$\Psi_1(x,s,q) = M\left(\frac{\alpha+\lambda}{2\delta(s,q)}, \frac{1}{2}, \frac{(\beta(s,q) - \delta(s,q)x)^2}{\delta(s,q)v^2(s,q)}\right) \tag{56}$$

and

$$\Psi_2(x,s,q) = U\left(\frac{\alpha+\lambda}{2\delta(s,q)}, \frac{1}{2}, \frac{(\beta(s,q) - \delta(s,q)x)^2}{\delta(s,q)v^2(s,q)}\right) \tag{57}$$

where we denote by

$$M(\varphi,\psi;z) = 1 + \sum_{k=1}^{\infty} \frac{(\varphi)_k}{(\psi)_k} \frac{z^k}{k!} \tag{58}$$

and

$$U(\varphi,\psi;z) = \frac{\Gamma(1-\psi)}{\Gamma(\varphi+1-\psi)} M(\varphi,\psi;z) + \frac{\Gamma(\psi-1)}{\Gamma(\varphi)} z^{1-\psi} M(\varphi+1-\psi, 2-\psi;z) \tag{59}$$

Kummer's confluent hypergeometric functions of the first and second kind, respectively, for $\psi \neq 0,-1,-2,\ldots$, $(\varphi)_k = \varphi(\varphi+1)\cdots(\varphi+k-1)$ and $(\psi)_k = \psi(\psi+1)\cdots(\psi+k-1)$, $k \in \mathbb{N}$. Note that the series in (58) converges under all $z > 0$ (see, e.g., Abramovitz and Stegun 1972, chp. XIII; Bateman and Erdély 1953, chp. VI), and $\Gamma$ denotes Euler's gamma function. Note that the functions in (58) and (59) admit the integral representations

$$M(\varphi,\psi;z) = \frac{\Gamma(\psi)}{\Gamma(\varphi)\Gamma(\psi-\varphi)} \int_0^1 e^{zv} v^{\varphi-1} (1-v)^{\psi-\varphi-1} dv, \tag{60}$$

for $\psi > \varphi > 0$ and all $z \in \mathbb{R}$, and

$$U(\varphi,\psi;z) = \frac{1}{\Gamma(\psi)} \int_0^{\infty} e^{-zv} v^{\varphi-1} (1+v)^{\psi-\varphi-1} dv, \tag{61}$$

for $\psi > 0$ and all $z > 0$, respectively (see, e.g., Abramovitz and Stegun 1972, chp. XIII and Bateman and Erdély 1953, chp. VI).

**Example 3.** Let $\mu(x,s,q) = (\beta(s,q) - v^2(s,q)/2)e^{-x} - \delta(s,q)$ and $\sigma(x,s,q) = v(s,q)e^{-x/2}$, for all $(x,s,q) \in E^3$ and some continuously differentiable functions $\beta(s,q)$, $\delta(s,q) \neq 0$, and $v(s,q) > 0$ such that $\beta(s,q) \geq v^2(s,q)/2$ on $[-\infty, \infty]^2$, so that the process $X$ from (1) represents locally the logarithm of a mean-reverting Feller square root diffusion process. In this case, we have

$$\Psi_1(x,s,q) = M\left(\frac{\alpha + \lambda}{\delta(s,q)}, \frac{2\beta(s,q)}{v^2(s,q)}, \frac{2\delta(s,q)e^x}{v^2(s,q)}\right) \tag{62}$$

and

$$\Psi_2(x,s,q) = U\left(\frac{\alpha + \lambda}{\delta(s,q)}, \frac{2\beta(s,q)}{v^2(s,q)}, \frac{2\delta(s,q)e^x}{v^2(s,q)}\right) \tag{63}$$

where the functions $M(\varphi, \psi; z)$ and $U(\varphi, \psi; z)$ are Kummer's confluent hypergeometric functions of the first and second kind given by (58) and (59) above, respectively.

## 4. The Result and Proof

Taking into account the facts proved above, we now formulate the main result of the paper, which extends the assertion of (Gapeev and Rodosthenous 2015, Theorem 4.1) to the case of the model with a random independent exponential time horizon and the $(X, S, Q)$-setting.

**Theorem 1.** *Suppose that the coefficients $\mu(x,s,q)$ and $\sigma(x,s,q)$ of the diffusion-type process $X$ given by (1)–(2) are continuously differentiable functions on $[-\infty, \infty]^3$ which are of at most linear growth in $x$ and uniformly bounded in $s$ and $q$. Let $\eta$ be a random time with the distribution $P(\eta > t) = e^{-\alpha t}$, for all $t \geq 0$ and some $\alpha > 0$ fixed, which is independent of the process $X$. Then, the joint Laplace transform $V_*(x,s,q)$ from (4) of the associated with $X$ random variables $\tau_a \wedge \eta$, $S_{\tau_a \wedge \eta}$, and $Q_{\tau_a \wedge \eta}$ such that $\tau_a < \zeta_b$ from (3), admits the representation*

$$V_*(x,s,q) = \begin{cases} V(x,s,q;\infty), & \text{if } q \leq s - a \leq x \leq q + b \leq s \\ V(x,s,q;a), & \text{if } q \leq s - a \leq x \leq s < q + b \\ V(x,s,q;b), & \text{if } s - a < q \leq x \leq q + b \leq s \\ V(x,s,q;0), & \text{if } s - a < q \leq x \leq s < q + b \end{cases} \tag{64}$$

*for any $a, b > 0$ fixed. Here, the function $V(x,s,q;\infty)$ takes the form of (30) with the coefficients $C_i(s,q;\infty)$, $i = 1,2$, given by (31)–(32), $V(x,s,q;a)$ takes the form of (33) with $C_i(s,q;a)$, $i = 1,2$, given by (34) and (40) (or (42) when $\mu(x,s,q) = \mu(x,s)$ and $\sigma(x,s,q) = \sigma(x,s)$ as well as $\kappa = 0$ and $b = \infty$) $V(x,s,q;b)$ takes the form of (43) with $C_i(s,q;b)$, $i = 1,2$, given by (44) and (50), and $V(x,s,q;0)$ takes the form of (52) with $C_i(s,q;0)$, $i = 1,2$, being a unique solution of the two-dimensional system of first-order partial differential equations in (28)–(29) and satisfying the conditions of (53)–(54) together with the property $C_2(r,r;0) \to 0$ as $r \downarrow -\infty$.*

**Proof.** In order to verify the assertion stated above, it remains to show that the function defined in (64) coincides with the value function in (6). For this purpose, let us denote by $V(x,s,q)$ the right-hand side of the expression in (64). Then, taking into account the fact that the function $V(x,s,q)$ is $C^{2,1,1}$ on $E^3$, by applying the change-of-variable formula from (Peskir 2007, Theorem 3.1) to $e^{-\lambda t} V(X_t, S_t, Q_t)$, we obtain that the expression

$$\begin{aligned} & e^{-\lambda(\tau_a \wedge \zeta_b \wedge t)} V(X_{\tau_a \wedge \zeta_b \wedge t}, S_{\tau_a \wedge \zeta_b \wedge t}, Q_{\tau_a \wedge \zeta_b \wedge t}) = V(x,s,q) + M_{\tau_a \wedge \zeta_b \wedge t} \\ & + \int_0^{\tau_a \wedge \zeta_b \wedge t} e^{-\lambda u} (\mathbb{L}V - (\alpha + \lambda)V + \alpha e^{-\theta S_u - \kappa Q_u})(X_u, S_u, Q_u) \, I(X_u \neq S_u, X_u \neq Q_u) \, du \\ & + \int_0^{\tau_a \wedge \zeta_b \wedge t} e^{-\lambda u} \partial_q V(X_u, S_u, Q_u) \, I(X_u = Q_u) \, dQ_u \\ & + \int_0^{\tau_a \wedge \zeta_b \wedge t} e^{-\lambda u} \partial_s V(X_u, S_u, Q_u) \, I(X_u = S_u) \, dS_u \end{aligned} \tag{65}$$

holds, for the stopping times $\tau_a$ and $\zeta_b$ given by (3), and all $t \geq 0$. Here, the process $M = (M_t)_{t \geq 0}$ defined by

$$M_t = \int_0^t e^{-\lambda u} \partial_x V(X_u, S_u, Q_u) I(X_u \neq S_u, X_u \neq Q_u) \sigma(S_u, Q_u) dB_u \tag{66}$$

is a continuous local martingale under $P_{x,s,q}$. Note that, since the time spent by the process $X$ at the hyperplanes $d_k^3$, $k = 1, 2$, is of Lebesgue measure zero, the indicators in the second line of the expression in (65) and in (66) can be ignored. Moreover, since the processes $S$ and $Q$ change their values only on the hyperplanes $d_1^3$ and $d_2^3$, respectively, the indicators appearing in the third and fourth lines of (65) can be set equal to one.

By virtue of straightforward calculations and the arguments of the previous section, it is verified that the function $V(x, s, q)$ solves the ordinary differential equation in (18) and satisfies the normal-reflection conditions in (21)–(22). Observe that the process $(M_{\tau_a \wedge \zeta_b \wedge t})_{t \geq 0}$ is a uniformly integrable martingale, since the derivative and the coefficient in (66) are bounded functions on the compact set $\{(x, s, q) \in \mathbb{R}^3 \mid a \vee q \leq x \leq s \wedge b\}$. Then, using the properties of the indicators mentioned above and taking the expectation with respect to $P_{x,s,q}$ in (65), by means of the optional sampling theorem (see, e.g., Liptser and Shiryaev [1977] 2001, chp. III, Theorem 3.6 or Karatzas and Shreve 1991, chp. I, Theorem 3.22), we get

$$E_{x,s,q}\left[e^{-\lambda(\tau_a \wedge \zeta_b \wedge t)} V(X_{\tau_a \wedge \zeta_b \wedge t}, S_{\tau_a \wedge \zeta_b \wedge t}, Q_{\tau_a \wedge \zeta_b \wedge t})\right] \tag{67}$$
$$= V(x, s, q) + E_{x,s,q}[M_{\tau_a \wedge \zeta_b \wedge t}] = V(x, s, q)$$

for all $(x, s, q) \in E^3$. Therefore, letting $t$ go to infinity and using the instantaneous-stopping conditions in (19)–(20) as well as the fact that $e^{-\lambda(\tau_a \wedge \zeta_b)} V(X_{\tau_a \wedge \zeta_b}, S_{\tau_a \wedge \zeta_b}, Q_{\tau_a \wedge \zeta_b}) = 0$ on $\{\tau_a \wedge \zeta_b = \infty\}$ ($P_{x,s,q}$-a.s.), we can apply the Lebesgue dominated convergence theorem for (67) to obtain the equalities

$$E_{x,s,q}\left[e^{-\lambda(\tau_a \wedge \zeta_b) - \theta S_{\tau_a \wedge \zeta_b} - \kappa Q_{\tau_a \wedge \zeta_b}} I(\tau_a < \zeta_b)\right] \tag{68}$$
$$= E_{x,s,q}\left[e^{-\lambda(\tau_a \wedge \zeta_b)} V(X_{\tau_a \wedge \zeta_b}, S_{\tau_a \wedge \zeta_b}, Q_{\tau_a \wedge \zeta_b})\right] = V(x, s, q)$$

for all $(x, s, q) \in E^3$, which directly implies the desired assertion. □

**Author Contributions:** P.V.G.: writing—original draft; N.R.: writing—review and editing; V.L.R.C.: writing—conceptualization.

**Funding:** This research was supported by a Small Grant from the Suntory and Toyota International Centres for Economics and Related Disciplines (STICERD) at the London School of Economics and Political Science.

**Acknowledgments:** The authors are grateful to Florin Avram and two anonymous referees for their valuable suggestions which helped to essentially improve the presentation of the paper.

**Conflicts of Interest:** The authors declare no conflict of interest.

## References

Abramovitz, Milton, and Irene A. Stegun. 1972. National Bureau of Standards. In *Handbook of Mathematical Functions with Formulas, Graphs, and Mathematical Tables*. New York: Wiley.

Agarwal, Ankush, Sandeep Juneja, and Ronnie Sircar. 2016. American options under stochastic volatility: Control variates, maturity randomization and multiscale asymptotics. *Quantitative Finance* 16: 17–30. [CrossRef]

Avram, Florin, A. E. Kyprianou, and M. Pistorius. 2004. Exit problems for spectrally negative Lévy processes and applications to (Canadized) Russian options. *Annals of Applied Probability* 14: 215–38.

Bakhach, Amer, Venkata L. Raju Chinthalapati, Edward P. K. Tsang, and Abdul R. El Sayed. 2018. Intelligent dynamic backlash agent: A trading strategy based on the directional change framework. *Algorithms* 11: 171. [CrossRef]

Bakhach, Amer, Edward P. K. Tsang, and Venkata L. Raju Chinthalapati. 2018. TSFDC: A trading strategy based on forecasting directional change. *Intelligent Systems in Accounting, Finance and Management* 25: 105–23. [CrossRef]

Bateman, Harry, and Arthur Erdélyi. 1953. *Higher Transcendental Functions*. New York: Mc Graw-Hill.

Borodin, Andrei N., and Paavo Salminen. 2002. *Handbook of Brownian Motion*, 2nd ed. Basel: Birkhäuser.

Carr, Peter. 1998. Randomization and the American put. *The Review of Financial Studies* 11: 597–626. [CrossRef]

Cont, Rama, and David-Antoine Fournié. 2013. Functional Itô calculus and stochastic integral representation of martingales. *Annals of Probability* 41: 109–33. [CrossRef]

Douady, R., A. N. Shiryaev, and M. Yor. 2000. On the probability characteristics of downfalls in a standard Brownian motion. *Theory of Probability and Its Applications* 44: 29–38. [CrossRef]

Dubins, Lester, Larry A. Shepp, and Albert Nikolaevich Shiryaev. 1993. Optimal stopping rules and maximal inequalities for Bessel processes. *Theory of Probability and Its Applications* 38: 226–61.

Dupire, Bruno. 1997. Pricing and hedging with smiles. In *The Volume Mathematics of Derivative Securities*. Edited by Dempster Michael A. H. and Pliska Stanley R. Cambridge: Cambridge University Press, pp. 103–11.

Dynkin, Eugene B. 1965. *Markov Processes. Volume I*. Berlin: Springer.

Forde, Martin. 2011. A diffusion-type process with a given joint law for the terminal level and supremum at an independent exponential time. *Stochastic Processes and Their Applications* 121: 2802–17. [CrossRef]

Forde, Martin, Andrey Pogudin, and Hongzhong Zhang. 2013. Hitting times, occupation times, tri-variate laws and the forward Kolmogorov equation for a one-dimensional diffusion with memory. *Advances in Applied Probability* 45: 860–75. [CrossRef]

Fournié, David-Antoine. 2010. Functional Itô Calculus and Applications. Ph.D. thesis, Columbia University, New York, NY, USA.

Gapeev, Pavel V., and Neofytos Rodosthenous. 2014. Optimal stopping problems in diffusion-type models with running maxima and drawdowns. *Journal of Applied Probability* 51: 799–817. [CrossRef]

Gapeev, Pavel V., and Neofytos Rodosthenous. 2015. On the drawdowns and drawups in diffusion-type models with running maxima and minima. *Journal of Mathematical Analysis and Applications* 434: 413–31. [CrossRef]

Gapeev, Pavel V., and Neofytos Rodosthenous. 2016. Perpetual American options in diffusion-type models with running maxima and drawdowns. *Stochastic Processes and their Applications* 126: 2038–61. [CrossRef]

Glattfelder, James, Alexandre Dupuis, and Richard Olsen. 2011. Patterns in high-frequency FX data: discovery of 12 empirical scaling laws. *Quantitative Finance* 11: 599–614. [CrossRef]

Glover, Kristoffer, Hardy Hulley, and Peskir, Goran. 2013. Three-dimensional Brownian motion and the golden ratio rule. *Annals of Applied Probability* 23: 895–922. [CrossRef]

Guo, Xin, and Larry A. Shepp. 2001. Some optimal stopping problems with nontrivial boundaries for pricing exotic options. *Journal of Applied Probability* 38: 647–58. [CrossRef]

Guo, Xin, and Mihail Zervos. 2010. $\pi$ options. *Stochastic Processes and their Applications* 120: 1033–59. [CrossRef]

Henry-Labordère, Pierre. 2009. Calibration of local stochastic volatility models to market smiles: A Monte-Carlo approach. *Risk Magazine*, September.

Jacka, S. D. 1991. Optimal stopping and best constants for Doob-like inequalities I: The case $p = 1$. *Annals of Probability* 19: 1798–821. [CrossRef]

Karatzas, Ioannis, and Steven E. Shreve. 1991. *Brownian Motion and Stochastic Calculus*, 2nd ed. New York: Springer.

Lehoczky, John P. 1977. Formulas for stopped diffusion processes with stopping times based on the maximum. *Annals of Probability* 5: 601–7. [CrossRef]

Liptser, Robert S., and Albert N. Shiryaev. 2001. *Statistics of Random Processes I*, 2nd ed. Berlin: Springer. First published 1977.

Magdon-Ismail, Malik, Amir F. Atiya, Amrit Pratap, and Yaser S. Abu-Mostafa. 2004. On the maximum drawdown of a Brownian motion. *Journal of Applied Probability* 41: 147–61. [CrossRef]

Mijatović, Aleksandar, and Martijn R. Pistorius. 2012. On the drawdown of completely asymmetric Lévy processes. *Stochastic Processes and Their Applications* 122: 3812–36. [CrossRef]

Pedersen, Jesper Lund. 2000. Discounted optimal stopping problems for the maximum process. *Journal of Applied Probability* 37: 972–83. [CrossRef]

Peskir, Goran. 1998. Optimal stopping of the maximum process: The maximality principle. *Annals of Probability* 26: 1614–40. [CrossRef]

Peskir, Goran. 2007. A change-of-variable formula with local time on surfaces. *Séminaire de Probabilité, Lecture Notes in Mathematics* 1899: 69–96.

Peskir, Goran. 2012. Optimal detection of a hidden target: The median rule. *Stochastic Processes and Their Applications* 122: 2249–63. [CrossRef]

Peskir, Goran. 2014. Quickest detection of a hidden target and extremal surfaces. *Annals of Applied Probability* 24: 2340–70. [CrossRef]

Peskir, Goran, and Albert N. Shiryaev. 2006. *Optimal Stopping and Free-Boundary Problems*. Basel: Birkhäuser.

Pospisil, Libor, Vecer Jan, and Olympia Hadjiliadis. 2009. Formulas for stopped diffusion processes with stopping times based on drawdowns and drawups. *Stochastic Processes and their Applications* 119: 2563–78. [CrossRef]

Ren, Yong, Dilip Madan, and Michael Qian Qian. 2007. Calibrating and pricing with embedded local volatility models. *Risk Magazine* 20: 138.

Rodosthenous, Neofytos, and Mihail Zervos. 2017. Watermark options. *Finance and Stochastics* 21: 157–86. [CrossRef]

Rodosthenous, Neofytos, and Hongzhong Zhang. 2018. Beating the Omega clock: An optimal stopping problem with random time-horizon under spectrally negative Lévy models. *Annals of Applied Probability* 28: 2105–40. [CrossRef]

Rogers, L. C. G., and David Williams. 1987. *Diffusions, Markov Processes and Martingales II. Itô Calculus*. New York: Wiley.

Shepp, Larry A., and Albert N. Shiryaev. 1993. The Russian option: Reduced regret. *Annals of Applied Probability* 3: 631–40. [CrossRef]

Shepp, Larry A., and Albert N. Shiryaev. 1994. A new look at the pricing of Russian options. *Theory Probability and Applications* 39: 103–19. [CrossRef]

Taylor, Howard M. 1975. A stopped Brownian motion formula. *Annals of Probability* 3: 234–46. [CrossRef]

Tsang, Edward P. K., Ran Tao, Antoaneta Serguieva, and Shuai Ma. 2017. Profiling high frequency equity price movements in directional changes. *Quantitative Finance* 17: 217–25. [CrossRef]

Zhang, Hongzhong. 2014. Occupation times, drawdowns, and drawups for one-dimensional regular diffusions. *Advances in Applied Probability* 47: 210–30. [CrossRef]

Zhang, Hongzhong. 2018. *Stochastic Drawdowns*. Singapore: World Scientific.

© 2019 by the authors. Licensee MDPI, Basel, Switzerland. This article is an open access article distributed under the terms and conditions of the Creative Commons Attribution (CC BY) license (http://creativecommons.org/licenses/by/4.0/).

Article

# A Review of First-Passage Theory for the Segerdahl-Tichy Risk Process and Open Problems

Florin Avram [1,*] and Jose-Luis Perez-Garmendia [2]

[1] Laboratoire de Mathématiques Appliquées, Université de Pau, 64000 Pau, France
[2] Centro de Investigación en Matemáticas, Guanajuato 36020, Mexico; jluis.garmendia@cimat.mx
* Correspondence: florin.avram@univ-Pau.fr

Received: 5 September 2019; Accepted: 13 November 2019; Published: 19 November 2019

**Abstract:** The Segerdahl-Tichy Process, characterized by exponential claims and state dependent drift, has drawn a considerable amount of interest, due to its economic interest (it is the simplest risk process which takes into account the effect of interest rates). It is also the simplest non-Lévy, non-diffusion example of a spectrally negative Markov risk model. Note that for both spectrally negative Lévy and diffusion processes, first passage theories which are based on identifying two "basic" monotone harmonic functions/martingales have been developed. This means that for these processes many control problems involving dividends, capital injections, etc., may be solved explicitly once the two basic functions have been obtained. Furthermore, extensions to general spectrally negative Markov processes are possible; unfortunately, methods for computing the basic functions are still lacking outside the Lévy and diffusion classes. This divergence between theoretical and numerical is strikingly illustrated by the Segerdahl process, for which there exist today six theoretical approaches, but for which almost nothing has been computed, with the exception of the ruin probability. Below, we review four of these methods, with the purpose of drawing attention to connections between them, to underline open problems, and to stimulate further work.

**Keywords:** Segerdahl process; affine coefficients; first passage; spectrally negative Markov process; scale functions; hypergeometric functions

## 1. Introduction and Brief Review of First Passage Theory

**Introduction.** The Segerdahl-Tichy Process Segerdahl (1955); Tichy (1984), characterized by exponential claims and state dependent drift, has drawn a considerable amount of interest—see, for example, Avram and Usabel (2008); Albrecher et al. (2013); Marciniak and Palmowski (2016), due to its economic interest (it is the simplest risk process which takes into account the effect of interest rates—see the excellent overview (Albrecher and Asmussen 2010, Chapter 8). It is also the simplest non-Lévy, non-diffusion example of a spectrally negative Markov risk model. Note that for both spectrally negative Lévy and diffusion processes, first passage theories which are based on identifying two "basic" monotone harmonic functions/martingales have been developed. This means that for these processes many control problems involving dividends, capital injections, etc., may be solved explicitly once the two basic functions have been obtained. Furthermore, extensions to general spectrally negative Markov processes are possible Landriault et al. (2017); Avram et al. (2018); Avram and Goreac (2019); Avram et al. (2019b). Unfortunately, methods for computing the basic functions are still lacking outside the Lévy and diffusion classes. This divergence between theoretical and numerical is strikingly illustrated by the Segerdahl process, for which there exist today six theoretical approaches, but for which almost nothing has been computed, with the exception of the ruin probability Paulsen and Gjessing (1997). Below, we review four of these methods (which apply also to certain generalizations provided in Avram and Usabel (2008); Czarna et al. (2017)), with the purpose of drawing attention to connections between them, to underline open problems, and to stimulate further work.

**Spectrally negative Markov processes with constant jump intensity.** To set the stage for our topic and future research, consider a spectrally negative jump diffusion on a filtered probability space $(\Omega, \{\mathcal{F}_t\}_{t\geq 0}, P)$, which satisfies the SDE:

$$dX_t = c(X_t)dt + \sigma(X_t)dB_t - dJ_t, \quad J_t = \sum_{i=1}^{N_\lambda(t)} C_i, \quad \forall X_t > 0 \tag{1}$$

and is absorbed or reflected when leaving the half line $(0, \infty)$. Here, $B_t$ is standard Brownian motion, $\sigma(x) > 0, c(x) > 0, \forall x > 0$, $N_\lambda(t)$ is a Poisson process of intensity $\lambda$, and $C_i$ are nonnegative random variables with distribution measure $F_C(dz)$ and finite mean. The functions $c(x)$, $a(x) := \frac{\sigma^2(x)}{2}$ and $\Pi(dz) = \lambda F_C(dz)$ are referred to as the Lévy-Khinchine characteristics of $X_t$. Note that we assume that all jumps go in the same direction and have constant intensity so that we can take advantage of potential simplifications of the first passage theory in this case.

**The Segerdahl-Tichy process** is the simplest example outside the spectrally negative Lévy and diffusion classes. It is obtained by assuming $a(x) = 0$ in (1), and $C_k$ to be exponential i.i.d random variables with density $f(x) = \mu e^{-\mu x}$ (see Segerdahl (1955) for the case $c(x) = c + rx, r > 0, c \geq 0$, and Tichy (1984) for nonlinear $c(x)$). Note that, for the case $c(x) = c + rx$, an explicit computation of the ruin probability has been provided (with some typos) in Paulsen and Gjessing (1997). See also Paulsen (2010) and see (Albrecher and Asmussen 2010, Chapter 8) for further information on risk processes with state dependent drift, and in particular the two pages of historical notes and references.

First passage theory concerns the first passage times above and below fixed levels. For any process $(X_t)_{t\geq 0}$, these are defined by

$$\begin{aligned} T_{b,+} &= T^X_{b,+} = \inf\{t \geq 0 : X_t > b\}, \\ T_{a,-} &= T^X_{a,-} = \inf\{t \geq 0 : X_t < a\}, \end{aligned} \tag{2}$$

with $\inf \emptyset = +\infty$, and the upper script $X$ typically omitted. Since $a$ is typically fixed below, we will write for simplicity $T$ instead of $T_{a,-}$.

First passage times are important in the control of reserves/risk processes. The rough idea is that when below low levels $a$, reserves processes should be replenished at some cost, and when above high levels $b$, they should be partly invested to yield income—see, for example, the comprehensive textbook Albrecher and Asmussen (2010).

The most important first passage functions are the solutions of the two-sided upward and downward exit problems from a bounded interval $[a, b]$:

$$\begin{cases} \overline{\Psi}^b_q(x, a) := E_x\left[e^{-qT_{b,+}} \mathbf{1}_{\{T_{b,+} < T_{a,-}\}}\right] = P_x\left[T_{b,+} < \min(T_{a,-}, \mathbf{e}_q)\right] \\ \Psi^b_q(x, a) := E_x\left[e^{-qT_{a,-}} \mathbf{1}_{\{T_{a,-} < T_{b,+}\}}\right] = P_x\left[T_{a,-} < \min(T_{b,+}, \mathbf{e}_q)\right] \end{cases} \quad q \geq 0, a \leq x \leq b, \tag{3}$$

where $\mathbf{e}_q$ is an independent exponential random variable of rate $q$. We will call them (killed) survival and ruin probabilities, respectively[1], but the qualifier killed will be usually dropped below. The absence of killing will be indicated by omitting the subindex $q$. Note that in the context of potential theory, (3) are called equilibrium potentials Blumenthal and Getoor (2007) (of the capacitors $\{b, a\}$ and $\{a, b\}$).

**Beyond ruin probabilities : scale functions, dividends, capital gains, etc.** Recall that for "completely asymmetric Lévy" processes, with jumps going all in the same direction, a large variety of first passage problems may be reduced to the computation of the two monotone "scale functions"

---

[1] See Ivanovs (2013) for a nice exposition of killing.

$W_q, Z_q$—see, for example, Suprun (1976), Bertoin (1997, 1998), Avram et al. (2004, 2007, 2015, 2016), Ivanovs and Palmowski (2012), Albrecher et al. (2016); Li and Palmowski (2016); Li and Zhou (2017); Avram and Zhou (2017), and see Avram et al. (2019a) for a recent compilation of more than 20 laws expressed in terms of $W_q, Z_q$.

For example, for spectrally negative Lévy processes, the Laplace transform/killed survival probability has a well known simple factorization[2]:

$$\overline{\Psi}_q^b(x,a) = \frac{W_q(x-a)}{W_q(b-a)}. \tag{4}$$

For a second example, the De-Finetti de Finetti (1957) discounted dividends fixed barrier objective for spectrally negative Lévy processes Avram et al. (2007) has a simple expression in terms of either the $W_q$ scale function or of its logarithmic derivative $\nu_q = \frac{W_q'}{W_q}$[3]:

$$V^b(x) = \begin{cases} \frac{W_q(x)}{W_q'(b)} = e^{-\int_x^b \nu_q(m)dm} \frac{1}{\nu_q(b)} & x \leq b \\ V^b(x) = x - b + V^b(b) & x > b \end{cases}. \tag{5}$$

Maximizing over the reflecting barrier $b$ is simply achieved by finding the roots of

$$W_q''(b) = 0 \Leftrightarrow \frac{\partial}{\partial b}\left[\frac{1}{\nu_q(b)}\right] = \frac{\partial}{\partial b}\left[V^b(b)\right] = 1. \tag{6}$$

**$W, Z$ formulas for first passage problems for spectrally negative Markov processes.** Since results for spectrally negative Lévy processes require often not much more than the strong Markov property, it is natural to attempt to extend them to the spectrally negative strong Markov case. As expected, everything worked out almost smoothly for "Lévy-type cases" like random walks Avram and Vidmar (2017), Markov additive processes Ivanovs and Palmowski (2012), etc. Recently, it was discovered that $W, Z$ formulas continue to hold a priori for spectrally negative Markov processes Landriault et al. (2017), Avram et al. (2018). The main difference is that in equations like Equation (4), $W_q(x-a)$ and the second scale function $Z_{q,\theta}(x-a)$ Avram et al. (2015); Ivanovs and Palmowski (2012) must be replaced by two-variable functions $W_q(x,a)$, $Z_{q,\theta}(x,a)$ (which reduces in the Lévy case to $W_q(x,y) = \widetilde{W}_q(x-y)$, with $\widetilde{W}_q$ being the scale function of the Lévy process). This unifying structure has lead to recent progress for the optimal dividends problem for spectrally negative Markov processes (see Avram and Goreac (2019)). However, since the computation of the two-variables scale functions is currently well understood only for spectrally negative Lévy processes and diffusions, AG could provide no example outside these classes. In fact, as of today, we are not aware of any explicit or numeric results on the control of the process (1) which have succeeded to exploit the $W, Z$ formalism.

**Literature review.** Several approaches may allow handling particular cases of spectrally negative Markov processes:

1. with phase-type jumps, there is Asmussen's embedding into a regime switching diffusion Asmussen (1995)—see Section 5, and the complex integral representations of Jacobsen and Jensen (2007), Jiang et al. (2019).
2. for Lévy driven Langevin-type processes, renewal equations have been provided in Czarna et al. (2017)—see Section 2
3. for processes with affine operator, an explicit integrating factor for the Laplace transform may be found in Avram and Usabel (2008)—see Section 3.

---

[2] The fact that the survival probability has the multiplicative structure (4) is equivalent to the absence of positive jumps, by the strong Markov property; this is the famous "gambler's winning" formula Kyprianou (2014).
[3] $\nu_q$ may be more useful than $\widetilde{W}_q$ in the spectrally negative Markov framework Avram and Goreac (2019).

4. for the Segerdahl process, the direct IDE solving approach is successful (Paulsen and Gjessing (1997)) —see Section 4.

We will emphasize here the third approach but use also the second to show how the third approach fits within it. The direct IDE solving approach is recalled for comparison, and Asmussen's approach is also recalled, for its generality.

Here is an example of an important problem we would like to solve:

**Problem 1.** *Find the de Finetti optimal barrier for the Segerdahl-Tichy process, extending the Equations (5) and (6).*

Contents. Section 2 reviews the recent approach based on renewal equations due to Czarna et al. (2017) (which needs still be justified for increasing premiums satisfying (8)). An important renewal (Equation (11)) for the "scale derivative" w is recalled here, and a new result relating the scale derivative to its integrating factor (16) is offered—see Theorem 1.

Section 3 reviews older computations of Avram and Usabel (2008) for more general processes with affine operator, and provides explicit formulas for the Laplace transforms of the survival and ruin probability (24), in terms of the same integrating factor (16) and its antiderivative.

Section 4 reviews the direct classic Kolmogorov approach for solving first passage problems with phase-type jumps. The discounted ruin probability ($q > 0$) for this process may be found explicitly (33) for the Segerdahl process by transforming the renewal equation (29) into the ODE (30), which is hypergeometric of order 2. This result due to Paulsen has stopped short further research for more general mixed exponential jumps, since it seems to require a separate "look-up" of hypergeometric solutions for each particular problem.

Section 5 reviews Asmussen's approach for solving first passage problems with phase-type jumps, and illustrates the simple structure of the survival and ruin probability of the Segerdahl-Tichy process, in terms of the scale derivative **w**. This approach yields quasi-explicit results when $q = 0$.

Section 6 checks that our integrating factor approach recovers various results for Segerdahl's process, when $q = 0$ or $x = 0$. Section 7 reviews necessary hypergeometric identities. Finally, Section 8 outlines further promising directions of research.

## 2. The Renewal Equation for the Scale Derivative of Lévy Driven Langevin Processes Czarna et al. (2017)

One tractable extension of the Segerdahl-Tichy process is provided by is the "Langevin-type" risk process defined by

$$X_t = x + \int_0^t c(X_s)\, ds + Y_t, \tag{7}$$

where $Y_t$ is a spectrally negative Lévy process, and $c(u)$ is a nonnegative premium function satisfying

$$\int_{x_0}^\infty \frac{1}{c(u)}\, dy = \infty, \quad c(u) > 0, \quad \forall x_0, u > 0. \tag{8}$$

The integrability condition above is necessary to preclude explosions. Indeed when $Y_t$ is a compound Poisson process, in between jumps (claims) the risk process (7) moves deterministically along the curves $x_t$ determined by the vector field

$$\frac{dx}{dt} = c(x) \Leftrightarrow t = \int_{x_0}^x \frac{du}{c(u)} := C(x; x_0), \forall x_0 > 0.$$

From the last equality, it may be noted that if $C(x; x_0)$ satisfies $\lim_{x \to \infty} C(x; x_0) < \infty$, then $x_t$ must explode, and the stochastic process $X_t$ may explode.

The case of Langevin processes has been tackled recently in Czarna et al. (2017), who provide the construction of the process (7) in the particular case of non-increasing functions $c(\cdot)$. This setup can be used to model dividend payments, and other mathematical finance applications.

Czarna et al. (2017) showed that the $W, Z$ scale functions which provide a basis for first passage problems of Lévy spectrally positive negative processes have two variables extensions $\mathcal{W}, \mathcal{Z}$ for the process (7), which satisfy integral equations. The equation for $\mathcal{W}$, obtained by putting $\phi(x) = c(a) - c(x)$ in (Czarna et al. 2017, eqn. (40)), is:

$$\mathcal{W}_q(x,a) = W_q(x-a) + \int_a^x (c(a) - c(z))W_q(x-z)\mathcal{W}_q'(z;a)dz, \tag{9}$$

where $W_q$ is the scale function of the Lévy process obtained by replacing $c(x)$ with $c(a)$.

It follows that the scale derivative

$$\mathbf{w}_q(x,a) = \frac{\partial}{\partial x}\mathcal{W}_q(x,a)$$

of the scale function of the process (7) satisfies a Volterra renewal equation (Czarna et al. 2017, eqn. (41)):

$$\mathbf{w}_q(x,a)\left(1 + (c(x) - c(a))W_q(0)\right) = w_q(x-a) + \int_a^x (c(a) - c(z))w_q(x-z)\mathbf{w}_q(z;a)dz, \tag{10}$$

where $w_q$ is the derivative of the scale function of the Lévy process $Y_t = Y_t^{(a)}$ obtained by replacing $c(x)$ with $c(a)$. This may further be written as:

$$w_q(x-a) + \int_a^x w_q(x-z)\mathbf{w}_q(z;a)(c(a)-c(z))dz = \begin{cases} \mathbf{w}_q(x,a), & Y_t \text{ of unbounded variation} \\ \mathbf{w}_q(x,a)\frac{c(x)}{c(a)}, & Y_t \text{ of bounded variation} \end{cases} \tag{11}$$

**Problem 2.** *It is natural to conjecture that the formula (11) holds for all drifts satisfying (8), but this is an open problem for now.*

**Remark 1.** *Note that renewal equations are a more appropriate tool than Laplace transforms for the general Langevin problem. Indeed, taking "shifted Laplace transform" $\mathcal{L}_a f(x) = \int_a^\infty e^{-s(y-a)} f(y) dy$ of (11), putting*

$$\begin{cases} \widehat{\mathbf{w}}_q(s,a) = \int_a^\infty e^{-s(y-a)} \mathbf{w}_q(y,a) dy, \\ \widehat{\mathbf{w}}_{q,c}(s,a) = \int_a^\infty e^{-s(y-a)} \mathbf{w}_q(y,a)c(y) dy, \\ \widehat{w}_q(s) = \int_0^\infty e^{-sy} w_q(y) dy \end{cases}$$

*and using*

$$\mathcal{L}_a\left[\int_a^x f(x-y)l(y)dy\right](s) = \mathcal{L}_0 f(s)\mathcal{L}_a l(s)$$

*yields equations with two unknowns:*

$$\widehat{w}_q(s)(1 + c(a)\widehat{\mathbf{w}}_q(s,a) - \widehat{\mathbf{w}}_{q,c}(s,a)) = \begin{cases} \widehat{\mathbf{w}}_q(s,a) & \text{unbounded variation case} \\ \frac{\widehat{\mathbf{w}}_{q,c}(s,a)}{c(a)} & \text{bounded variation case} \end{cases}, \tag{12}$$

*whose solution is not obvious.*

*The Linear Case $c(x) = rx + c$*

To get explicit Laplace transforms, we will turn next to Ornstein-Uhlenbeck type processes[4] $X(\cdot)$, with $c(x) = c(a) + r(x-a)$, which implies

$$\widehat{w}_{q,c}(s,a) = \int_a^\infty e^{-s(y-a)} w_q(y,a)(r(y-a) + c(a))dy = -r\widehat{w}_q'(s,a) + c(a)\widehat{w}_q(s,a). \tag{13}$$

Equation (12) simplify then to:

$$\widehat{w}_q(s)(1 + r\widehat{w}_q'(s,a)) = \begin{cases} \widehat{w}_q(s,a) & \text{unbdd variation case} \\ \widehat{w}_q(s,a) - \frac{r}{c(a)}\widehat{w}_q'(s,a) & \text{bdd variation case} \end{cases} \tag{14}$$

**Remark 2.** *Note that the only dependence on a in this equation is via $c(a)$, and via the shifted Laplace transform. Since a is fixed, we may and will from now on simplify by assuming w.l.o.g. $a = 0$, and write $c = c(a)$.*

Let now

$$\kappa(s) = \alpha_0 s^2 + cs - s\widehat{\overline{\Pi}}(s) - q, \alpha_0 > 0,$$

denote the Laplace exponent or symbol of the Lévy process $Y_t = \sqrt{2\alpha_0}B_t - J_t + ct$, and recall that

$$w_q(s) = \begin{cases} \frac{s}{\kappa(s)} & \text{unbdd variation case} \\ \frac{s}{\kappa(s)} - \frac{1}{c} & \text{bdd variation case} \end{cases}$$

(where we have used that $W_q(0) = 0(\frac{1}{c})$ in the two cases, respectively).

We obtain now from (14) the following ODE

$$r\widehat{w}_q'(s,a) - \frac{\kappa(s)}{s}\widehat{w}_q(s,a) = -1 + \frac{\kappa(s)}{s}W_q(0) = \begin{cases} -1 & \text{unbdd variation case} \\ -1 + \frac{\kappa(s)}{cs} := -\frac{h(s)}{c} & \text{bdd variation case} \end{cases}, \tag{15}$$

where

$$h(s) = \widehat{\overline{\Pi}}(s) + \frac{q}{s}.$$

**Remark 3.** *The Equation (15) is easily solved multiplying by an integrating factor*

$$I_q(s, s_0) = e^{-\int_{s_0}^s \frac{\kappa(z)/z}{r} dz} = e^{-\int_{s_0}^s \frac{\alpha_0 z + c - \widehat{\overline{\Pi}}(z) - q/z}{r} dz}, \tag{16}$$

*where $s_0 > 0$ is an arbitrary integration limit chosen so that the integral converges (the formula (16) appeared first in Avram and Usabel (2008)). To simplify, we may choose $s_0 = 0$ to integrate the first part $\alpha_0 z + c - \widehat{\overline{\Pi}}(z)$, and a different lower bound $s_0 = 1$ to integrate $q/z$. Putting $\tilde{q} = \frac{q}{r}, \tilde{c} = \frac{c}{r}, \tilde{\alpha}_0 = \frac{\alpha_0}{r}$, we get that*

$$I_q(s) = e^{-\int_\cdot^s \frac{\kappa(z)/z}{r} dz} = s^{\tilde{q}} e^{-\left[\left(\frac{\tilde{\alpha}_0}{2}\right)s^2 + \tilde{c}s\right] + \frac{1}{r}\int_0^s \widehat{\overline{\Pi}}(z) dz} := s^{\tilde{q}} I(s) := e^{-\tilde{c}s} i_q(s), \tag{17}$$

*where we replaced $s_0$ by $\cdot$ to indicate that two different lower bounds are in fact used, and we put $I(s) = I_0(s)$ (the subscript 0 will be omitted when $q = 0$).*

Solving (15) yields:

---

[4] For some background first passage results on these processes, see for example Borovkov and Novikov (2008); Loeffen and Patie (2010).

**Theorem 1.** *Fix a and put $\bar{I}_q(s) = \int_s^\infty I_q(y)dy$. Then, the Laplace transform of the scale derivative of an Ornstein-Uhlenbeck type process* (7) *satisfies:*

$$\widehat{w}_q(s,a) = \frac{\bar{I}_q(s)}{rI_q(s)} - W_q(0) = \begin{cases} \frac{\bar{I}_q(s)}{rI_q(s)}, & \text{in the unbounded variation case} \\ \frac{\bar{I}_q(s)}{rI_q(s)} - \frac{1}{c}, & \text{in the bounded variation case} \end{cases}. \quad (18)$$

**Proof.** In the unbounded variation case, applying the integrating factor to (15) yields immediately:

$$\widehat{w}_q(s,a)I_q(s) = r^{-1}\int_s^\infty I_q(y)dy = r^{-1}\bar{I}_q(s).$$

In the bounded variation case, we observe that

$$i'_q(s) = \frac{h(s)}{r}i_q(s),$$

where $i_q$ is defined in (17). An integration by parts now yields

$$\widehat{w}_q(s,a)I_q(s) = \int_s^\infty \frac{h(y)}{cr}I_q(y)dy = \int_s^\infty \frac{h(y)}{cr}e^{-\tilde{c}y}i_q(y)dy$$
$$= c^{-1}\int_s^\infty e^{-\tilde{c}y}i'_q(y)dy = c^{-1}(-I_q(s) + \tilde{c}\int_s^\infty e^{-\tilde{c}y}i_q(y)dy) = r^{-1}\bar{I}_q(s)dy - c^{-1}I_q(s).$$

□

**Remark 4.** *The result* (18) *is quite similar to the Laplace transform for the survival and ruin probability (Gerber-Shiu functions) derived in (Avram and Usabel 2008, p. 470)—see* (23), (24) *below; the main difference is that in that case additional effort was needed for finding the values* $\overline{\Psi}(a,a), \Psi(a,a)$.

## 3. The Laplace transform-Integrating Factor Approach for Jump-Diffusions with Affine Operator Avram and Usabel (2008)

We summarize now for comparison the results of Avram and Usabel (2008) for the still tractable, more general extension of the Segerdahl-Tichy process provided by jump-diffusions with affine premium and volatility

$$\begin{cases} c(x) = rx + c \\ \frac{v^2(x)}{2} = \alpha_1 x + \alpha_0, \ \alpha_1, \alpha_0 \geq 0. \end{cases} \quad (19)$$

Besides Ornstein-Uhlenbeck type processes, (19) includes another famous particular case, Cox-Ingersoll-Ross (CIR) type processes, obtained when $\alpha_1 > 0$.

Introduce now a combined ruin-survival expected payoff at time $t$

$$V(t,u) = \mathbb{E}_{X_0=u}\left[w(X_T)\,1_{\{T<t\}} + p(X_t)\,1_{\{T\geq t\}}\right] \quad (20)$$

where $w, p$ represent, respectively:

- A penalty $w(X_T)$ at a stopping time $T$, $w: \mathbb{R} \to \mathbb{R}$
- A reward for survival after $t$ years: $p(X_t)$, $p: \mathbb{R} \to \mathbb{R}^+$.

Some particular cases of interest are the survival probability for $t$ years, obtained with

$$w(X_T) = 0, \ p(X_t) = 1_{\{X_t \geq 0\}}$$

and the ruin probability with deficit larger in absolute value than $y$, obtained with

$$w(X_T) = 1_{\{X_T < -y\}}, \quad p(X_t) = 0$$

Let

$$V_q(x) = \int_0^\infty q e^{-qt} V(t,x) dt = E_x \left[ w(X_T) 1_{\{T < e_q\}} + p(X_{e_q}) 1_{\{T \geq e_q\}} \right], \quad (21)$$

denote a "Laplace-Carson"/"Gerber Shiu" discounted penalty/pay-off.

**Proposition 1.** (*Avram and Usabel 2008, Lem. 1, Thm. 2*) (a) *Consider the process* (19). *Let $V_q(x)$ denote the corresponding Gerber-Shiu function* (21), *let $w_\Pi(x) = \int_x^\infty w(x-u)\Pi(du)$ denote the expected payoff at ruin, and let $g(x) := w_\Pi(x) + qp(x), \hat{g}(s)$ denote the combination of the two payoffs and its Laplace transform; note that the particular cases*

$$\hat{g}(s) = \frac{q}{s}, \quad \hat{g}(s) = \lambda \overline{F}(s)$$

*correspond to the survival and ruin probability, respectively* Avram and Usabel (2008).
*Then, the Laplace transform of the derivative*

$$V_*(x) = \int_0^\infty e^{-sx} dV_q(x) = s\widehat{V_q}(s) - V_q(0)$$

satisfies the ODE

$$(\alpha_1 s + r) V_*(s)' - \left(\frac{\kappa(s)}{s} - \alpha_1\right) V_*(s) = -h(s) V_q(0) - \alpha_0 V_q'(0) + \hat{g}(s) \Longrightarrow$$

$$V_*(s) I_q(s) = \int_s^\infty I_q(y) \frac{h(y) V_q(0) + \alpha_0 V_q'(0) - \hat{g}(y)}{r + \alpha_1 y} dy, \quad (22)$$

where $h(s) = \widehat{\overline{\Pi}}(s) + \frac{q}{s}$ (this corrects a typo in (Avram and Usabel 2008, eqn. (9))), and where the integrating factor is obtained from (16) by replacing $c$ with $c - \alpha_1$ (Avram and Usabel 2008, eqn. (11))). Equivalently,

$$r\left(s\widehat{V_q}(s)\right)' - \frac{\kappa(s)}{s} s\widehat{V_q}(s) = -(c + \alpha_0 s) V_q(0) - \alpha_0 V_q'(0) + \hat{g}(s) \Longrightarrow$$

$$s\widehat{V_q}(s) I_q(s) = \int_s^\infty I_q(y) \frac{(c + \alpha_0 s) V_q(0) + \alpha_0 V_q'(0) - \hat{g}(y)}{r + \alpha_1 y} dy. \quad (23)$$

(b) *If $\alpha_0 = 0 = \alpha_1$ and $q > 0$, the survival probability satisfies*

$$\overline{\Psi}_q(0) = \frac{\tilde{q}\bar{I}_{q-1}(0)}{\tilde{c}\bar{I}_q(0)} \quad (24)$$

$$s\widehat{\overline{\Psi}}_q(s) I_q(s) = \int_s^\infty I_q(y)\left(\tilde{c}\overline{\Psi}_q(0) - \frac{\tilde{q}}{y}\right) dy = \tilde{c}\overline{\Psi}_q(0)\bar{I}_q(s) - \tilde{q}\bar{I}_{q-1}(s) = \tilde{q}\left(\frac{\bar{I}_{q-1}(0)}{\bar{I}_q(0)}\bar{I}_q(s) - \bar{I}_{q-1}(s)\right)$$

**Proof.** (b) The survival probability follow from (a), by plugging $\hat{g}(y) = \frac{q}{y}$. Indeed, the Equation (23) becomes for the survival probability

$$s\widehat{\overline{\Psi}}_q(s) I_q(s) = \int_s^\infty I_q(y)\left(\tilde{c}\overline{\Psi}_q(0) - \frac{\tilde{q}}{y}\right) dy = \tilde{c}\overline{\Psi}_q(0)\bar{I}_q(s) - \tilde{q}\bar{I}_{q-1}(s).$$

Letting $s \to 0$ in this equation yields $\overline{\Psi}_q(0) = \frac{\tilde{q}\bar{I}_{q-1}(0)}{\tilde{c}\bar{I}_q(0)}$.

As a check, let us verify also Equation (23) for the ruin probability, by plugging $\hat{g}(y) = \lambda \overline{F}(y)$:

$$s\widehat{\Psi}_q(s) I_q(s) = \int_s^\infty I_q(y)(\tilde{c}\Psi_q(0) - \lambda \overline{F}(y)) dy = \tilde{c}\Psi_q(0) \overline{I}_q(y) - J(y),$$

$$J(y) = \int_s^\infty y^{\tilde{q}} e^{-\tilde{c}y} j'(y) dy, \quad j(y) := e^{\tilde{\lambda} \int_0^y \overline{F}(z) dz}.$$

Integrating by parts, $J(y) = -I_q(s) + \tilde{c} \overline{I}_q(s) - \tilde{q} \overline{I}_{q-1}(s)$. Finally,

$$s\widehat{\Psi}_q(s) I_q(s) = \tilde{c}(1 - \Psi_q(0)) \overline{I}_q(s) - \left( - I_q(s) + \tilde{c} \overline{I}_q(s) - \tilde{q} \overline{I}_{q-1}(s) \right) =$$

$$I_q(s) + \tilde{q} \overline{I}_{q-1}(s) - \tilde{c} \overline{\Psi}_q(0) \overline{I}_q(s) = I_q(s) - s\widehat{\overline{\Psi}}_q(s) I_q(s). \tag{25}$$

☐

*Segerdahl's Process via the Laplace Transform Integrating Factor*

We revisit now the particular case of Segerdahl's process with exponential claims of rate $\mu$ and $\alpha_0 = \alpha_1 = 0$. Using $\overline{\Pi}(y) = \lambda F_C(y) dy = \frac{\lambda}{y+\mu}$ we find that for Segerdahl's process the integrand in the exponent is

$$\frac{\kappa(s)}{rs} = \tilde{c} - \tilde{\lambda}/(s+\mu) - \tilde{q}/s,$$

and the integrating factor (17) may be taken as

$$I_q(x) = x^{\tilde{q}} e^{-\tilde{c}x} (1 + x/\mu)^{\tilde{\lambda}}.$$

The antiderivative $\overline{I}_q(x)$ is not explicit, except for:

1. $x = 0$, when it holds that
$$\overline{I}_q(0) = \mu^{\tilde{q}+1} U(\tilde{q}+1, \tilde{q}+\tilde{\lambda}+2, \tilde{c}\mu),$$

where (Abramowitz and Stegun 1965, 13.2.5)[5]

$$U[a, a+c, z] = \frac{1}{\Gamma[a]} \int_0^\infty e^{-zt} t^{a-1} (t+1)^{c-1} dt, \; Re[z] > 0, Re[a] > 0.$$

2. for $q = 0$, when it holds that

$$I(x) = e^{-\tilde{c}x}(1 + x/\mu)^{\tilde{\lambda}}, \; \overline{I}(x) = \int_x^\infty I(y) dy = \frac{e^{\tilde{c}\mu}(\tilde{c}\mu)^{-\tilde{\lambda}} \Gamma(\tilde{\lambda}+1, \tilde{c}(x+\mu))}{\tilde{c}}.$$

However, the Laplace transforms of the integrating factor $I_q(x)$ and its primitive are explicit:

$$\widehat{I}_q(s) = \int_0^\infty e^{-(s+\tilde{c})x} x^{\tilde{q}} (1+x/\mu)^{\tilde{\lambda}} = \Gamma(\tilde{q}+1) U(\tilde{q}+1, \tilde{q}+\tilde{\lambda}+2, \mu(\tilde{c}+s)),$$

$$\widehat{\overline{I}}_q(s) = \Gamma(\tilde{q}+1) \frac{U(\tilde{q}+1, \tilde{q}+\tilde{\lambda}+2, \mu\tilde{c}) - U(\tilde{q}+1, \tilde{q}+\tilde{\lambda}+2, \mu(\tilde{c}+s))}{s}. \tag{26}$$

Finally, we may compute:

---

[5] Note that when $c = 1$, this function reduces to a power: $U(a, a+1, z) = \frac{\int_0^\infty t^{a-1} e^{-zt} dt}{\Gamma(a)} = z^{-a}$.

$$\overline{\Psi}_q(0) = \frac{\tilde{q}\tilde{I}_{q-1}(0)}{\tilde{c}\tilde{I}_q(0)} = \frac{\tilde{q}U(\tilde{q}, \tilde{q} + \tilde{\lambda} + 1, \tilde{c}\mu)}{\tilde{c}\mu U(\tilde{q} + 1, \tilde{q} + \tilde{\lambda} + 2, \tilde{c}\mu)}$$

$$\Psi_q(0) = 1 - \overline{\Psi}_q(0) = 1 - \frac{\tilde{q}U(\tilde{q}, \tilde{q} + \tilde{\lambda} + 1, \tilde{c}\mu)}{\tilde{c}\mu U(\tilde{q} + 1, \tilde{q} + \tilde{\lambda} + 2, \tilde{c}\mu)}$$

$$= \frac{\tilde{c}\mu U(\tilde{q} + 1, \tilde{q} + \tilde{\lambda} + 2, \tilde{c}\mu) - \tilde{q}U(\tilde{q}, \tilde{q} + \tilde{\lambda} + 1, \tilde{c}\mu)}{\tilde{c}\mu U(\tilde{q} + 1, \tilde{q} + \tilde{\lambda} + 2, \tilde{c}\mu)}$$

$$= \left(\frac{\lambda}{c\mu}\right) \frac{U(\tilde{q} + 1, \tilde{q} + 1 + \tilde{\lambda}, \mu \tilde{c})}{U(\tilde{q} + 1, \tilde{q} + \tilde{\lambda} + 2, \mu \tilde{c})}, \tag{27}$$

where we used the identity (Abramowitz and Stegun 1965, 13.4.18)

$$U[a - 1, b, z] + (b - a)U[a, b, z] = zU[a, b + 1, z], a > 1, \tag{28}$$

with $a = \tilde{q} + 1, b = \tilde{q} + \tilde{\lambda} + 1$. This checks the (corrected) Paulsen result (38) for $x = 0$.

**Remark 5.** *We can now numerically answer Problem 1: (a) obtain the antiderivative $\tilde{I}_q(x)$ by numerical integration; (b) compute the Laplace transform of the scale derivative by (18); c) Invert the Laplace transform.*

The example above raises the question:

**Problem 3.** *Is it possible to compute explicitly the Laplace transforms of the integrating factor $I_q(x)$ and its primitive for affine processes with phase-type jumps?*

## 4. Direct Conversion to an Ode of Kolmogorov'S Integro-Differential Equation for the Discounted Ruin Probability with Phase-Type Jumps

One may associate to the process (1) a Markovian semi-group with generator

$$\mathcal{G}h(x) = a(x)h''(x) + c(x)h'(x) + \int_{(0,\infty)} [h(x-y) - h(x)]\Pi(dy)$$

acting on $h \in C^2_{(0,\infty)}$, up to the minimum between its explosion and exit time $T_{0,-}$.

The classic approach for computing the ruin, survival, optimal dividends, and other similar functions starts with the well-known Kolmogorov integro-differential equations associated to this operator. With phase-type jumps, one may remove the integral term in Kolmogorov's equation above by applying to it the differential operator $n(D)$ given by the denominator of the Laplace exponent $\kappa(D)$. For example, with exponential claims, we would apply the operator $\mu + D$.

### 4.1. Paulsen's Result for Segerdahl's Process with Exponential Jumps Paulsen and Gjessing (1997), ex. 2.1

When $a(x) = 0$ and $C_k$ in (1) are exponential i.i.d random variables with density $f(x) = \mu e^{-\mu x}$, the Kolmogorov integro-differential equation for the ruin probability is:

$$c(x)\Psi_q(x,a)' + \lambda\mu \int_a^x e^{-\mu(x-z)}\Psi_q(z,a)dz - (\lambda + q)\Psi_q(x,a) + \lambda e^{-\mu x} = 0, \ \Psi_q(b,a) = 1, \Psi_q(x,a) = 0, x < a. \tag{29}$$

To remove the convolution term $\Psi_q * f_C$, apply the operator $\mu + D$, which replaces the convolution term by $\lambda \mu \Psi_q(x)$[6] yielding finally

$$\left( c(x) D^2 + (c'(x) + \mu c(x) - (\lambda + q)) D - q\mu \right) \Psi_q(x) = 0$$

When $c(x) = c + rx, a = 0, b = \infty$, the ruin probability satisfies:

$$\left[ (\tilde{c} + x) D^2 + (1 + \mu(\tilde{c} + x) - \tilde{q} - \tilde{\lambda}) D - \mu \tilde{q} \right] \Psi_q(x) = 0,$$
$$(-cD + \lambda + q)\Psi_q(0) = \lambda^7, \quad \Psi_q(\infty) = 0 \tag{30}$$

see (Paulsen and Gjessing 1997, (2.14),(2.15)), where $\tilde{\lambda} = \frac{\lambda}{r}, \tilde{q} = \frac{q}{r}$, and $-\tilde{c} := -\frac{c}{r}$ is the absolute ruin level.

Changing the origin to $-\tilde{c}$ by $z = \mu(x + \tilde{c}), \Psi_q(x) = y(z)$ brings this to the form

$$zy''(z) + (z + 1 - n) y'(z) - \tilde{q} y(z) = 0, \ n = \tilde{\lambda} + \tilde{q}, \tag{31}$$

(we corrected here two wrong minuses in Paulsen and Gjessing (1997)), which corresponds to the process killed at the absolute ruin, with claims rate $\mu = 1$. Note that the (Sturm-Liouville) Equation (31) intervenes also in the study of the squared radial Ornstein-Uhlenbeck diffusion (also called Cox-Ingersoll-Ross process) (Borodin and Salminen 2012, p. 140, Chapter II.8).

Let $K_i(z) = K_i(\tilde{q}, n, z), i = 1, 2, n = \tilde{q} + \tilde{\lambda}$ denote the (unique up to a constant) increasing/decreasing solutions for $z \in (0, \infty)$ of the confluent hypergeometric Equation (31). The solution of (31) is thus

$$c_1 K_1(\tilde{q}, n, z) + c_2 K_2(\tilde{q}, n, z) = c_1 z^n e^{-z} M(\tilde{q} + 1, n + 1, z) + c_2 z^n e^{-z} U(\tilde{q} + 1, n + 1, z), \tag{32}$$

where (Abramowitz and Stegun 1965, 13.2.5) $U[a, a+c, z] = \frac{1}{\Gamma[a]} \int_0^\infty e^{-zt} t^{a-1} (t+1)^{c-1} dt, Re[z] > 0, Re[a] > 0$ is Tricomi's decreasing hypergeometric U function and $M(a, a+c, z) = {}_1F_1(a, a+c; z)$ is Kummer's increasing nonnegative confluent hypergeometric function of the first kind.[8]

The fact that the killed ruin probability must decrease to 0 implies the absence of the function $K_1$. The next result shows that the function $K_2(\mu(x + c(a)/r))$ is proportional to the ruin probability on an arbitrary interval $[a, \infty), a > -\tilde{c}$, and determines the proportionality constant. $K_1$ yields the absolute survival probability (and scale function) on $[-\tilde{c}, \infty)$, but over an arbitrary interval we must use a combination of $K_1$ and $K_2$.

**Theorem 2.** *(a) Put $z(x) = \mu(\tilde{c} + x), \tilde{c} = c(a)/r$. The ruin probability on $[a, \infty)$ is*

$$\Psi_q(x, a) = E_x[e^{-qT_{a,-}}] = \frac{\tilde{\lambda}}{\tilde{c}\mu} \frac{e^{-\mu x}(1 + x/\tilde{c})^{(\tilde{q}+\tilde{\lambda})} U(1 + \tilde{q}, 1 + \tilde{q} + \tilde{\lambda}, \mu(\tilde{c} + x))}{U(1 + \tilde{q}, 2 + \tilde{q} + \tilde{\lambda}, \mu \tilde{c})} \tag{33}$$

---

[6] More generally, for any phase-type jumps $C_i$ with Laplace transform $\hat{f}_C(s) = \frac{a(s)}{b(s)}$, it may be checked that $\Psi_q * f_C = \hat{f}_C(D)\Psi_q$ in the sense that $b(D)\Psi_q * f_C = a(D)\Psi_q$, thus removing the convolution by applying the denominator $b(D)$.
[7] this is implied by the Kolmogorov integro-differential equation $(\mathcal{G} - \lambda - q)\Psi_q(x) + \lambda \overline{F}(x) = 0, x \geq 0$
[8] $M(a, b, z)$ and $U(a, b, z)$ are the increasing/decreasing solutions of the to Weiler's canonical form of Kummer equation $zf''(z) + (b-z)f'(z) - a f(z) = 0$, which is obtained via the substitution $y(z) = e^{-z} z^n f(z)$, with $a = \tilde{q} + 1, b = n + 1$. Some computer systems use instead of $M$ the Laguerre function defined by $M(a, b, z) = L_{-a}^{b-1}(z) \frac{\Gamma(1-a)\Gamma(b)}{\Gamma(b-a)}$, which yields for natural $-a$ the Laguerre polynomial of degree $-a$.

(when $q = 0$, $K_2(0, n, z) = \Gamma(\tilde{\lambda}, z)$ and (33) reduces to $\frac{\Gamma(\tilde{\lambda}, \mu(\tilde{c}+x))}{\Gamma(\tilde{\lambda}+1, \mu\tilde{c})}$).[9]

(b) The **scale** function $W_q(x, a)$ on $[a, \infty)$ is up to a proportionality constant

$$K_1(\tilde{q}, \tilde{\lambda}, z(x)) - kK_2(\tilde{q}, \tilde{\lambda}, z(x)),$$

with k defined in (40).

**Proof.** (a) Following (Paulsen and Gjessing 1997, ex. 2.1), note that the limit $\lim_{z \to \infty} U(z) = 0$ implies

$$\Psi_q(x) = k\, K_2(z) = k e^{-z} z^{\tilde{q}+\tilde{\lambda}} U(\tilde{q}+1, \tilde{q}+\tilde{\lambda}+1, z), \quad z = \mu(x+\tilde{c}).$$

The proportionality constant $k$ is obtained from the boundary condition (30). Putting $G_b[h](x) := [c(x)(h)'(x) - (\lambda+q)h(x)]_{x=0}$,

$$G_b[\Psi_q](x) + \lambda = 0 \implies k = \frac{\lambda}{-G_b[K_2](z(x))},$$

$$= -e^{-z} z^{\tilde{q}+\tilde{\lambda}+1} U(\tilde{q}+1, \tilde{q}+\tilde{\lambda}+2, z)$$

Putting $z_0 = \mu c$, we find

$$-G_b[K_2](z(x)) = z_0 e^{-z_0} z_0^{\tilde{q}+\tilde{\lambda}-1} U(\tilde{q}, \tilde{q}+\tilde{\lambda}, z_0) + (\tilde{q}+\tilde{\lambda})e^{-z_0} z_0^{\tilde{q}+\tilde{\lambda}} U(\tilde{q}+1, \tilde{q}+\tilde{\lambda}+1, z_0)$$
$$= e^{-z_0} z_0^{\tilde{q}+\tilde{\lambda}} (U(\tilde{q}, \tilde{q}+\tilde{\lambda}, z_0) + (\tilde{q}+\tilde{\lambda}) U(\tilde{q}+1, \tilde{q}+\tilde{\lambda}+1, z_0)),$$

where we have used the identity (Borodin and Salminen 2012, p. 640)

$$K_2'(z) = -e^{-z} z^{\tilde{q}+\tilde{\lambda}-1} U(\tilde{q}, \tilde{q}+\tilde{\lambda}, z). \tag{34}$$

This may be further simplified to

$$-G_b[K_2](z(x)) = e^{-z_0} z_0^{\tilde{q}+\tilde{\lambda}+1} U(\tilde{q}+1, \tilde{q}+\tilde{\lambda}+2, z_0)),$$

by using the identity

$$U[a, b, z] + bU[a+1, b+1, z] = zU[a+1, b+2, z], a > 1, \tag{35}$$

which is itself a consequence of the identities (Abramowitz and Stegun 1965, 13.4.16, 13.4.18)

$$(b-a)U[a, b, z] + zU[a, 2+b, z] = (z+b)U[a, 1+b, z] \tag{36}$$
$$U[a, b, z] + (b-a-1)U[a+1, b, z] = zU[a+1, b+1, z] \tag{37}$$

(replace $a$ by $a+1$ in the first identity, and subtract the second).

---

[9] Note that we have corrected Paulsen's original denominator by using the identity (Abramowitz and Stegun 1965, 13.4.18) $U[a-1, b, z] + (b-a)U[a, b, z] = zU[a, b+1, z], a > 1$.

Finally, we obtain:

$$\Psi_q(x) = \left(\frac{\tilde{\lambda}}{\tilde{c}\mu}\right) \frac{e^{-\mu x}(1+\frac{x}{\tilde{c}})^{\tilde{q}+\tilde{\lambda}} U\left(\tilde{q}+1,\tilde{q}+1+\tilde{\lambda},\mu(x+\tilde{c})\right)}{U\left(\tilde{q}+1,\tilde{q}+1+\tilde{\lambda}+1,\mu\tilde{c}\right)}$$

$$= \left(\frac{\tilde{\lambda}}{\mu}\right) \frac{\int_x^\infty (s-x)^{\tilde{q}}(s+\tilde{c})^{\tilde{\lambda}-1} e^{-\mu s} ds}{\int_0^\infty s^{\tilde{q}}(s+\tilde{c})^{\tilde{\lambda}} e^{-\mu s} ds},$$

and

$$\Psi_q(0) = \left(\frac{\lambda}{c\mu}\right) \frac{U\left(\tilde{q}+1,\tilde{\lambda}+\tilde{q}+1,\tilde{c}\mu\right)}{U\left(\tilde{q}+1,\tilde{\lambda}+\tilde{q}+2,\tilde{c}\mu\right)} = \left(\frac{\lambda}{c\mu}\right) \frac{\int_0^\infty t^{\tilde{q}}(1+t)^{\tilde{\lambda}-1} e^{-\tilde{c}\mu t} dt}{\int_0^\infty t^{\tilde{q}}(1+t)^{\tilde{\lambda}} e^{-\tilde{c}\mu t} dt}$$

$$= \left(\frac{\tilde{\lambda}}{\mu}\right) \frac{\int_0^\infty s^{\tilde{q}}(s+\tilde{c})^{\tilde{\lambda}-1} e^{-\mu s} ds}{\int_0^\infty s^{\tilde{q}}(s+\tilde{c})^{\tilde{\lambda}} e^{-\mu s} ds} = \left(\frac{\tilde{\lambda}}{\tilde{c}}\right) \frac{\int_0^\infty t^{\tilde{q}}(t+\mu)^{\tilde{\lambda}-1} e^{-\tilde{c}t} dt}{\int_0^\infty t^{\tilde{q}}(t+\mu)^{\tilde{\lambda}} e^{-\tilde{c}t} dt}.$$

For $(a,\infty), a > -\tilde{c}$, the same proof works after replacing $c, z(0)$ by $c(a), z(a)$.

(b) On $(0,\infty)$, we must determine, up to proportionality, a linear combination $W_q(x) = K_1(\tilde{q},\tilde{\lambda},z(x)) - kK_2(\tilde{q},\tilde{\lambda},z(x))$ satisfying the boundary condition

$$G_b W_q(x) = 0 \Longrightarrow k = \frac{G_b[K_1](0)}{G_b[K_2](0)}, G_b h(x) := [c(x)(h)'(x) - (\lambda+q)h(x)]_{x=0}. \quad (38)$$

Recall we have already computed $G_b[K_2](z(0)) = -e^{-z_0} z_0^{\tilde{q}+\tilde{\lambda}+1} U\left(\tilde{q}+1,\tilde{q}+\tilde{\lambda}+2,z_0\right)$ in the proof of (a). Similarly, using (Borodin and Salminen 2012, p. 640) (reproduced for convenience in (55) below)

$$-G_b[K_1](z(0)) = (\tilde{\lambda}+\tilde{q})e^{-z_0} z_0^{\tilde{q}+\tilde{\lambda}} \left[M\left(\tilde{q}+1,\tilde{q}+\tilde{\lambda}+1,z_0\right) - M\left(\tilde{q},\tilde{q}+\tilde{\lambda},z_0\right)\right].$$

Hence[10]

$$k = \frac{\tilde{\lambda}+\tilde{q}}{z_0} \frac{M\left(\tilde{q}+1,\tilde{q}+\tilde{\lambda}+1,z_0\right) - M\left(\tilde{q},\tilde{q}+\tilde{\lambda},z_0\right)}{U(\tilde{q}+1,\tilde{q}+\tilde{\lambda}+2,z_0)}. \quad (40)$$

□

**Remark 6.** *Note that on $(-\tilde{c},\infty)$, choosing $-\tilde{c}$ as the origin yields $z_0 = 0 = k$ (since $M\left(\tilde{q},\tilde{q}+\tilde{\lambda},0\right) = 1$ (Abramowitz and Stegun 1965, 13.1.2)) and the scale function is proportional to $K_1(z)$, which follows also from the uniqueness of the nondecreasing solution.*

**Corollary 1.** *(a) Differentiating the scale function yields*

$$w_q(x) = e^{-z} z^{\tilde{q}+\tilde{\lambda}-1} \left((\tilde{q}+\tilde{\lambda})M(\tilde{q},\tilde{q}+\tilde{\lambda},z) + kU(\tilde{q},\tilde{q}+\tilde{\lambda},z)\right).$$

*(b) The scale function is increasing.*

---

[10] Putting $M_{++} = M\left(\tilde{q}+2,\tilde{q}+\tilde{\lambda}+2,\mu\tilde{c}\right), U_{++} = U\left(\tilde{q}+2,\tilde{q}+\tilde{\lambda}+2,\mu\tilde{c}\right)$, we must solve the equation

$$M - lU = \frac{\tilde{q}+1}{\tilde{q}+\tilde{\lambda}+1} M_{++} + l(\tilde{q}+1)U_{++}$$

$$\Leftrightarrow l = \frac{(\tilde{q}+\tilde{\lambda}+1)M - (\tilde{q}+1)M_{++}}{(\tilde{q}+\tilde{\lambda}+1)(U+(\tilde{q}+1)U_{++})} = \frac{\tilde{\lambda}}{\tilde{q}+\tilde{\lambda}+1} \frac{M_+}{U_+}, \quad (39)$$

where we put $M_+ = M\left(\tilde{q}+1,\tilde{q}+\tilde{\lambda}+2,\mu\tilde{c}\right), U_+ = U\left(\tilde{q}+1,\tilde{q}+\tilde{\lambda}+2,\mu\tilde{c}\right)$, and applied the identities (Abramowitz and Stegun 1965, 13.4.3, 13.4.4).

## 5. Asmussen's Embedding Approach for Solving Kolmogorov's Integro-Differential Equation with Phase-Type Jumps

One of the most convenient approaches to get rid of the integral term in (29) is a probabilistic transformation which gets rid of the jumps as in Asmussen (1995), when the downward phase-type jumps have a survival function

$$\bar{F}_C(x) = \int_x^\infty f_C(u)du = \vec{\beta}e^{Bx}\mathbf{1},$$

where $B$ is a $n \times n$ stochastic generating matrix (nonnegative off-diagonal elements and nonpositive row sums), $\vec{\beta} = (\beta_1,\ldots,\beta_n)$ is a row probability vector (with nonnegative elements and $\sum_{j=1}^n \beta_j = 1$), and $\mathbf{1} = (1,1,\ldots,1)$ is a column probability vector.

The density is $f_C(x) = \vec{\beta}e^{-Bx}b$, where $b = (-B)\mathbf{1}$ is a column vector, and the Laplace transform is

$$\hat{b}(s) = \vec{\beta}(sI - B)^{-1}b.$$

Asmussen's approach Asmussen (1995); Asmussen et al. (2002) replaces the negative jumps by segments of slope $-1$, embedding the original spectrally negative Lévy process into a continuous Markov modulated Lévy process. For the new process we have auxiliary unknowns $A_i(x)$ representing ruin or survival probabilities (or, more generally, Gerber-Shiu functions) when starting at $x$ conditioned on a phase $i$ with drift downwards (i.e., in one of the "auxiliary stages of artificial time" introduced by changing the jumps to segments of slope $-1$). Let $\mathbf{A}$ denote the column vector with components $A_1,\ldots,A_n$. The Kolmogorov integro-differential equation turns then into a system of ODE's, due to the continuity of the embedding process.

$$\begin{pmatrix} \Psi_q'(x) \\ \mathbf{A}'(x) \end{pmatrix} = \begin{pmatrix} \frac{\lambda+q}{c(x)} & -\frac{\lambda}{c(x)}\vec{\beta} \\ b & B \end{pmatrix} \begin{pmatrix} \Psi_q(x) \\ \mathbf{A}(x) \end{pmatrix}, \quad x \geq 0. \tag{41}$$

For the ruin probability with exponential jumps of rate $\mu$ for example, there is only one downward phase, and the system is:

$$\begin{pmatrix} \Psi_q'(x) \\ A'(x) \end{pmatrix} = \begin{pmatrix} \frac{\lambda+q}{c(x)} & -\frac{\lambda}{c(x)} \\ \mu & -\mu \end{pmatrix} \begin{pmatrix} \Psi_q(x) \\ A(x) \end{pmatrix} \quad x \geq 0. \tag{42}$$

For survival probabilities, one only needs to modify the boundary conditions—see the following section.

### 5.1. Exit Problems for the Segerdahl-Tichy process, with $q = 0$

Asmussen's approach is particular convenient for solving exit problems for the Segerdahl-Tichy process.

**Example 1.** *The eventual ruin probability.* When $q = 0$, the system for the ruin probabilities with $x \geq 0$ is:

$$\begin{cases} \Psi'(x) = \frac{\lambda}{c(x)}(\Psi(x) - A(x)), & \Psi(\infty) = A(\infty) = 0 \\ A'(x) = \mu(\Psi(x) - A(x)), & A(0) = 1 \end{cases} \tag{43}$$

This may be solved by subtracting the equations. Putting

$$K(x) = e^{-\mu x + \int_0^x \frac{\lambda}{c(v)}dv},$$

we find:
$$\begin{cases} \Psi(x) - A(x) &= (\Psi(0) - A(0))K(x), \\ A(x) &= \mu(A(0) - \Psi(0))\int_x^\infty K(v)dv, \end{cases} \quad (44)$$

whenever $K(v)$ is integrable at $\infty$.

The boundary condition $A(0) = 1$ implies that $1 - \Psi(0) = \frac{1}{\mu \int_0^\infty K(v)dv}$ and

$$A(x) = \mu(1 - \Psi(0))\int_x^\infty K(v)dv = \frac{\int_x^\infty K(v)dv}{\int_0^\infty K(v)dv},$$

$$\Psi(x) - A(x) = -\frac{K(x)}{\mu \int_0^\infty K(v)dv}.$$

Finally,

$$\Psi(x) = A(x) + (\Psi(x) - A(x)) = \frac{\mu \int_x^\infty K(v)dv - K(x)}{\mu \int_0^\infty K(v)dv},$$

and for the survival probability $\overline{\Psi}$,

$$\overline{\Psi}(x) = \frac{\mu \int_0^x K(v)dv + K(x)}{\mu \int_0^\infty K(v)dv} := \overline{\Psi}(0)\mathbf{W}(x) = \frac{\mathbf{W}(x)}{\mathbf{W}(\infty)}, \quad (45)$$

$$\mathbf{W}(x) = \mu \int_0^x K(v)dv + K(x),$$

where $\overline{\Psi}(0) = \frac{1}{\mathbf{W}(\infty)}$ by plugging $\mathbf{W}(0) = 1$ in the first and last terms in (45).

We may also rewrite (45) as:

$$\overline{\Psi}(x) = \frac{1 + \int_0^x \mathbf{w}(v)dv}{1 + \int_0^\infty \mathbf{w}(v)dv} \Leftrightarrow \Psi(x) = \frac{\int_x^\infty \mathbf{w}(v)dv}{1 + \int_0^\infty \mathbf{w}(v)dv}, \mathbf{w}(x) := \mathbf{W}'(x) = \frac{\lambda K(x)}{c(x)} \quad (46)$$

Note that $\mathbf{w}(x) > 0$ implies that the scale function $\mathbf{W}(x)$ is nondecreasing.

**Example 2.** *For the two sided exit problem on $[a, b]$, a similar derivation yields the scale function*

$$\mathbf{W}(x, a) = \mu \int_a^x \frac{K(v)}{K(a)}dv + \frac{K(x)}{K(a)} = 1 + \frac{1}{K(a)} \int_a^x \mathbf{w}(y)dy,$$

*with scale derivative derivative* $\mathbf{w}(x, a) = \frac{1}{K(a)}\mathbf{w}(x)$, *where* $\mathbf{w}(x)$ *given by (46) does not depend on a.*

*Indeed, the analog of (44) is:*

$$\begin{cases} \overline{\Psi}^b(x, a) - A^b(x) = \overline{\Psi}^b(a, a)\frac{K(x)}{K(a)}, \\ A^b(x) = \mu \overline{\Psi}^b(a, a)\int_a^x \frac{K(v)}{K(a)}dv, \end{cases}$$

implying by the fact that $\overline{\Psi}^b(b,a) = 1$ that

$$\overline{\Psi}^b(x,a) = \overline{\Psi}^b(a,a)\left(\frac{K(x)}{K(a)} + \mu\int_a^x \frac{K(v)}{K(a)}dv\right) = \frac{W(x,a)}{W(b,a)} = \frac{1 + \frac{1}{K(a)}\int_a^x \mathbf{w}(u)du}{1 + \frac{1}{K(a)}\int_a^b \mathbf{w}(u)du} \Leftrightarrow$$

$$\Psi^b(x,a) = \frac{\int_x^b \mathbf{w}(u)du}{K(a) + \int_a^b \mathbf{w}(u)du} \Leftrightarrow \qquad (47)$$

$$\psi^b(x,a) := -(\Psi^b)'(x,a) = \frac{\mathbf{w}(x)}{K(a) + \int_a^b \mathbf{w}(u)du} = \mathbf{w}(x,a)\frac{\overline{\Psi}(a,a)}{\overline{\Psi}(b,a)}.$$

**Remark 7.** *The definition adopted in this section for the scale function* $W(x,a)$ *uses the normalization* $W(a,a) = 1$, *which is only appropriate in the absence of Brownian motion.*

**Problem 4.** *Extend the equations for the survival and ruin probability of the Segerdahl-Tichy process in terms of the scale derivative* $\mathbf{w}_q$, *when* $q > 0$. *Essentially, this requires obtaining*

$$T_q(x) = E_x\left[e^{-q[T_{a,-} \min T_{b,+}]}\right]$$

## 6. Revisiting Segerdahl's Process via the Scale Derivative/Integrating Factor Approach, When $q = 0$

Despite the new scale derivative/integrating factor approach, we were not able to produce further explicit results beyond (33), due to the fact that neither the scale derivative, nor the integral of the integrating factor are explicit when $q > 0$ (this is in line with Avram et al. (2010)). (33) remains thus for now an outstanding, not well-understood exception.

**Problem 5.** *Are there other explicit first passage results for Segerdahl's process when* $q > 0$?

In the next subsections, we show that via the scale derivative/integrating factor approach, we may rederive well-known results for $q = 0$.

### 6.1. Laplace Transforms of the Eventual Ruin and Survival Probabilities

For $q = 0$, both Laplace transforms and their inverses are explicit, and several classic results may be easily checked. The scale derivative may be obtained using Proposition 1 and $\Gamma(\tilde{\lambda}+1,v) = e^{-v}v^{\tilde{\lambda}} + \lambda\Gamma(\tilde{\lambda},v)$ with $v = \tilde{c}(s+\mu)$. We find

$$\widehat{\mathbf{w}}(s,a) = \frac{e^{\tilde{c}\mu}(\tilde{c}\mu)^{-\tilde{\lambda}}\Gamma(\tilde{\lambda}+1,\tilde{c}(s+\mu))}{e^{-\tilde{c}s}(1+s/\mu)^{\tilde{\lambda}}} - 1 = 1 + \lambda e^v(v)^{-\tilde{\lambda}}\Gamma(\tilde{\lambda},v) - 1 = \tilde{\lambda}U(1,1+\tilde{\lambda},\tilde{c}(s+\mu))$$

$$\Longrightarrow \mathbf{w}(x,a) = \frac{\tilde{\lambda}}{\tilde{c}}\left(1+\frac{x}{\tilde{c}}\right)^{\tilde{\lambda}-1}e^{-\mu x}, \qquad (48)$$

which checks (46). Using again $\widehat{\mathbf{w}}(s) = \tilde{c}\frac{\overline{I}(y)}{I(y)} - 1$ yields the ruin and survival probabilities:

$$s\widehat{\overline{\Psi}}(s) = \frac{\int_s^\infty \tilde{c}\overline{\Psi}(0)I(y)dy}{I(s)} = \overline{\Psi}(0)(\widehat{\mathbf{w}}(s)+1)$$

$$s\widehat{\Psi}(s) = \frac{\int_s^\infty (\tilde{c}\overline{\Psi}(0) - \frac{\tilde{\lambda}}{y+\mu})I(y)dy}{I(s)} = \Psi(0)(\widehat{\mathbf{w}}(s)+1) - \widehat{\mathbf{w}}(s).$$

Letting $s \to 0$ yields

$$\Psi(0) = \frac{\widehat{w}(0)}{\widehat{w}(0)+1} = \frac{\tilde{\lambda} U(1,1+\tilde{\lambda},\mu\tilde{c})}{\mu\tilde{c}\, U(1,2+\tilde{\lambda},\mu\tilde{c})} = \frac{\tilde{\lambda}\Gamma(\tilde{\lambda},\tilde{c}\mu)}{\Gamma(\tilde{\lambda}+1,\tilde{c}\mu)} \Leftrightarrow$$

$$\overline{\Psi}(0) = \frac{\lim_{s\to 0} s\widehat{\overline{\Psi}}(s)}{\widehat{w}(0)+1} = \frac{\overline{\Psi}(\infty)}{1+\tilde{\lambda}\, U(1,1+\tilde{\lambda},\mu\tilde{c})} = \frac{1}{\mu\tilde{c}\, U(1,2+\tilde{\lambda},\mu\tilde{c})} \tag{49}$$

For the survival probability, we finally find

$$s\widehat{\overline{\Psi}}(s) = \overline{\Psi}(0)(1+\widehat{w}(s)) = \frac{1+\tilde{\lambda}U(1,1+\tilde{\lambda},\mu(\tilde{c}+s))}{1+\tilde{\lambda}U(1,1+\tilde{\lambda},\mu\tilde{c})} = \frac{\tilde{c}(\mu+s)U(1,2+\tilde{\lambda},\mu(\tilde{c}+s))}{\tilde{c}\mu U(1,2+\tilde{\lambda},\mu\tilde{c})},$$

which checks with the Laplace transform of the Segerdahl result (53).

### 6.2. The Eventual Ruin and survival probabilities

These may also be obtained directly by integrating the explicit scale derivative $w(x,a) = \frac{\tilde{\lambda}}{\tilde{c}}\left(1+\frac{x}{\tilde{c}}\right)^{\tilde{\lambda}-1} e^{-\mu x}$ (48) Indeed,

$$\int_u^\infty w(x)dx = \int_u^\infty \frac{\tilde{\lambda}}{\tilde{c}}\left(1+\frac{x}{\tilde{c}}\right)^{\tilde{\lambda}-1} e^{-\mu x} dx = \tilde{\lambda} e^{\mu\tilde{c}} \int_{1+\frac{u}{\tilde{c}}}^\infty y^{\tilde{\lambda}-1} e^{\mu\tilde{c}y} dy$$

$$= \tilde{\lambda} e^{\mu\tilde{c}} \frac{1}{(\mu\tilde{c})^{\tilde{\lambda}}} \int_{\mu(\tilde{c}+u)}^\infty t^{\tilde{\lambda}-1} e^{-t} dt = \tilde{\lambda} e^{\mu\tilde{c}} (\mu\tilde{c})^{-\tilde{\lambda}} \Gamma(\tilde{\lambda},\mu(\tilde{c}+u)),$$

where $\Gamma(\eta,x) = \int_x^\infty t^{\eta-1} e^{-t} dt$ is the incomplete gamma function. The ruin probability is Segerdahl (1955), (Paulsen and Gjessing 1997, ex. 2.1):

$$\Psi(x) = \tilde{\lambda} \frac{\exp(\mu\tilde{c})(\mu\tilde{c})^{-\tilde{\lambda}}\Gamma(\tilde{\lambda},\mu(\tilde{c}+x))}{1+\tilde{\lambda}\exp(\mu\tilde{c})(\mu\tilde{c})^{-\tilde{\lambda}}\Gamma(\tilde{\lambda},\mu\tilde{c})} = \tilde{\lambda}\frac{e^{-\mu x}(1+x/\tilde{c})^{\tilde{\lambda}} U(1,1+\tilde{\lambda},\mu(\tilde{c}+x))}{1+\tilde{\lambda}U(1,1+\tilde{\lambda},\mu\tilde{c})}$$

$$= \frac{\tilde{\lambda}}{\mu\tilde{c}} \frac{e^{-\mu x}(1+x/\tilde{c})^{\tilde{\lambda}} U(1,1+\tilde{\lambda},\mu(\tilde{c}+x))}{U(1,2+\tilde{\lambda},\mu\tilde{c})} = \frac{\tilde{\lambda}\Gamma(\tilde{\lambda},\mu(\tilde{c}+x))}{\Gamma(\tilde{\lambda}+1,\tilde{c}\mu)}, \tag{50}$$

where we used

$$U(1,1+\tilde{\lambda},v) = e^v v^{-\tilde{\lambda}} \Gamma(\tilde{\lambda},v) \tag{51}$$

and

$$1+\tilde{\lambda}U(1,1+\tilde{\lambda},v) = vU(1,2+\tilde{\lambda},v), \tag{52}$$

which holds by integration by parts.

A simpler formula holds for the rate of ruin $\psi(x)$ and its Laplace transform

$$\psi(x) = -\Psi'(x) = \frac{w(x)}{1+\int_0^\infty w(x)dx} = \frac{\tilde{\lambda}}{\Gamma(\tilde{\lambda}+1,\tilde{c}\mu)} \mu(\mu(\tilde{c}+x))^{\tilde{\lambda}-1} e^{-\mu(\tilde{c}+x)} = e^{-\mu\tilde{c}} \gamma_{\tilde{\lambda},\mu}(x+\tilde{c}) \Leftrightarrow$$

$$\widehat{\psi}(s) = \overline{\Psi}(0)\widehat{w}(s) = \begin{cases} \frac{\tilde{\lambda}U(1,1+\tilde{\lambda},\tilde{c}(s+\mu))}{\tilde{c}\mu U(1,2+\tilde{\lambda},\tilde{c}\mu)}, & c > 0 \\ (1+s/\mu)^{-\tilde{\lambda}}, & c = 0 \end{cases}, \tag{53}$$

where $\gamma$ denotes a (shifted) Gamma density. Of course, the case $c > 0$ simplifies to a Gamma density when moving the origin to the "absolute ruin" point $-\tilde{c} = -\frac{c}{r}$, i.e., by putting $y = x + \tilde{c}, Y_t = X_t + \tilde{c}$, where the process $Y_t$ has drift rate $rY_t$.

**Problem 6.** *Find a relation between the ruin derivative $\psi_q(x) = -\Psi'_q(x)$ and the scale derivative $w_q(x)$ when $q > 0$.*

## 7. Further Details on the Identities Used in the Proof of Theorem 2

We recall first some continuity and differentiation relations needed here Abramowitz and Stegun (1965)

**Proposition 2.** *Using the notation $M = M(a,b,z)$, $M(a+) = M(a+1,b,z)$, $M(+,+) = M(a+1,b+1,z)$, and so on, the Kummer and Tricomi functions satisfy the following identities:*

$$bM + (a-b)M(b+) = aM(a+) \tag{13.4.3}$$

$$b(M(a+) - M) = zM(+,+) \tag{13.4.4}$$

$$(b-a)U + zU(b+2) = (z+b)U(b+1) \tag{13.4.16}$$

$$U + aU(+,+) = U(b+) \tag{13.4.17}$$

$$U + (b-a-1)U(a+1) = zU(+,+) \tag{13.4.18}$$

(see corresponding equations in Abramowitz and Stegun (1965)).

$$U' = -aU(+,+), \quad M' = \frac{a}{b}M(+,+). \tag{54}$$

**Proposition 3.** *The functions $K_i(\tilde{q}, \tilde{\lambda}, z)$ defined by (32) satisfy the identities*

$$K'_1(\tilde{q}, n, z) = (\tilde{q} + \tilde{\lambda})e^{-z}z^{\tilde{q}+\tilde{\lambda}-1} M(\tilde{q}, \tilde{q} + \tilde{\lambda}, z) = (\tilde{q} + \tilde{\lambda})K_1(\tilde{q} - 1, \tilde{\lambda}, z) \tag{55}$$

$$K'_2(\tilde{q}, n, z) = -e^{-z}z^{\tilde{q}+\tilde{\lambda}-1} U(\tilde{q}, \tilde{q} + \tilde{\lambda}, z) = -K_2(\tilde{q} - 1, n, z) \tag{56}$$

$$K_2(\tilde{q}, n, z) = \int_z^\infty (y - z)^{\tilde{q}} (y)^{n - \tilde{q} - 1} e^{-y} dy \tag{57}$$

**Proof:** For the first identity, note, using (Abramowitz and Stegun 1965, 13.4.3, 13.4.4), that

$$\frac{e^z}{z^{\tilde{q}+\tilde{\lambda}-1}}K'_1(z) = (\tilde{q} + \tilde{\lambda} - z)M(\tilde{q} + 1, \tilde{q} + 1 + \tilde{\lambda}, z) + z\frac{\tilde{q}+1}{\tilde{q}+\tilde{\lambda}+1}M(\tilde{q} + 2, \tilde{q} + 2 + \tilde{\lambda}, z)$$

$$= (\tilde{q} + \tilde{\lambda})M(\tilde{q} + 1, \tilde{q} + 1 + \tilde{\lambda}, z)$$
$$+ \frac{z}{\tilde{q}+\tilde{\lambda}+1}\left((\tilde{q}+1)M(\tilde{q}+2, \tilde{q}+2+\tilde{\lambda}, z) - (\tilde{q}+\tilde{\lambda}+1)M(\tilde{q}+1, \tilde{q}+1+\tilde{\lambda}, z)\right)$$

$$= (\tilde{q} + \tilde{\lambda})M(\tilde{q}+1, \tilde{q}+1+\tilde{\lambda}, z) - \frac{z}{\tilde{q}+\tilde{\lambda}+1}\tilde{\lambda}M(\tilde{q}+1, \tilde{q}+2+\tilde{\lambda}, z)$$

$$= (\tilde{q} + \tilde{\lambda})M(\tilde{q}+1, \tilde{q}+1+\tilde{\lambda}, z) - \tilde{\lambda}\left(M(\tilde{q}+1, \tilde{q}+1+\tilde{\lambda}, z) - M(\tilde{q}, \tilde{q}+1+\tilde{\lambda}, z)\right)$$

$$= \tilde{q}M(\tilde{q}+1, \tilde{q}+1+\tilde{\lambda}, z) + \tilde{\lambda}M(\tilde{q}, \tilde{q}+1+\tilde{\lambda}, z).$$

The second formula may be derived similarly using 13.4.17, or by considering the function

$$_z\tilde{U}(\tilde{q}+1, \tilde{q}+1+\tilde{\lambda}, \mu) := \Gamma(q+1)K_2(z) = \int_z^\infty (s-z)^{\tilde{q}} (s)^{\tilde{\lambda}-1} e^{-\mu s} ds$$

appearing in the numerator of the last form of (57). An integration by parts yields

$$_z\tilde{U}'(\tilde{q}+1,\tilde{q}+1+\tilde{\lambda},1) = \int_z^\infty (s-z)^{\tilde{q}} \frac{d}{dz}[(s)^{\tilde{\lambda}-1}e^{-s}]ds$$
$$= (\tilde{\lambda}-1)_z\tilde{U}(\tilde{q}+1,\tilde{q}+\tilde{\lambda},1) -_z \tilde{U}(\tilde{q}+1,\tilde{q}+\tilde{\lambda}+1,1), \implies$$
$$K_2'(\tilde{q}+1,\tilde{\lambda},z) = e^{-z}z^{\tilde{q}+\tilde{\lambda}-1}\left((\tilde{\lambda}-1)U(\tilde{q}+1,\tilde{q}+\tilde{\lambda},z) - U(\tilde{q}+1,\tilde{q}+\tilde{\lambda}+1,z)\right)$$

and the result follows by (Abramowitz and Stegun 1965, 13.4.18.)[11]

The third formula is obtained by the substitution $y = z(t+1)$.

## 8. Conclusions and Future Work

Two promising fundamental functions have been proposed for working with generalizations of Segerdahl's process: (a) the scale derivative **w** Czarna et al. (2017) and (b) the integrating factor $I$ Avram and Usabel (2008), and they are shown to be related via Thm. 1.

Segerdahl's process per se is worthy of further investigation. A priori, many risk problems (with absorbtion/reflection at a barrier $b$ or with double reflection, etc.) might be solved by combinations of the hypergeometric functions $U$ and $M$.

However, this approach leads to an impasse for more complicated jump structures, which will lead to more complicated hypergeometric functions. In that case, we would prefer answers expressed in terms of the fundamental functions **w** or $I$.

We conclude by mentioning two promising numeric approaches, not discussed here. One due to Jacobsen and Jensen (2007) bypasses the need to deal with high-order hypergeometric solutions by employing complex contour integral representations. The second one uses Laguerre-Erlang expansions—see Abate, Choudhury and Whitt (1996); Avram et al. (2009); Zhang and Cui (2019). Further effort of comparing their results with those of the methods discussed above seems worthwhile.

**Author Contributions:** The authors had equal contributions for conceptualization, methodology, validation, formal analysis, investigation, writing—original draft preparation, review and editing, and project administration.

**Funding:** This research received no external funding.

**Conflicts of Interest:** The authors declare no conflict of interest.

## References

Abramowitz, Milton, and Irene Ann Stegun. 1965. *Handbook of Mathematical Functions: With Formulas, Graphs and Mathematical Tables*. Mineola: Dover Publications, vol. 55.

Hansjörg, Albrecher, and Sören Asmussen. 2010. *Ruin Probabilities*. Singapore: World Scientific, vol. 14.

Hansjörg, Albrecher, Jevgenijs Ivanovs, and Xiaowen Zhou. 2016. Exit identities for Lévy processes observed at Poisson arrival times. *Bernoulli* 22: 1364–82. [CrossRef]

Hansjörg, Albrecher, Corina Constantinescu, Zbigniew Palmowski, Georg Regensburger, and Markus Rosenkranz. 2013. Exact and asymptotic results for insurance risk models with surplus-dependent premiums. *SIAM Journal on Applied Mathematics* 73: 47–66. [CrossRef]

Asmussen, Søren. 1995. Stationary distributions for fluid flow models with or without brownian noise. *Communications in Statistics Stochastic Models* 11: 21–49. [CrossRef]

Asmussen, Soren, Florin Avram, and Miguel Usabel. 2002. Erlangian approximations for finite-horizon ruin probabilities. *ASTIN Bulletin: The Journal of the IAA* 32: 267–81. [CrossRef]

Abate, Joseph, Choudhury Gagan, and Whitt Whitt. 1996. On the Laguerre method for numerically inverting Laplace transforms. *INFORMS Journal on Computing* 8: 413–27. [CrossRef]

---

[11] See also (Borodin and Salminen 2012, p. 640), where however the first formula has a typo.

Florin, Avram, José F Cariñena, and Javier de Lucas. 2010. A lie systems approach for the first passage-time of piecewise deterministic processes. *Mathematics* arXiv:1008.2625.

Florin, Avram, and Dan Goreac. 2019. A pontryaghin maximum principle approach for the optimization of dividends/consumption of spectrally negative markov processes, until a generalized draw-down time. *Scandinavian Actuarial Journal* 2019: 1–25.

Florin, Avram, Danijel Grahovac, and Ceren Vardar-Acar. 2019a. The $W, Z$ scale functions kit for first passage problems of spectrally negative Lévy processes and applications to the optimization of dividends. *Mathematics* arXiv:1706.06841.

Florin, Avram, Danijel Grahovac, and Ceren Vardar-Acar. 2019b. The $w, z/\nu, \delta$ paradigm for the first passage of strong markov processes without positive jumps. *Risks* 7: 18. [CrossRef]

Florin, Avram, Andreas Kyprianou, and Martijn Pistorius. 2004. Exit problems for spectrally negative Lévy processes and applications to (Canadized) Russian options. *The Annals of Applied Probability* 14: 215–38.

Florin, Avram, Bin Li, and Shu Li. 2018. A unified analysis of taxed draw-down spectrally negative markov processes. *Risks* 7: 18.

Florin, Avram, Zbigniew Palmowski, and Martijn Pistorius. 2007. On the optimal dividend problem for a spectrally negative Lévy process. *The Annals of Applied Probability* 17: 156–80. [CrossRef]

Florin, Avram, Nikolai Leonenko, and Landy Rabehasaina. 2009. Series Expansions for the First Passage Distribution of Wong–Pearson Jump-Diffusions. *Stochastic Analysis and Applications* 27: 770–96. [CrossRef]

Avram, Florin, Zbigniew Palmowski, and Martijn Pistorius. 2015. On Gerber–Shiu functions and optimal dividend distribution for a Lévy risk process in the presence of a penalty function. *The Annals of Applied Probability* 25: 1868–935. [CrossRef]

Florin, Avram, José Luis Pérez, and Kazutoshi Yamazaki. 2016. Spectrally negative Lévy processes with parisian reflection below and classical reflection above. *Mathematics* arXiv:1604.01436.

Florin, Avram, and Miguel Usabel. 2008. The gerber-shiu expected discounted penalty-reward function under an affine jump-diffusion model. *Astin Bulletin* 38: 461–81.

Florin, Avram, and Matija Vidmar. 2017. First passage problems for upwards skip-free random walks via the $phi, w, z$ paradigm. *Mathematics* arXiv:1708.06080.

Florin, Avram, and Xiaowen Zhou. 2017. On fluctuation theory for spectrally negative Lévy processes with parisian reflection below and applications. *Theory of Probability and Mathematical Statistics* 95: 17–40.

Bertoin, Jean. 1997. Exponential decay and ergodicity of completely asymmetric Lévy processes in a finite interval. *The Annals of Applied Probability* 7: 156–69. [CrossRef]

Bertoin, Jean. 1998. *Lévy Processes*. Cambridge: Cambridge University Press, vol. 121.

Blumenthal, Robert McCallum, and Ronald Kay Getoor. 2007. *Markov Processes and Potential Theory*. Chelmsford: Courier Corporation.

Borodin, Andrei N., and Paavo Salminen. 2012. *Handbook of Brownian Motion-Facts and Formulae*. Basel: Birkhäuser.

Borovkov, Konstantin, and Alexander Novikov. 2008. On exit times of levy-driven ornstein–uhlenbeck processes. *Statistics & Probability Letters* 78: 1517–25.

Czarna, Irmina, José-Luis Pérez Tomasz Rolski, and Kazutoshi Yamazaki. 2017. Fluctuation theory for level-dependent Lévy risk processes. *Mathematics* arXiv:1712.00050.

de Finetti, B. 1957. Su un'impostazione alternativa della teoria collettiva del rischio. In *Transactions of the XVth International Congress of Actuaries*. New York: Mallon, vol. 2, pp. 433–43.

Ivanovs, Jevgenijs. 2013. A note on killing with applications in risk theory. *Insurance: Mathematics and Economics* 52: 29–34. [CrossRef]

Ivanovs, Jevgenijs, and Zbigniew Palmowski. 2012. Occupation densities in solving exit problems for Markov additive processes and their reflections. *Stochastic Processes and Their Applications* 122: 3342–60. [CrossRef]

Jacobsen, Martin, and Anders Tolver Jensen. 2007. Exit times for a class of piecewise exponential markov processes with two-sided jumps. *Stochastic Processes and Their Applications* 117: 1330–56. [CrossRef]

Jiang, Pingping, Bo Li, and Yongjin Wang. 2019. Exit times, undershoots and overshoots for reflected cir process with two-sided jumps. *Methodology and Computing in Applied Probability* 2019: 1–18. [CrossRef]

Kyprianou, Andreas. 2014. *Fluctuations of Lévy Processes with Applications: Introductory Lectures*. Berlin/Heidelberg: Springer Science & Business Media.

Landriault, David, Bin Li, and Hongzhong Zhang. 2017. A unified approach for drawdown (drawup) of time-homogeneous markov processes. *Journal of Applied Probability* 54: 603–26. [CrossRef]

Li, Bo, and Zbigniew Palmowski. 2016. Fluctuations of omega-killed spectrally negative Lévy processes. *Mathematics* arXiv:1603.07967.

Li, Bo, and Xiaowen Zhou. 2017. On weighted occupation times for refracted spectrally negative Lévy processes. *Mathematics* arXiv:1703.05952.

Loeffen, Ronnie Lambertus, and Pierre Patie. 2010. Absolute ruin in the ornstein-uhlenbeck type risk model. *arXiv* arXiv:1006.2712.

Marciniak, Ewa, and Zbigniew Palmowski. 2016. On the optimal dividend problem for insurance risk models with surplus-dependent premiums. *Journal of Optimization Theory and Applications* 168: 723–42. [CrossRef]

Paulsen, Jostein, and Hakon Gjessing. 1997. Ruin theory with stochastic return on investments. *Advances in Applied Probability* 29: 965–85. [CrossRef]

Paulsen, Jostein. 2010. Ruin models with investment income. In *Encyclopedia of Quantitative Finance*. Hoboken: Wiley Online Library.

Segerdahl, C.-O. 1955. When does ruin occur in the collective theory of risk? *Scandinavian Actuarial Journal* 38: 22–36. [CrossRef]

Suprun, V. N. 1976. Problem of destruction and resolvent of a terminating process with independent increments. *Ukrainian Mathematical Journal* 28: 39–51. [CrossRef]

Tichy, Roland. 1984. Uber eine zahlentheoretische Methode zur numerischen Integration und zur Behandlung von Integralgleichungen. *Osterreichische Akademie der Wissenschaften Mathematisch-Naturwissenschaftliche Klasse Sitzungsberichte II* 193: 329–58.

Zhang, Zhimin, and Cui, Zhenyu. 2019. Laguerre series expansion for scale functions and applications in risk theory. Preprint.

© 2019 by the authors. Licensee MDPI, Basel, Switzerland. This article is an open access article distributed under the terms and conditions of the Creative Commons Attribution (CC BY) license (http://creativecommons.org/licenses/by/4.0/).

*Article*

# Ruin Probability Approximations in Sparre Andersen Models with Completely Monotone Claims

Hansjörg Albrecher and Eleni Vatamidou *

Department of Actuarial Science, Faculty of Business and Economics, University of Lausanne, Quartier UNIL-Chamberonne Bâtiment Extranef, 1015 Lausanne, Switzerland; hansjoerg.albrecher@unil.ch
* Correspondence: eleni.vatamidou@unil.ch

Received: 20 September 2019; Accepted: 8 October 2019; Published: 14 October 2019

**Abstract:** We consider the Sparre Andersen risk process with interclaim times that belong to the class of distributions with rational Laplace transform. We construct error bounds for the ruin probability based on the Pollaczek–Khintchine formula, and develop an efficient algorithm to approximate the ruin probability for completely monotone claim size distributions. Our algorithm improves earlier results and can be tailored towards achieving a predetermined accuracy of the approximation.

**Keywords:** Sparre Andersen model; heavy tails; completely monotone distributions; error bounds; hyperexponential distribution

---

## 1. Introduction

The Sparre Andersen model is a classical object of study in insurance risk theory, see e.g., Labbé and Sendova (2009); Li and Garrido (2005); Temnov (2004, 2014); Willmot (2007); and Asmussen and Albrecher (2010) for an overview. In this model, claims occur according to a renewal process, which generalises the Cramér–Lundberg model, where claims arrive according to a Poisson process. Ruin probabilities in such a general setting are typically expressed as solutions of defective renewal equations, differential equations, the so-called Wiener–Hopf factorisation, etc., but the latter are typically inadequate to be used for numerical computations. However, if either the interclaim times or the claim sizes belong to the class of phase-type distributions, then ruin-related quantities can be found in an explicit form; see, e.g., Albrecher and Boxma (2005); Dickson (1998); Li and Garrido (2005) and Landriault and Willmot (2008), respectively.

However, in many relevant situations in practice, the behaviour of the claim sizes is better captured by heavy-tailed distributions (Embrechts et al. 1997); however, in that case, explicit expressions are hard or impossible to evaluate even in terms of Laplace transforms. Under a heavy-tailed setting, a standard approach is hence to seek for asymptotic approximations (Albrecher et al. 2012; Dong and Liu 2013; Wei et al. 2008), for initial capital levels being very large. At the same time, this capital level typically has to be very large, so as to be reasonably accurate, when actual magnitudes matter. One mathematically appealing solution is then to look for higher-order approximations (see e.g., Albrecher et al. 2010); but, then an actual error bound for fixed values also cannot be given. Another alternative is to approximate the actual heavy-tailed claim distribution by a tractable light-tailed one and control the introduced error in some way. Spectral approximations in this spirit were recently developed in Vatamidou et al. (2014) for the classical Cramér–Lundberg model.

The present paper proposes an extension of techniques in Vatamidou et al. (2014) to the more general Sparre Andersen model, and at the same time improves the bound derived there and the efficiency of the algorithm to establish it. Using the geometric compound tail representation of the ruin probability, we derive our error bound in terms of the ladder height distribution, which is explicitly available when the distribution of the interclaim times has a rational Laplace transform. We focus on heavy-tailed claim sizes, where numerical evaluations of ruin probabilities are typically challenging, and we develop an algorithm for the class of completely monotone distributions. Concretely, we approximate the ladder height distribution by a hyperexponential distribution, and we are able to prescribe the number of required phases for a desired resulting accuracy for the ruin probability.

The rest of the paper is organised as follows. In Section 2, we introduce the model and provide the exact formula for the ladder height distribution. As a next step, we derive, in Section 3, the error bound for the ruin probability, and we construct our approximation algorithm. In Section 4, we compare our approximations with existing asymptotic approximations. In Section 5, we then perform an extensive numerical analysis to check the tightness of the bound and the quality of the derived approximations. Finally, we conclude in Section 6.

## 2. Model Description

Consider the Sparre Andersen risk model for an insurance surplus process defined as

$$R(t) = u + ct - \sum_{i=1}^{N(t)} X_i, \quad t \geq 0, \tag{1}$$

where $u \geq 0$ is the initial capital, $c > 0$ is the constant premium rate and the i.i.d. positive random variables $\{X_i\}_{i \geq 1}$ with distribution function $F_X$ represent the claim sizes. The counting process $\{N(t), t \geq 0\}$ denotes the number of claims within $[0, t]$ and is defined as $N(t) = \max\{n \in \mathbb{N} : W_1 + W_2 + \ldots W_n \leq t\}$, where the interclaim times $W_i$ are assumed to be i.i.d. with common distribution function $K$, independent of the claim sizes; see, e.g., Asmussen and Albrecher (2010). We also assume $c\mathbb{E}W > \mathbb{E}X$, providing a positive safety loading condition.

Now, let $T = \inf\{t \geq 0 : R(t) < 0\}$ be the time of ultimate ruin. Then, the ruin probability is defined as

$$\psi(u) = \mathbb{P}(T < \infty \mid R(0) = u). \tag{2}$$

The ruin probability satisfies the defective renewal equation

$$\psi(u) = \phi \int_0^u \psi(u - x) dH(x) + \phi \overline{H}(u), \quad u \geq 0, \tag{3}$$

where $\phi = \psi(0)$, $H(u)$ is the distribution of the ascending ladder height associated with the surplus process $S(t) := u - R(t)$ and $\overline{H}(u) = 1 - H(u)$, for $u \geq 0$; see, e.g., Willmot et al. (2001). The solution to Equation (3) is the Pollaczek–Khintchine-type formula

$$\psi(u) = \sum_{n=1}^{\infty} (1 - \phi) \phi^n \overline{H}^{*n}(u), \tag{4}$$

i.e., $\psi(u)$ is a geometric compound tail with geometric parameter $\phi$; see Section 1.2.3 in Willmot and Woo (2017) for details.

Although Equation (4) provides a closed-form formula for the ruin probability, it is impractical, because the ladder height distribution $H(u)$ is not available in most cases of interest. However, when the distribution $K$ of the interclaim times has a rational Laplace transform, $H(u)$ has an explicit form (Li and Garrido 2005), which we recall in the next subsection. In the sequel, we will then use this as a starting point for developing highly accurate approximations for $\psi(u)$, which is of particular interest for heavy-tailed claim sizes.

*The Ladder Height Distribution with Interclaim Times of Rational Laplace Transform*

We assume now that the Laplace transform of the interclaim times is a rational function of the form

$$\tilde{k}(s) = \frac{\mu^* + s\beta(s)}{\prod_{n=1}^{N}(s+\mu_n)}, \qquad (5)$$

where $\mu_n > 0$, $\forall n = 1,\ldots,N$, $\mu^* = \prod_{n=1}^{N}\mu_n$ and $\beta(s)$ is a polynomial of degree $N-2$ or less. Obviously, $\mathbb{E}W = -\tilde{k}'(0) = \sum_{n=1}^{N}\frac{1}{\mu_n} - \frac{\beta(0)}{\mu^*}$. If $\tilde{f}_X(s) = \int_0^{+\infty} e^{-sx} dF_X(x)$ is the Laplace–Stieltjes transform (LST) of the claim sizes, it is shown in Li and Garrido (2005) that the generalised Lundberg equation

$$\frac{\prod_{n=1}^{N}(\mu_n - cs)}{\mu^* - cs\beta(-cs)} = \tilde{f}_X(s), \qquad s \in \mathbb{C}, \qquad (6)$$

has exactly $N$ roots $\rho_1, \rho_2,\ldots, \rho_N$, with $\rho_N = 0$ and $\Re(\rho_n) > 0$, $n = 1,2,\ldots,N-1$. These roots play an important role in the evaluation of the ladder height distribution and the geometric parameter $\phi$. Denote with $\overline{F}_X(x)$ the complementary cumulative distribution function (ccdf) of the claim sizes, and consider the Dickson–Hipp operator

$$\mathcal{T}_r f(x) := \int_x^\infty e^{-r(y-x)} f(y) dy = \int_0^\infty e^{-ry} f(y+x) dy, \qquad (7)$$

for a function $f(x)$ (see Dickson and Hipp 2001). Moreover, let $\rho^* = \prod_{n=1}^{N-1} \rho_n$. Then, as shown in Li and Garrido (2005), the ccdf of the ascending ladder heights is calculated via the formula

$$\overline{H}(u) = \frac{1}{\phi c^N} \sum_{n=1}^{N} \frac{\mu^* - c\rho_n \beta(-c\rho_n)}{\prod_{\substack{k=1\\k\neq n}}^{N}(\rho_k - \rho_n)} \mathcal{T}_{\rho_n} \overline{F}_X(u), \qquad (8)$$

where

$$\phi = 1 - \frac{\mu^*(c\mathbb{E}W - \mathbb{E}X)}{\rho^*} < 1. \qquad (9)$$

Although the ladder height distribution in this model has an explicit formula, it is difficult to evaluate $\psi(u)$ either via Equation (4) or by taking Laplace transforms (an equivalent formula to the Pollaczek–Khinchine in the Cramér–Lundberg model). In particular, this is the case when the claim sizes follow a heavy-tailed distribution, as already mentioned in Section 1. As a result, in such cases, opting for approximations seems a natural solution.

In the next section, we will study error bounds for $\psi(u)$ when the ladder height distribution is approximated by a phase-type distribution. In particular, we will provide an efficient algorithm to construct approximations for $\psi(u)$ when approximating $H(u)$ by the subclass of hyperexponential distributions.

## 3. Spectral Approximation for the Ruin Probability

The starting point for the approximation of $\psi(u)$ is its geometric compound tail representation in Equation (4). Note that this representation is similar to the Pollaczek–Khintchine formula for $\psi(u)$ in the Cramér–Lundberg model where $\phi$ is replaced by the average amount of claim per unit time $\rho < 1$ and the ladder height distribution is equal to the stationary excess claim size distribution. Therefore, following the reasoning in Vatamidou et al. (2014), we will approximate the ladder height distribution by a hyperexponential distribution (which has a rational Laplace transform), to construct approximations for the ruin probability.

## 3.1. Error Bound for the Ruin Probability

Let $\hat{H}(u)$ be an approximation of the ladder height distribution $H(u)$ and $\hat{\psi}(u)$ be the exact result we obtain from (4) when we use $\hat{H}(u)$. From Equation (4) and the triangle inequality, the error between the ruin probability and its approximation then is

$$|\psi(u) - \hat{\psi}(u)| \leq \sum_{n=1}^{\infty} (1-\phi)\phi^n \left|\overline{H}^{*n}(u) - \hat{\overline{H}}^{*n}(u)\right|. \tag{10}$$

If we define the sup norm distance between two distribution functions $F_1$ and $F_2$ as $\mathcal{D}(F_1, F_2) := \sup_x |F_1(x) - F_2(x)|$, $x \geq 0$ (also referred to as Kolmogorov metric), the following result holds.

**Theorem 1.** *A bound for the approximation error of the ruin probability is*

$$|\psi(u) - \hat{\psi}(u)| \leq \frac{\mathcal{D}(H, \hat{H})(1-\phi)\phi}{(1 - \phi H(u))(1 - \phi \hat{H}(u))}, \qquad \forall u > 0.$$

**Proof.** The result is a direct application of Theorem 4.1 of Peralta et al. (2018) by (i) choosing the functions $\hat{F}_1$ and $\hat{F}_2$ to be $H$ and $\hat{H}$, respectively; (ii) taking $\rho = \phi$; and (iii) recognising that $\sup_{y<u} \{|H(y) - \hat{H}(y)|\} \leq \mathcal{D}(H, \hat{H})$. □

**Remark 1.** *As $\lim_{u \to +\infty} H(u) = \lim_{u \to +\infty} \hat{H}(u) = 1$, it is immediately obvious that the bound converges to $\mathcal{D}(H, \hat{H})\phi/(1-\phi)$, which means that the bound is asymptotically uniform in $u$.*

To sum up, when the ladder height distribution is approximated with some desired accuracy, a bound for the ruin probability is guaranteed by Theorem 1. Although this result holds for any approximation $\hat{H}$ of $H$, we will in the sequel focus on hyperexponential approximations, as these lead to very tractable expressions and at the same time are sufficiently accurate for the purpose. Consequently, our next goal is to construct an algorithm to approximate the ladder height distribution by a hyperexponential distribution.

## 3.2. Completely Monotone Claim Sizes

We are mostly interested in evaluating ruin probabilities when the claim sizes follow a heavy-tailed distribution, such as Pareto or Weibull. These two distributions belong to the class of completely monotone distributions.

**Definition 1.** *A pdf $f$ is said to be completely monotone (c.m.) if all derivatives of $f$ exist and if*

$$(-1)^n f^{(n)}(u) \geq 0 \text{ for all } u > 0 \text{ and } n \geq 1.$$

Completely monotone distributions can be approximated arbitrarily closely by hyperexponentials; see, e.g., Feldmann and Whitt (1998). Here, we provide a method to approximate a completely monotone ladder height distribution with a hyperexponential one to achieve any desired accuracy for the ruin probability. The following result is standard; see Feller (1971).

**Theorem 2.** *A ccdf $\overline{F}$ is completely monotone if and only if it is the Laplace–Stieltjes transform of some probability distribution $S$ defined on the positive half-line, i.e.,*

$$\overline{F}(u) = \int_0^{\infty} e^{-yu} dS(y). \tag{11}$$

We call $S$ the spectral cdf.

**Remark 2.** *With a slight abuse of terminology, we will say that a function S is the spectral cdf of a distribution if it is the spectral cdf of its ccdf.*

Note that Theorem 2 also extends to the case where $S(y)$ is not a distribution but simply a finite measure on the positive half-line, i.e., a function $f$ is completely monotone if and only if it can be expressed as the Laplace–Stieltjes integral of such a finite measure $S(y)$. We will show that under the assumption that the claim size distribution is c.m. and the ladder height distribution is c.m. too. We first need the following intermediate result.

**Lemma 1.** *If the ccdf $\overline{F}_X(u)$ is c.m., then $\mathcal{T}_{\rho_n}\overline{F}_X(u)$ is a c.m. function, $\forall n = 1, \ldots, N$.*

**Proof.** Assume that the claim sizes are completely monotone, i.e., $\overline{F}_X(u) = \int_0^\infty e^{-uy} dS(y)$, for some spectral cdf $S(y)$. In this case, it holds that

$$\mathcal{T}_{\rho_n}\overline{F}_X(u) = \int_{t=0}^\infty e^{-\rho_n t}\overline{F}_X(t+u)dt = \int_{t=0}^\infty e^{-\rho_n t} \int_{y=0}^\infty e^{-(t+u)y} dS(y) dt$$

$$= \int_{y=0}^\infty e^{-uy} dS(y) \int_{t=0}^\infty e^{-(y+\rho_n)t} dt = \int_0^\infty e^{-uy} \frac{dS(y)}{y+\rho_n} = \int_0^\infty e^{-uy} dS_{\mathcal{T}_{\rho_n}}(y),$$

where $dS_{\mathcal{T}_{\rho_n}}(y) = \frac{dS(y)}{y+\rho_n}$, $n = 1, \ldots, N$, is a finite measure on the positive half-line with $\tilde{S}_{\mathcal{T}_{\rho_n}}(+\infty) = (1 - \tilde{f}_X(\rho_n))/\rho_n$, $n = 1, \ldots, N-1$, and $S_{\mathcal{T}_0}(+\infty) = \mathbb{E}X$. □

We can now state the following result.

**Proposition 1.** *If the ccdf $\overline{F}_X(u)$ is c.m., i.e., $\overline{F}_X(u) = \int_0^\infty e^{-uy} dS(y)$, for some spectral cdf $S(y)$, then the ladder height distribution is c.m. too, i.e., $\overline{H}(u) = \int_0^\infty e^{-uy} dS_H(y)$, where $S_H(y)$ is a spectral cdf such that*

$$dS_H(y) = \frac{1}{\phi c N} \sum_{n=1}^N \frac{\mu^* - c\rho_n \beta(-c\rho_n)}{(y+\rho_n)\prod_{\substack{k=1 \\ k \neq n}}^N (\rho_k - \rho_n)} dS(y).$$

**Proof.** It was proven in Chiu and Yin (2014) that the ascending ladder height distribution in the Sparre Andersen model is c.m. if the claim size distribution is c.m, meaning that $\overline{H}(u)$ can be represented as the Laplace–Stieltjes transform of some spectal cdf $S_H(y)$. Due to the uniqueness of Laplace transforms, it, therefore, suffices to find the formula of the spectral cdf $S_H(y)$ by applying Lemma 1 to (8). □

We show in the next section how to utilise the above results to construct approximations for the ruin probability $\psi(u)$ that have a guaranteed error bound given by Theorem 1.

### 3.3. Approximation Algorithm

Following the proof of Lemma 2 in Vatamidou et al. (2014), we can directly deduce the following result.

**Lemma 2.** *Let $S_H$ be the spectral cdf of the c.m. ladder height distribution $H$ and $\hat{S}_H$ a step function such that $\mathcal{D}(S_H, \hat{S}_H) \leq \epsilon$. Consequently, $\mathcal{D}(H, \hat{H}) \leq \epsilon$, where $\hat{H}$ is the c.m. approximate ladder height distribution with spectral cdf $\hat{S}_H$.*

The above lemma states that if we want to approximate a c.m. ladder height distribution with a hyperexponential one with some fixed accuracy $\epsilon$, it suffices to approximate its spectral cdf with a step function with the same accuracy. As pointed out in Remark 1 of Vatamidou et al. (2014), we could approximate $S_H$ with a step function having $k$ jumps that occur at the quantiles $\lambda_i$, such that $S_H(\lambda_i) = i/(k+1)$, $i = 1, \ldots, k$ and are all of size $1/k$ to achieve $\mathcal{D}(H, \hat{H}) \leq \epsilon = 1/(k+1)$.

Another possibility is to use the step function in Step 4d of our Algorithm 1; see also Figure 1 for a graphical representation of the approximate step function and its corresponding hyperexponential distribution. Clearly, this new step function leads to $\mathcal{D}(H, \hat{H}) \leq \epsilon = 1/2(k-1)$.

The error bound for the approximate ruin probability $\hat{\psi}(u)$ can be calculated afterwards through Theorem 1. An interesting question in this context is how many phases $k$ for the approximate ladder height distribution suffice to guarantee an error bound $|\psi(u) - \hat{\psi}(u)| \leq \delta$ for some predetermined $\delta > 0$. We answer this question in the next lemma.

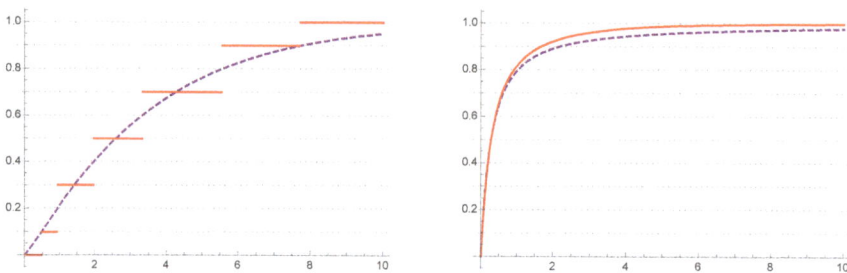

**Figure 1.** Approximating the ladder height distribution with a hyperexponential one with 6 phases to achieve accuracy $\epsilon = 0.1$, under Pareto(2,3) claim sizes. On the **left** graph, the purple dashed line corresponds to the spectral cdf $S_H$ and the red solid line to its approximate step function $\hat{S}_H$, whereas on the **right** graph we see $H$ and $\hat{H}$, respectively.

**Lemma 3.** *To achieve $|\psi(u) - \hat{\psi}(u)| \leq \delta$ for some predetermined $\delta > 0$, the ladder height distribution $H(u)$ must be approximated by a hyperexponential one with at least $k$ phases, such that*

$$k = k(u) = \left\lceil \min\left\{ \frac{\phi\left(1 - \phi + \delta(1 - \phi H(u))\right)}{2\delta(1 - \phi H(u))^2}, \frac{\phi}{2\delta(1 - \phi)} \right\} \right\rceil + 1, \qquad (12)$$

*where $\lceil x \rceil$ is the integer that is greater than or equal to x but smaller than $x + 1$.*

**Proof.** Observe that the error bound in Theorem 1 depends on the approximate hyperexponential distribution $\hat{H}(u)$, which means that one should first determine $\hat{H}(u)$ and then calculate the error bound. However, when $\mathcal{D}(H, \hat{H}) \leq \epsilon$, this translates to $H(u) - \epsilon \leq \hat{H}(u) \leq H(u) + \epsilon$. Therefore, the worst-case scenario for the bound is when $\hat{H}(u) = H(u) + \epsilon$ and consequently $\mathcal{D}(H, \hat{H}) = \epsilon$. As a result, if we want to achieve $|\psi(u) - \hat{\psi}(u)| \leq \delta$ for all possible scenarios of $\hat{H}(u)$, we should solve the inequality

$$\frac{\epsilon(1-\phi)\phi}{(1 - \phi H(u))(1 - \phi H(u) - \phi\epsilon)} \leq \delta,$$

with respect to $\epsilon$. By substituting $\epsilon = 1/2(k-1)$, we calculate

$$k \geq \frac{\phi\left(1 - \phi + \delta(1 - \phi H(u))\right)}{2\delta(1 - \phi H(u))^2} + 1.$$

In addition, the bound is asymptotically equal to $\epsilon\phi/(1-\phi)$ according to Remark 1. Consequently, it must also hold that

$$\frac{\epsilon\phi}{1-\phi} \leq \delta \quad \Rightarrow \quad k \geq \frac{\phi}{2\delta(1-\phi)} + 1.$$

Finally, as the number of phases $k$ must be an integer, the smallest possible integer that satisfies at least one of the inequalities is the one described in Equation (12). □

After this, we present our algorithm under the setting that we fix the desired accuracy $\delta$ for the approximation of the ruin probability $\hat{\psi}(u)$.

---
**Algorithm 1. Spectral Approximation**
**Steps:**

1. Calculate the roots $\rho_n$, $n = 1, \ldots, N-1$ using Equation (6).
2. Find the spectral cdf $S(y)$ of $\overline{F}_X(x)$.
3. Use Proposition 1 to calculate the spectral cdf $S_H(y)$ of $\overline{H}(u)$.
4. Approximate $\overline{H}(u)$ by a hyperexponential distribution with $k$ phases.

    (a) Choose the accuracy of the ruin probability $\delta$ for a fixed $u > 0$.
    (b) Calculate $k$ required to achieve this accuracy using Lemma 3 and set $\epsilon = \dfrac{1}{2(k-1)}$.
    (c) Define $k$ quantiles such that $S_H(\lambda_1) = \epsilon$, $S_H(\lambda_i) = 2(i-1)\epsilon$, $i = 2, \ldots, k-1$, and $S_H(\lambda_k) = 1 - \epsilon$.
    (d) Approximate the spectral cdf $S_H(y)$ with the step function

    $$\hat{S}_H(y) = \begin{cases} 0, & y \in [0, \lambda_1), \\ \epsilon, & y \in [\lambda_1, \lambda_2), \\ (2i-1)\epsilon, & y \in [\lambda_i, \lambda_{i+1}), \ i = 2, \ldots, k-1, \\ 1, & y \geq \lambda_k. \end{cases}$$

    (e) Find the ladder height distribution $\hat{\overline{H}}(u) = \left[ e^{-\lambda_1 u} + 2 \sum_{i=2}^{k-1} e^{-\lambda_i u} + e^{-\lambda_k u} \right] / 2(k-1)$ and calculate its Laplace transform $\mathcal{L}\left\{ \hat{\overline{H}}(u) \right\}(s) = \dfrac{1}{2(k-1)} \left[ \dfrac{1}{s+\lambda_1} + 2 \sum_{i=2}^{k-1} \dfrac{1}{s+\lambda_i} + \dfrac{1}{s+\lambda_k} \right]$.

5. Calculate the Laplace transform of the ruin probability as $\mathcal{L}\{\hat{\psi}(u)\}(s) = \dfrac{\phi \mathcal{L}\left\{ \hat{\overline{H}}(u) \right\}(s)}{\phi s \mathcal{L}\left\{ \hat{\overline{H}}(u) \right\}(s) + 1 - \phi}$.

6. Use simple fraction decomposition to determine positive real numbers $R_i$, $\eta_i$, $i = 1, \ldots, k$, with $\sum_{i=1}^k R_i = 1$, such that $\mathcal{L}\{\hat{\psi}(u)\}(s) = \phi \sum_{i=1}^k R_i \dfrac{1}{s + \eta_i}$.

7. Invert the previous Laplace transform to find $\hat{\psi}(u) = \phi \sum_{i=1}^k R_i e^{-\eta_i u}$, $u \geq 0$.

8. The accuracy for $\hat{\psi}(u)$ is then $\dfrac{\mathcal{D}(H, \hat{H})(1-\phi)\phi}{(1 - \phi H(u))(1 - \phi \hat{H}(u))}$, $\forall u > 0$.

---

**Remark 3.** The decomposition of $\mathcal{L}\{\hat{\psi}(u)\}(s)$ at Step 6 is guaranteed by Asmussen and Rolski (1992), who showed that the ruin probability in the Sparre Andersen model has a phase-type representation when the claim sizes are phase-type. Moreover, the particular hyperexponential representation of $\hat{\psi}(u)$ at Step 7 occurs because the poles of $\mathcal{L}\{\hat{\psi}(u)\}(s)$ are exactly the roots of the polynomial function $P_\phi(s) = \prod_{i=1}^k (s + \lambda_i) - \phi \Big( \prod_{i=1}^k (s + \lambda_i) - s \big( \prod_{i=1}^k (2 - \delta_{i1} - \delta_{ik})(s + \lambda_i) \big)' / 2(k-1) \Big)$, where $\delta_{ij}$ is the Kronecker delta. It is immediate from perturbation theory that $P_\phi(s)$ has exactly $k$ simple roots analytic in $\phi$; see Baumgärtel (1985) for details.

**Remark 4.** The above algorithm is an extension of the one developed for the Cramér–Lundberg model in Vatamidou et al. (2014), to which we refer for further details on technical implementation.

## 4. Asymptotic Approximation

In many cases, it is of importance to investigate the asymptotic behaviour of the ruin probability when the initial risk reserve tends to infinity. This question is particularly interesting in the case of heavy-tailed claim sizes. Towards this direction, when the claim sizes belong to the class of subexponential distributions $\mathcal{S}$ (Teugels 1975), e.g., Pareto, Weibull, Lognormal, etc., the following asymptotic approximation is classical (see, e.g., Embrechts and Veraverbeke 1982):

**Theorem 3.** *Suppose in the general Sparre Andersen model that the claim sizes and interclaim times have both finite means $\mathbb{E}X$ and $\mathbb{E}W$, respectively, such that $c\mathbb{E}W > \mathbb{E}X$. If $\frac{1}{\mathbb{E}X}\int_0^u \overline{F}_X(x)dx \in \mathcal{S}$, then*

$$\psi(u) \sim \psi_S(u) := \frac{1}{c\mathbb{E}W - \mathbb{E}X}\int_u^{+\infty} \overline{F}_X(x)dx, \quad \text{as } u \to +\infty.$$

Note that the heavy-tail approximation $\psi_S(u)$ holds for any interclaim time distribution. However, further modifications have been attained in Willmot (1999), when the Laplace transform of the interclaim times is a rational function of the form (5) with $\beta(s) = \beta$ and $F_X$ belongs to the subclass of regularly varying distributions, i.e., $\overline{F}_X(u) \sim L(u)u^{-\alpha-1}e^{-\gamma u}$, $u \to +\infty$, where $L(u)$ a slowly varying function and $\alpha > 0$, $\gamma \geq 0$. For example, the Pareto$(a,b)$ distribution (see Section 5.2.1) belongs to the class of regularly varying distributions with $L(u) = (b+1/u)^{-a}$, $\alpha = a-1$ and $\gamma = 0$, and its modified asymptotic approximation is then given by

$$\psi(u) \sim \psi_M(u) := \frac{L(u)u^{-\alpha}}{\alpha(c\mathbb{E}W - \mathbb{E}X)} = \frac{(1+bu)^{-a+1}}{(a-1)(b+\frac{1}{u})(c\mathbb{E}W - \mathbb{E}X)},$$

which is smaller than $\psi_S(u)$ by a factor $\frac{bu}{bu+1}$ that converges to 1 as $u \to +\infty$; see Willmot (1999) for details.

Clearly, the heavy-tail approximation admits a simple formula whenever the expectations of the interclaim times and claim sizes are finite; however, it has a drawback that occurs when $c\mathbb{E}W \approx \mathbb{E}X$ the approximation is useful only for extremely large values of $u$.

In the next section, we compare the accuracy of the spectral approximation to the accuracy of the heavy tail one, i.e., $\psi_S(u)$. An interesting observation is that the spectral approximation converges faster to zero than any heavy-tailed distribution due to the exponential decay rate of the former. Thus, the heavy-tail approximation is expected to outperform the spectral approximation in the far tail, but for medium values, this new approximation can be very competitive.

## 5. Numerical Analysis

The goal of this section is to implement our algorithm in order to check the accuracy of the spectral approximation and the tightness of its accompanying bound, which is given in Theorem 1. To perform the numerical examples, we need to make a selection for the distribution $K$ of the interclaim times as well as the claim size distribution $F_X$.

### 5.1. Interclaims Times

We choose a hyperexponential distribution with two phases, i.e., $K \sim H_2(\theta, 1-\theta; \nu_1, \nu_2)$, such that $\tilde{k}(s) = \frac{\nu_1\nu_2 + s(\theta\nu_1 + (1-\theta)\nu_2)}{(s+\nu_1)(s+\nu_2)}$. As $N = 2$, it is evident that there exists only one positive and real root $\rho_1$ to the generalised Lundberg equation of Equation (6). Therefore, given also that $\beta(s) = \theta\nu_1 + (1-\theta)\nu_2$, the ladder height distribution takes the form

$$\overline{H}(u) = \frac{1}{\phi c^2}\left(\frac{\nu_1\nu_2 - c\rho_1(\theta\nu_1 + (1-\theta)\nu_2)}{-\rho_1}T_{\rho_1}\overline{F}_X(u) + \frac{\nu_1\nu_2}{\rho_1}T_0\overline{F}_X(u)\right),$$

which is in accordance with Li and Garrido (2005).

## 5.2. Claim Sizes

For the claim sizes, we consider the Pareto$(a, b)$ distribution with shape parameter $a > 0$ and scale parameter $b > 0$ and the Weibull$(c, a)$ distribution with $c$ and $a$ positive shape and scale parameters, respectively.

### 5.2.1. Pareto

This distribution is c.m., as its ccdf $\bar{F}_X(x) = (1 + bu)^{-a}$ can be written as the LST of the Gamma distribution with shape and scale parameters $a$ and $b$, respectively, i.e.,

$$(1 + bu)^{-a} = \int_0^{+\infty} e^{-uy} \frac{y^{a-1}}{\Gamma(a) b^a} e^{-y/b} dy.$$

The $n$th moment of the Pareto distribution exists if and only if the shape parameter is greater than $n$. As we are interested in comparing the spectral approximation to the asymptotic approximation of Section 4, it is necessary to have a finite first moment for the claim sizes. Therefore, the shape parameter $a$ must be chosen to be greater than 1.

Using Proposition 1, we can easily verify that

$$dS_H(y) = \frac{1}{\phi c^2} \left( \frac{v_1 v_2 - c\rho_1(\theta v_1 + (1-\theta)v_2)}{-(y+\rho_1)\rho_1} + \frac{v_1 v_2}{y\rho_1} \right) \frac{y^{a-1}}{\Gamma(a) b^a} e^{-y/b} dy.$$

### 5.2.2. Weibull

It can be verified that the ccdf $\bar{F}_X(x) = e^{-(u/a)^c}$ with fixed shape parameter $c = 1/2$ arises as a c.m. distribution (Jewell 1982), where the mixing measure (measure of the spectral function) $S$ is given by

$$dS(y) = \frac{e^{-\frac{1}{4ay}}}{2\sqrt{a\pi y^3}} dy.$$

Similarly, we can find using Proposition 1 that

$$dS_H(y) = \frac{1}{\phi c^2} \left( \frac{v_1 v_2 - c\rho_1(\theta v_1 + (1-\theta)v_2)}{-(y+\rho_1)\rho_1} + \frac{v_1 v_2}{y\rho_1} \right) \frac{e^{-\frac{1}{4ay}}}{2\sqrt{a\pi y^3}} dy.$$

## 5.3. Numerical Results

The goal of this section is to implement our algorithm to check the accuracy of the spectral approximation and the tightness of its accompanying bound, which is given in Theorem 1.

For Pareto claim sizes, we choose $a = 2$, $b = 3$, $c = 1$, $\theta = 0.4$, $v_1 = 1$ and $v_2 = 5$, and we obtain $\mathbb{E}W = 0.52$, $\mathbb{E}X = 0.33$ and $\phi = 0.72897$. For Weibull claim sizes, we choose $a = 3$, $c = 1$, $\theta = 0.2$, $v_1 = 1$ and $v_2 = 1/9$, and we obtain $\mathbb{E}W = 7.4$, $\mathbb{E}X = 6$ and $\phi = 0.83184$. Note that we performed extensive numerical experiments for various combinations of parameters, but we chose to present only these two cases since the qualitative conclusions are comparable among all cases. Our experiments are illustrated below.

- Impact of phases. It is intuitively true that the spectral approximation becomes more accurate as the number of phases increases. To test this hypothesis, we compare three different spectral approximations with number of phases 10, 30 and 100, respectively, with the exact value of the ruin probability (which we obtain through simulation). We display our results in Table 1 only for Pareto claim sizes. The conclusion is that, indeed, a more accurate spectral approximation is achieved, as the number of phases increases for every fixed initial capital $u$, which is in line with expectations.

**Table 1.** The spectral approximation for different number of phases, under Pareto(2,3) claim sizes. The numbers in the brackets correspond to the confidence intervals of the exact ruin probability.

| $u$ | Simulation | sa 10 Phases | sa 30 Phases | sa 100 Phases |
|---|---|---|---|---|
| 0 | 0.72888 (±0.00016) | 0.72897 | 0.72897 | 0.72897 |
| 1 | 0.42859 (±0.00018) | 0.42505 | 0.42828 | 0.42859 |
| 2 | 0.30991 (±0.00017) | 0.29972 | 0.30877 | 0.30984 |
| 5 | 0.16095 (±0.00014) | 0.13236 | 0.15608 | 0.15996 |
| 10 | 0.08189 (±0.00010) | 0.04214 | 0.07216 | 0.08017 |
| 15 | 0.05240 (±0.00008) | 0.01463 | 0.03978 | 0.05025 |

- Quality of the bound. A compelling question regarding the bound is if it is strict or pessimistic, i.e., how far it is from the true error of the spectral approximation. To answer this question, we first need to determine the accuracy $\delta$ we would like to guarantee for the ruin probability. Using Lemma 3, we present, in Figure 2, the number of phases required in order to guarantee $\delta = 0.02$ under Pareto(2,3) claim sizes and $\delta = 0.05$ under Weibull(0.5, 3) claim sizes as a function of $u$. For $u = 30$, the required number of phases is equal to $k = 67$ in the Pareto case. Similarly, we find that $k = 11$ for $u = 17$ in the Weibull case. We generate the spectral approximations with 67 and 11 phases, respectively, and compare in Figure 3 the true error (difference between simulation and spectral approximation) with the predicted error bound of Theorem 1 (green dotted line). The dashed cyan line in the left graph represents the worst-case scenario for the bound that was used in the proof of Lemma 3 to calculate the optimal number of phases to guarantee an error of at most $\delta = 0.02$ up to $u = 30$.

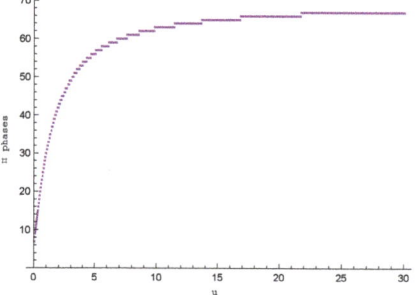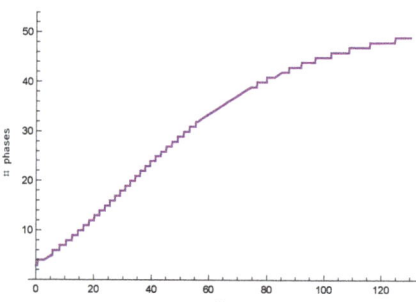

**Figure 2.** Number of phases required to guarantee for each initial capital $u$ an error bound (i) $\delta = 0.02$ under Pareto(2,3) claim sizes (**left graph**) and (ii) $\delta = 0.05$ under Weibull(0.5, 3) claim sizes (**right graph**).

As we can observe in Figure 3, the true error is significantly smaller than the predicted error bound for small values of $u$, under Pareto(2,3) claim sizes. This may be because, for small values of $u$, a smaller number of phases $k$ is enough to guarantee $\delta = 0.02$; see also Figure 2. Afterwards, the true error increases to the error bound by reaching its maximum value close to $u = 40$, and then drops to zero as $u \to \infty$, whereas the predicted bound remains constant. A similar behaviour is recognised under Weibull(0.5, 3) claim sizes, where now the true error is close to the predicted error bound for small values of $u$, as $k = 11$ is already a small number itself.

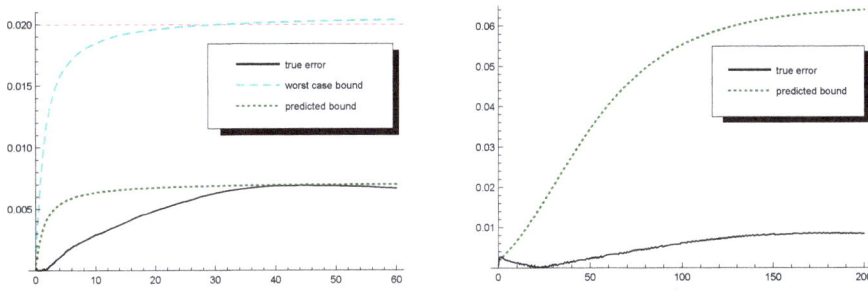

**Figure 3.** Comparison between the error bound and the true error, under Pareto(2, 3) (**left graph**) and Weibull(0.5, 3) (**right graph**) claim sizes. The dashed cyan line in the left graph corresponds to the worst-case scenario for the bound that was used to determine the number of phases in the spectral approximation that guarantee $\delta = 0.02$ up to $u = 30$.

Finally, notice that the predicted error bound is almost 4 times smaller than $\delta = 0.02$ in the Pareto case. This happens because $\mathcal{D}(H, \hat{H})$ could be a lot smaller than $\epsilon$; see also Figure 1 where $\mathcal{D}(H, \hat{H}) < 0.1$. However, most importantly, the true error is close to the predicted bound, and thus we can say that Lemma 3 provides a good proxy for the necessary number of phases $k$ to achieve it.

- Comparison between spectral and heavy-tail approximations. As we pointed out in Section 4, the spectral approximation is expected to underestimate both the exact ruin probability and the asymptotic approximation $\psi_S(u)$ in Theorem 3 for large $u$, due to its exponential decay rate. It is of interest to see the magnitude of $u$ for which the asymptotic approximation outperforms the spectral approximation.

We select the spectral approximations with $k = 67$ phases for Pareto(2, 3) claim sizes and $k = 11$ phases for Weibull(0.5, 3) claim sizes, as in the previous experiment, and present the distributions in a graph. The pink shadow in Figure 4 enfolding the spectral approximation represents its bound. We observe that for small values of $u$, the spectral approximation is more accurate than the heavy-tail approximation, where the second provides a rough estimate of the ruin probability. On the other hand, the heavy-tail approximation is slightly more accurate than the spectral approximation in the tail, i.e., for $u > 25$, under Pareto claim sizes. However, for the Weibull case, we observe that, even for values of $u$ around 300, the spectral approximation still outperforms the heavy-tail approximation.

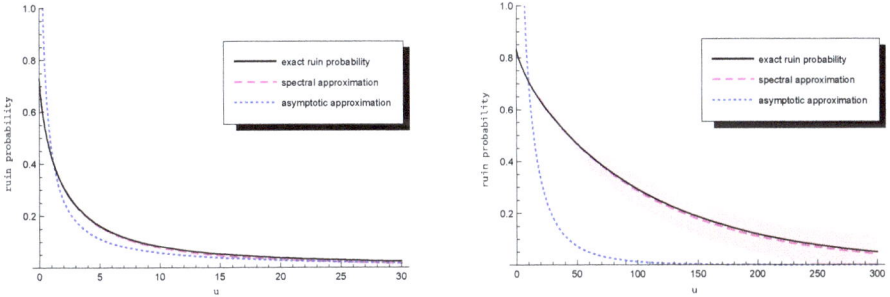

**Figure 4.** Comparison between the spectral approximation with $k = 67$ under Pareto(2, 3) claim sizes (**left graph**) and with $k = 11$ under Weibull(0.5, 3) claim sizes (**right graph**) and the heavy-tail approximation.

## 6. Conclusions

In this paper, we considered the ruin probability of the Sparre Andersen model with heavy-tailed claim sizes and interclaim times with rational Laplace transform. Using the geometric random sum representation, we developed an explicit bound and also constructed a spectral approximation by approximating the c.m. ladder height distribution with a hyperexponential one. Our spectral approximation algorithm advances on the algorithm established in Vatamidou et al. (2014) in various aspects. We provide below a summary of our conclusions both for the spectral approximation and the bound.

- When comparing with the technique proposed in Vatamidou et al. (2014), the strategic selection of the quantiles in Step 4d reduces the number of phases to almost a half, to guarantee a certain accuracy for the ladder height distribution.
- As the bound depends on the initial capital, we were able to focus on one area and optimise the required number of phases to achieve a desired accuracy, e.g., we would need 110 phases for $u = 5$ and 132 phases for $u = 30$ to guarantee accuracy of at most $\delta = 0.01$ in our example.
- The step function is constructed to guarantee $\mathcal{D}(H, \hat{H}) \leq \epsilon$, but in most applications $\mathcal{D}(H, \hat{H})$ is a lot smaller than $\epsilon$. Thus, the use of $\mathcal{D}(H, \hat{H})$ in the bound makes it tighter.

To sum up, the spectral approximation is highly accurate for all values of $u$ as opposed to the heavy-tail approximation, which fails to provide a good fit for small values. Moreover, it is accompanied by a rather tight bound.

Finally, note that the results of this paper are also valid for the risk model with two-sided jumps, i.e.,

$$\check{R}(t) = u + ct + \sum_{j=1}^{N_+(t)} Y_j - \sum_{i=1}^{N_-(t)} X_i, \quad t \geq 0, \tag{13}$$

where $u$, $c$ and $X_i$ are defined as before, whereas $N_+(t)$ and $N_-(t)$ are independent Poisson processes with intensities $\lambda_+$ and $\lambda_-$, respectively; see, e.g., Albrecher et al. (2010). In addition, the sequence $\{Y_j\}_{j \geq 1}$ of i.i.d. r.v.'s, independent of $\{X_i\}_{i \geq 1}$, $N_+(t)$ and $N_-(t)$, and having the common d.f. $G_Y$ that belongs to the class of distributions with rational Laplace transform, are the sizes of premium payments. The positive security loading condition in this model becomes $c + \lambda_+ \mathbb{E} Y > \lambda_- \mathbb{E} X$.

Let $\tau_n$ be the time when the $n$th claim occurs with $\tau_0 = 0$. As ruin occurs only at the epochs when claims occur, we define the discrete time process $\check{R} = \{\check{R}_n : n = 0, 1, 2, \dots\}$, where $\check{R}_0 = 0$ and $\check{R}_n = \check{R}(\tau_n)$, which denotes the surplus immediately after the $n$th claim, i.e.,

$$\check{R}_n = u + c\tau_n + \sum_{j=1}^{N_+(\tau_n)} Y_j - \sum_{i=1}^{n} X_i = u + c\check{\tau}_n - \sum_{i=1}^{n} X_i, \quad n = 0, 1, 2, \dots, \tag{14}$$

where $\check{\tau}_n = \tau_n + \sum_{j=1}^{N_+(\tau_n)} Y_j/c$ with $\check{\tau}_0 = 0$. Equation (14) corresponds to the discrete-time embedded process of the Sparre Andersen risk model (1), and the counting process $N(t)$ denotes the number of claims up to time $t$ with the modified interclaim times $W_i = \check{\tau}_j - \check{\tau}_{j-1}$. Clearly,

$$\check{k}(s) = \frac{\lambda_-}{\lambda_- + s + \lambda_+(1 - \check{g}_Y(s/c))}, \tag{15}$$

where $\check{g}_Y(s) = \int_0^{+\infty} e^{-sx} dG_Y(x)$ is the Laplace transform of the premium payments; see Dong and Liu (2013). Let now $\check{T} = \inf\{t \geq 0 \mid \check{R}(t) < 0\}$ and $\check{\psi}(u) = \mathbb{P}(\check{T} < \infty \mid \check{R}(0) = u)$. Obviously, $\check{\psi}(u) = \psi(u)$.

**Author Contributions:** Both authors contributed equally.

**Funding:** Financial support from the Swiss National Science Foundation Project 200021_168993 is gratefully acknowledged.

**Conflicts of Interest:** The authors declare no conflict of interest.

## References

Albrecher, Hansjörg, and Onno J. Boxma. 2005. On the discounted penalty function in a Markov-dependent risk model. *Insurance: Mathematics and Economics* 37: 650–72. [CrossRef]

Albrecher, Hansjörg, Corina Constantinescu, and Enrique Thomann. 2012. Asymptotic results for renewal risk models with risky investments. *Stochastic Processes and Their Applications* 122: 3767–89. [CrossRef]

Albrecher, Hansjörg, Hans U. Gerber, and Hailiang Yang. 2010. A direct approach to the discounted penalty function. *North American Actuarial Journal* 14: 420–34. [CrossRef]

Albrecher, Hansjörg, Christian Hipp, and Dominik Kortschak. 2010. Higher-order expansions for compound distributions and ruin probabilities with subexponential claims. *Scandinavian Actuarial Journal* 2010: 105–35. [CrossRef]

Asmussen, Søren, and Hansjörg Albrecher. 2010. *Ruin Probabilities*, 2nd ed. Advanced Series on Statistical Science & Applied Probability, 14. Singapore: World Scientific.

Asmussen, Søren, and Tomasz Rolski. 1992. Computational methods in risk theory: A matrix-algorithmic approach. *Insurance: Mathematics and Economics* 10: 259–74. [CrossRef]

Baumgärtel, Hellmut. 1985. *Analytic Perturbation Theory for Matrices and Operators*. Basel: Birkhäuser.

Chiu, Sung Nok, and Chuancun Yin. 2014. On the complete monotonicity of the compound geometric convolution with applications in risk theory. *Scandinavian Actuarial Journal* 2014: 116–24. [CrossRef]

Dickson, David C. M. 1998. On a class of renewal risk processes. *North American Actuarial Journal* 2: 60–68. [CrossRef]

Dickson, David C. M., and Christian Hipp. 2001. On the time to ruin for erlang (2) risk processes. *Insurance: Mathematics and Economics* 29: 333–44. [CrossRef]

Dong, Hua, and Zaiming Liu. 2013. The ruin problem in a renewal risk model with two-sided jumps. *Mathematical and Computer Modelling* 57: 800–11. [CrossRef]

Embrechts, Paul, Claudia Klüppelberg, and Thomas Mikosch. 1997. *Modelling Extremal Events: For Insurance and Finance, Volume 33 of Applications of Mathematics*. Berlin: Springer.

Embrechts, Paul, and Noël Veraverbeke. 1982. Estimates for the probability of ruin with special emphasis on the possibility of large claims. *Insurance: Mathematics and Economics* 1: 55–72. [CrossRef]

Feldmann, Anja, and Ward Whitt. 1998. Fitting mixtures of exponentials to long-tail distributions to analyze network performance models. *Performance Evaluation* 31: 245–79. [CrossRef]

Feller, William. 1971. *An Introduction to Probability Theory and Its Applications, Volume II*, 2nd ed. New York: John Wiley & Sons Inc.

Jewell, Nicholas P. 1982. Mixtures of exponential distributions. *The Annals of Statistics* 10: 479–84. [CrossRef]

Labbé, Chantal, and Kristina P. Sendova. 2009. The expected discounted penalty function under a risk model with stochastic income. *Applied Mathematics and Computation* 215: 1852–67. [CrossRef]

Landriault, David, and Gordon Willmot. 2008. On the gerber–shiu discounted penalty function in the sparre andersen model with an arbitrary interclaim time distribution. *Insurance: Mathematics and Economics* 42: 600–8. [CrossRef]

Li, Shuanming, and José Garrido. 2005. On a general class of renewal risk process: Analysis of the Gerber-Shiu function. *Advances in Applied Probability* 37: 836–56. [CrossRef]

Peralta, Oscar, Leonardo Rojas-Nandayapa, Wangyue Xie, and Hui Yao. 2018. Approximation of ruin probabilities via erlangized scale mixtures. *Insurance: Mathematics and Economics* 78: 136–56. [CrossRef]

Temnov, Gregory. 2004. Risk process with random income. *Journal of Mathematical Sciences* 123: 3780–94. [CrossRef]

Temnov, Gregory. 2014. Risk models with stochastic premium and ruin probability estimation. *Journal of Mathematical Sciences* 196: 84–96. [CrossRef]

Teugels, Jozef L. 1975. The class of subexponential distributions. *The Annals of Probability* 3: 1000–11. [CrossRef]

Vatamidou, Eleni, Ivo Jean Baptiste François Adan, Maria Vlasiou, and Bert Zwart. 2014. On the accuracy of phase-type approximations of heavy-tailed risk models. *Scandinavian Actuarial Journal* 2014: 510–34. [CrossRef]

Wei, Jiaqin, Rongming Wang, and Dingjun Yao. 2008. The asymptotic estimate of ruin probability under a class of risk model in the presence of heavy tails. *Communications in Statistics–Theory and Methods* 37: 2331–41. [CrossRef]

Willmot, Gordon E. 1999. A laplace transform representation in a class of renewal queueing and risk processes. *Journal of Applied Probability* 36: 570–84. [CrossRef]

Willmot, Gordon E. 2007. On the discounted penalty function in the renewal risk model with general interclaim times. *Insurance: Mathematics and Economics* 41: 17–31. [CrossRef]

Willmot, Gordon E., X. Sheldon Lin, and X. Sheldon Lin. 2001. *Lundberg Approximations for Compound Distributions with Insurance Applications*. New York: Springer, vol. 156.

Willmot, Gordon E., and Jae-Kyung Woo. 2017. *Surplus Analysis of Sparre Andersen Insurance Risk Processes*. Cham: Springer.

© 2019 by the authors. Licensee MDPI, Basel, Switzerland. This article is an open access article distributed under the terms and conditions of the Creative Commons Attribution (CC BY) license (http://creativecommons.org/licenses/by/4.0/).

Article

# Logarithmic Asymptotics for Probability of Component-Wise Ruin in a Two-Dimensional Brownian Model

Krzysztof Dębicki [1,†], Lanpeng Ji [2,*] and Tomasz Rolski [1,†]

1. Mathematical Institute, University of Wrocław, 50-137 Wrocław, Poland
2. School of Mathematics, University of Leeds, Woodhouse Lane, Leeds LS2 9JT, UK
* Correspondence: l.ji@leeds.ac.uk
† These authors contributed equally to this work.

Received: 14 June 2019; Accepted: 29 July 2019; Published: 1 August 2019

**Abstract:** We consider a two-dimensional ruin problem where the surplus process of business lines is modelled by a two-dimensional correlated Brownian motion with drift. We study the ruin function $P(u)$ for the component-wise ruin (that is both business lines are ruined in an infinite-time horizon), where $u$ is the same initial capital for each line. We measure the goodness of the business by analysing the adjustment coefficient, that is the limit of $-\ln P(u)/u$ as $u$ tends to infinity, which depends essentially on the correlation $\rho$ of the two surplus processes. In order to work out the adjustment coefficient we solve a two-layer optimization problem.

**Keywords:** adjustment coefficient; logarithmic asymptotics; quadratic programming problem; ruin probability; two-dimensional Brownian motion

## 1. Introduction

In classical risk theory, the surplus process of an insurance company is modelled by the compound Poisson risk model. For both applied and theoretical investigations, calculation of ruin probabilities for such model is of particular interest. In order to avoid technical calculations, *diffusion approximation* is often considered e.g., (Asmussen and Albrecher 2010; Grandell 1991; Iglehart 1969; Klugman et al. 2012), which results in tractable approximations for the interested finite-time or infinite-time ruin probabilities. The basic premise for the approximation is to let the number of claims grow in a unit time interval and to make the claim sizes smaller in such a way that the risk process converges to a Brownian motion with drift. Precisely, the Brownian motion risk process is defined by

$$R(t) = x + pt - \sigma B(t), \quad t \geq 0,$$

where $x > 0$ is the *initial capital*, $p > 0$ is the *net profit rate* and $\sigma B(t)$ models the net loss process with $\sigma > 0$ the volatility coefficient. Roughly speaking, $\sigma B(t)$ is an approximation of the total claim amount process by time $t$ minus its expectation, the latter is usually called the *pure premium* amount and calculated to cover the average payments of claims. The net profit, also called *safety loading*, is the component which protects the company from large deviations of claims from the average and also allows an accumulation of capital. Ruin related problems for Brownian models have been well studied; see, for example, Asmussen and Albrecher (2010); Gerber and Shiu (2004).

*Risks* **2019**, *7*, 83; doi:10.3390/risks7030083 www.mdpi.com/journal/risks

In recent years, multi-dimensional risk models have been introduced to model the surplus of multiple business lines of an insurance company or the suplus of collaborating companies (e.g., insurance and reinsurance). We refer to Asmussen and Albrecher (2010) [Chapter XIII 9] and Avram and Loke (2018); Avram and Minca (2017); Avram et al. (2008a, 2008b); Albrecher et al. (2017); Azcue and Muler (2018); Azcue et al. (2019); Foss et al. (2017); Ji and Robert (2018) for relevant recent discussions. It is concluded in the literature that in comparison with the well-understood 1-dimensional risk models, study of multi-dimensional risk models is much more challenging. It was shown recently in Delsing et al. (2019) that multi-dimensional Brownian model can serve as approximation of a multi-dimensional classical risk model in a Markovian environment. Therefore, obtained results for multi-dimensional Brownian model can serve as approximations of the multi-dimensional classical risk models in a Markovian environment; ruin probability approximation has been used in the aforementioned paper. Actually, multi-dimensional Brownian models have drawn a lot of attention due to its tractability and practical relevancy.

A $d$-dimensional Brownian model can be defined in a matrix form as

$$R(t) = x + pt - X(t), \quad t \geq 0, \quad \text{with } X(t) = AB(t),$$

where $x = (x_1, \cdots, x_d)^\top, p = (p_1, \cdots, p_d)^\top \in (0, \infty)^d$ are, respectively, (column) vectors representing the initial capital and net profit rate, $A \in \mathbb{R}^{d \times d}$ is a non-singular matrix modelling dependence between different business lines and $B(t) = (B_1(t), \ldots, B_d(t))^\top, t \geq 0$ is a standard $d$-dimensional Brownian motion (BM) with independent coordinates. Here $\top$ is the transpose sign. In what follows, vectors are understood as column vectors written in bold letters.

Different types of ruin can be considered in multi-dimensional models, which are relevant to the probability that the surplus of one or more of the business lines drops below zero in a certain time interval $[0, T]$ with $T$ either a finite constant or infinity. One of the commonly studied is the so-called *simultaneous ruin probability* defined as

$$Q_T(x) := \mathbb{P}\left\{\exists_{t \in [0,T]} \bigcap_{i=1}^{d} \{R_i(t) < 0\}\right\},$$

which is the probability that at a certain time $t \in [0, T]$ all the surpluses become negative. Here for $T < \infty$, $Q_T(x)$ is called finite-time simultaneous ruin probability, and $Q_\infty(x)$ is called infinite-time simultaneous ruin probability. Simultaneous ruin probability, which is essentially the hitting probability of $R(t)$ to the orthant $\{y \in \mathbb{R}^d : y_i < 0, i = 1, \ldots, d\}$, has been discussed for multi-dimensional Brownian models in different contexts; see Dębicki et al. (2018); Garbit and Raschel (2014). In Garbit and Raschel (2014), for fixed $x$ the asymptotic behaviour of $Q_T(x)$ as $T \to \infty$ has been discussed. Whereas, in Dębicki et al. (2018), the asymptotic behaviour, as $u \to \infty$, of the infinite-time ruin probability $Q_\infty(x)$, with $x = \alpha u = (\alpha_1 u, \alpha_2 u, \ldots, \alpha_d u)^\top, \alpha_i > 0, 1 \leq i \leq d$ has been obtained. Note that it is common in risk theory to derive the later type of asymptotic results for ruin probabilities; see, for example, Avram et al. (2008a); Embrechts et al. (1997); Mikosch (2008).

Another type of ruin probability is the *component-wise (or joint) ruin probability* defined as

$$P_T(x) := \mathbb{P}\left\{\bigcap_{i=1}^{d} \{\exists_{t \in [0,T]} R_i(t) < 0\}\right\} = \mathbb{P}\left\{\bigcap_{i=1}^{d} \{\sup_{t_i \in [0,T]} (X_i(t_i) - p_i t_i) > x_i\}\right\}, \tag{1}$$

which is the probability that all surpluses get below zero but possibly at different times. It is this possibility that makes the study of $P_T(x)$ more difficult.

The study of joint distribution of the extrema of multi-dimensional BM over finite-time interval has been proved to be important in quantitative finance; see, for example, He et al. (1998); Kou and Zhong (2016).

We refer to Delsing et al. (2019) for a comprehensive summary of related results. Due to the complexity of the problem, two-dimensional case has been the focus in the literature and for this case some explicit formulas can be obtained by using a PDE approach. Of particular relevance to the ruin probability $P_T(x)$ is a result derived in He et al. (1998) which shows that

$$\mathbb{P}\left\{\sup_{t\in[0,T]}(X_1(t)-p_1t)\leq x_1,\ \sup_{s\in[0,T]}(X_2(s)-p_2s)\leq x_2\right\}$$
$$=e^{a_1x_1+a_2x_2+bT}f(x_1,x_2,T),$$

where $a_1, a_2, b$ are known constants and $f$ is a function defined in terms of infinite-series, double-integral and Bessel function. Using the above formula one can derive an expression for $P_T(x)$ in two-dimensional case as follows

$$P_T(x) = 1 - \mathbb{P}\left\{\sup_{t\in[0,T]}(X_1(t)-p_1t)\leq x_1\right\} - \mathbb{P}\left\{\sup_{s\in[0,T]}(X_2(s)-p_2s)\leq x_2\right\} \quad (2)$$
$$+\mathbb{P}\left\{\sup_{t\in[0,T]}(X_1(t)-p_1t)\leq x_1,\ \sup_{s\in[0,T]}(X_2(s)-p_2s)\leq x_2\right\},$$

where the expression for the distribution of single supremum is also known; see He et al. (1998). Note that even though we have obtained explicit expression of $P_T(x)$ in (2) for the two-dimensional case, it seems difficult to derive the explicit form of the corresponding infinite-time ruin probability $P_\infty(x)$ by simply putting $T \to \infty$ in (2).

By assuming $x = \alpha u = (\alpha_1 u, \alpha_2 u, \ldots, \alpha_d u)^\top, \alpha_i > 0, 1 \leq i \leq d$, we aim to analyse the asymptotic behaviour of the infinite-time ruin probability $P_\infty(x)$ as $u \to \infty$. Applying Theorem 1 in Dębicki et al. (2010) we arrive at the following logarithmic asymptotics

$$-\frac{1}{u}\ln P_\infty(x) \sim \frac{1}{2}\inf_{t>0}\inf_{v\geq a+pt} v^\top \Sigma_t^{-1} v, \quad \text{as } u \to \infty \quad (3)$$

provided $\Sigma_t$ is non-singular, where $pt := (p_1t_1, \cdots, p_dt_d)^\top$, inequality of vectors are meant component-wise, and $\Sigma_t^{-1}$ is the inverse matrix of the covariance function $\Sigma_t$ of $(X_1(t_1), \cdots, X_d(t_d))$, with $t = (t_1, \cdots, t_d)^\top$ and $0 = (0, \cdots, 0)^\top \in \mathbb{R}^d$. Let us recall that conventionally for two given positive functions $f(\cdot)$ and $h(\cdot)$, we write $f(x) \sim h(x)$ if $\lim_{x\to\infty} f(x)/h(x) = 1$.

For more precise analysis on $P_\infty(x)$, it seems crucial to first solve the two-layer optimization problem in (3) and find the optimization points $t_0$. As it can be recognized in the following, when dealing with $d$-dimensional case with $d > 2$ the calculations become highly nontrivial and complicated. Therefore, in this contribution we only discuss a tractable two-dimensional model and aim for an explicit logarithmic asymptotics by solving the minimization problem in (3).

In the classical ruin theory when analysing the compound Poisson model or Sparre Andersen model, the so-called *adjustment coefficient* is used as a measure of goodness; see, for example, Asmussen and Albrecher (2010) or Rolski et al. (2009). It is of interest to obtain the solution of the minimization problem in (3) from a practical point of view, as it can be seen as an analogue of the adjustment coefficient and thus we could get some insights about the risk that the company is facing. As discussed in Asmussen and Albrecher (2010) and Li et al. (2007) it is also of interest to know how the dependence between different risks influences the joint ruin probability, which can be easily analysed through the obtained logarithmic asymptotics; see Remark 2.

The rest of this paper is organised as follows. In Section 2, we formulate the two-dimensional Brownian model and give the main results of this paper. The main lines of proof with auxiliary lemmas are displayed in Section 3. In Section 4 we conclude the paper. All technical proofs of the lemmas in Section 3 are presented in Appendix A.

## 2. Model Formulation and Main Results

Due to the fact that component-wise ruin probability $P_\infty(x)$ does not change under scaling, we can simply assume that the volatility coefficient for all business lines is equal to 1. Furthermore, noting that the timelines for different business lines should be distinguished as shown in (1) and (3), we introduce a two-parameter extension of correlated two-dimensional BM defined as

$$(X_1(t), X_2(s)) = \left(B_1(t), \rho B_1(s) + \sqrt{1-\rho^2} B_2(s)\right), \quad t, s \geq 0,$$

with $\rho \in (-1,1)$ and mutually independent Brownian motions $B_1, B_2$. We shall consider the following two dependent insurance risk processes

$$R_i(t) = u + \mu_i t - X_i(t), \quad t \geq 0, \quad i = 1, 2,$$

where $\mu_1, \mu_2 > 0$ are net profit rates, $u$ is the initial capital (which is assumed to be the same for both business lines, as otherwise, the calculations become rather complicated). We shall assume without loss of generality that $\mu_1 \leq \mu_2$. Here, $\mu_i$ is different from $p_i$ (see (1)) in the sense that it corresponds to the (scaled) model with volatility coefficient standardized to be 1.

In this contribution, we shall focus on the logarithmic asymptotics of

$$P(u) := P_\infty(u(1,1)^\top) = \mathbb{P}\{\{\exists_{t \geq 0} R_1(t) < 0\} \cap \{\exists_{s \geq 0} R_2(s) < 0\}\} \tag{4}$$

$$= \mathbb{P}\left\{\sup_{t \geq 0}(X_1(t) - \mu_1 t) > u, \sup_{s \geq 0}(X_2(s) - \mu_2 s) > u\right\}, \quad \text{as } u \to \infty.$$

Define

$$\hat{\rho}_1 = \frac{\mu_1 + \mu_2 - \sqrt{(\mu_1+\mu_2)^2 - 4\mu_1(\mu_2-\mu_1)}}{4\mu_1} \in [0, \tfrac{1}{2}), \quad \hat{\rho}_2 = \frac{\mu_1+\mu_2}{2\mu_2} \tag{5}$$

and let

$$t^* = t^*(\rho) = s^* = s^*(\rho) := \sqrt{\frac{2(1-\rho)}{\mu_1^2 + \mu_2^2 - 2\rho\mu_1\mu_2}}. \tag{6}$$

The following theorem constitutes the main result of this contribution.

**Theorem 1.** *For the joint infinite-time ruin probability (4) we have, as $u \to \infty$,*

$$-\frac{\log(P(u))}{u} \sim \begin{cases} 2(\mu_2 + (1-2\rho)\mu_1), & \text{if } -1 < \rho \leq \hat{\rho}_1; \\ \frac{\mu_1 + \mu_2 + 2/t^*}{1+\rho}, & \text{if } \hat{\rho}_1 < \rho < \hat{\rho}_2; \\ 2\mu_2, & \text{if } \hat{\rho}_2 \leq \rho < 1. \end{cases}$$

**Remark 2.** *(a) Following the classical one-dimensional risk theory we can call quantities on the right hand side in Theorem 1 as adjustment coefficients. They serve sometimes as a measure of goodness for a risk business.*

(b) One can easily check that adjustment coefficient as a function of $\rho$ is continuous, strictly decreasing on $(-1, \hat{\rho}_2]$ and it is constant, equal to $2\mu_2$ on $[\hat{\rho}_2, 1)$. This means that as the two lines of business becomes more positively correlated the risk of ruin becomes larger, which is consistent with the intuition.

Define
$$g(t,s) := \inf_{\substack{x \geq 1 + \mu_1 t \\ y \geq 1 + \mu_2 s}} (x,y) \Sigma_{ts}^{-1} (x,y)^\top, \quad t,s > 0, \tag{7}$$

where $\Sigma_{ts}^{-1}$ is the inverse matrix of $\Sigma_{ts} = \begin{pmatrix} t & \rho\, t \wedge s \\ \rho\, t \wedge s & s \end{pmatrix}$, with $t \wedge s = \min(t,s)$ and $\rho \in (-1, 1)$.

The proof of Theorem 1 follows from (3) which implies that the logarithmic asymptotics for $P(u)$ is of the form
$$-\frac{1}{u} \ln P(u) \sim \frac{g(t_0)}{2}, \quad u \to \infty, \tag{8}$$

where
$$g(t_0) = \inf_{(t,s) \in (0,\infty)^2} g(t,s), \tag{9}$$

and Proposition 3 below, wherein we list dominating points $t_0$ that optimize the function $g$ over $(0, \infty)^2$ and the corresponding optimal values $g(t_0)$.

In order to solve the two-layer minimization problem in (9) (see also (7)) we define for $t, s > 0$ the following functions:
$$g_1(t) = \frac{(1+\mu_1 t)^2}{t}, \quad g_2(s) = \frac{(1+\mu_2 s)^2}{s},$$
$$g_3(t,s) = (1+\mu_1 t, 1+\mu_2 s) \Sigma_{ts}^{-1} (1+\mu_1 t, 1+\mu_2 s)^\top.$$

Since $t \wedge s$ appears in the above formula, we shall consider a partition of the quadrant $(0, \infty)^2$, namely
$$(0,\infty)^2 = A \cup L \cup B, \quad A = \{s < t\}, \; L = \{s = t\}, \; B = \{s > t\}. \tag{10}$$

For convenience we denote $\overline{A} = \{s \leq t\} = A \cup L$ and $\overline{B} = \{s \geq t\} = B \cup L$. Hereafter, all sets are defined on $(0, \infty)^2$, so $(t,s) \in (0, \infty)^2$ will be omitted.

Note that $g_3(t,s)$ can be represented in the following form:
$$g_3(t,s) = \begin{cases} g_A(t,s) := \frac{(1+\mu_2 s)^2}{s} + \frac{((1+\mu_1 t) - \rho(1+\mu_2 s))^2}{t - \rho^2 s}, & \text{if } (t,s) \in \overline{A} \\ g_B(t,s) := \frac{(1+\mu_1 t)^2}{t} + \frac{((1+\mu_2 s) - \rho(1+\mu_1 t))^2}{s - \rho^2 t}, & \text{if } (t,s) \in \overline{B}. \end{cases} \tag{11}$$

Denote further
$$g_L(s) := g_A(s,s) = g_B(s,s) = \frac{(1+\mu_1 s)^2 + (1+\mu_2 s)^2 - 2\rho(1+\mu_1 s)(1+\mu_2 s)}{(1-\rho^2)s}, \quad s > 0. \tag{12}$$

In the next proposition we identify the so-called dominating points, that is, points $t_0$ for which function defined in (7) achieves its minimum. This identification might also be useful for deriving a more subtle asymptotics for $P(u)$.

**Notation:** In the following, in order to keep the notation consistent, $\rho \leq \mu_1/\mu_2$ is understood as $\rho < 1$ if $\mu_1 = \mu_2$.

**Proposition 3.**

(i) Suppose that $-1 < \rho < 0$.
For $\mu_1 < \mu_2$ we have

$$g(t_0) = g_A(t_A, s_A) = 4(\mu_2 + (1-2\rho)\mu_1),$$

where, $(t_A, s_A) = (t_A(\rho), s_A(\rho)) := \left(\frac{1-2\rho}{\mu_1}, \frac{1}{\mu_2 - 2\mu_1\rho}\right)$ is the unique minimizer of $g(t,s)$, $(t,s) \in (0,\infty)^2$.
For $\mu_1 = \mu_2 =: \mu$ we have

$$g(t_0) = g_A(t_A, s_A) = g_B(t_B, s_B) = 8(1-\rho)\mu,$$

where $(t_A, s_A) = \left(\frac{1-2\rho}{\mu}, \frac{1}{(1-2\rho)\mu}\right) \in A$, $(t_B, s_B) := \left(\frac{1}{(1-2\rho)\mu}, \frac{1-2\rho}{\mu}\right) \in B$ are the only two minimizers of $g(t,s)$, $(t,s) \in (0,\infty)^2$.

(ii) Suppose that $0 \leq \rho < \hat{\rho}_1$. We have

$$g(t_0) = g_A(t_A, s_A) = 4(\mu_2 + (1-2\rho)\mu_1),$$

where $(t_A, s_A) \in A$ is the unique minimizer of $g(t,s)$, $(t,s) \in (0,\infty)^2$.

(iii) Suppose that $\rho = \hat{\rho}_1$. We have

$$g(t_0) = g_A(t_A, s_A) = 4(\mu_2 + (1-2\rho)\mu_1),$$

where $(t_A, s_A) = (t_A(\hat{\rho}_1), s_A(\hat{\rho}_1)) = (t^*(\hat{\rho}_1), s^*(\hat{\rho}_1)) \in L$, is the unique minimizer of $g(t,s)$, $(t,s) \in (0,\infty)^2$, with $(t^*, s^*)$ defined in (6).

(iv) Suppose that $\hat{\rho}_1 < \rho < \hat{\rho}_2$. We have

$$g(t_0) = g_A(t^*, s^*) = g_L(t^*) = \frac{2}{1+\rho}(\mu_1 + \mu_2 + 2/t^*),$$

where $(t^*, s^*) \in L$ is the unique minimizer of $g(t,s)$, $(t,s) \in (0,\infty)^2$.

(v) Suppose that $\rho = \hat{\rho}_2$. We have $t^*(\hat{\rho}_2) = s^*(\hat{\rho}_2) = 1/\mu_2$ and

$$g(t_0) = g_A(1/\mu_2, 1/\mu_2) = g_L(1/\mu_2) = g_2(1/\mu_2) = 4\mu_2,$$

where the minimum of $g(t,s)$, $(t,s) \in (0,\infty)^2$ is attained at $(1/\mu_2, 1/\mu_2)$, with $g_3(1/\mu_2, 1/\mu_2) = g_2(1/\mu_2)$ and $1/\mu_2$ is the unique minimizer of $g_2(s)$, $s \in (0,\infty)$.

(vi) Suppose that $\hat{\rho}_2 < \rho < 1$. We have

$$g(t_0) = g_2(1/\mu_2) = 4\mu_2,$$

where the minimum of $g(t,s)$, $(t,s) \in (0,\infty)^2$ is attained when $g(t,s) = g_2(s)$.

**Remark 4.** In case that $\mu_1 = \mu_2$, we have $\hat{\rho}_1 = 0, \hat{\rho}_2 = 1$ and thus scenarios (ii) and (vi) do not apply.

## 3. Proofs of Main Results

As discussed in the previous section, Proposition 3 combined with (8), straightforwardly implies the thesis of Theorem 1. In what follows, we shall focus on the proof of Proposition 3, for which we need to find the dominating points $t_0$ by solving the two-layer minimization problem (9).

The solution of quadratic programming problem of the form (7) (inner minimization problem of (9)) has been well understood; for example, Hashorva (2005); Hashorva and Hüsler (2002) (see also Lemma 2.1 of Dębicki et al. (2018)). For completeness and for reference, we present below Lemma 2.1 of Dębicki et al. (2018) for the case where $d = 2$.

We introduce some more notation. If $I \subset \{1,2\}$, then for a vector $a \in \mathbb{R}^2$ we denote by $a_I = (a_i, i \in I)$ a sub-block vector of $a$. Similarly, if further $J \subset \{1,2\}$, for a matrix $M = (m_{ij})_{i,j \in \{1,2\}} \in \mathbb{R}^{2 \times 2}$ we denote by $M_{IJ} = M_{I,J} = (m_{ij})_{i \in I, j \in J}$ the sub-block matrix of $M$ determined by $I$ and $J$. Further, write $M_{II}^{-1} = (M_{II})^{-1}$ for the inverse matrix of $M_{II}$ whenever it exists.

**Lemma 5.** *Let $M \in \mathbb{R}^{2 \times 2}$ be a positive definite matrix. If $b \in \mathbb{R}^2 \setminus (-\infty, 0]^2$, then the quadratic programming problem*

$$P_M(b) : \text{Minimise } x^\top M^{-1} x \text{ under the linear constraint } x \geq b$$

*has a unique solution $\widetilde{b}$ and there exists a unique non-empty index set $\widetilde{I} \subseteq \{1,2\}$ such that*

$$\widetilde{b}_I = b_I \neq 0_I, \quad M_{II}^{-1} b_I > 0_I,$$
$$\text{and} \quad \text{if } I^c = \{1,2\} \setminus I \neq \emptyset, \text{ then } \widetilde{b}_{I^c} = M_{I^c I} M_{II}^{-1} b_I \geq b_{I^c}.$$

*Furthermore,*

$$\min_{x \geq b} x^\top M^{-1} x = \widetilde{b}^\top M^{-1} \widetilde{b} = b_I^\top M_{II}^{-1} b_I > 0,$$
$$x^\top M^{-1} \widetilde{b} = x_I^\top M_{II}^{-1} \widetilde{b}_I = x_I^\top M_{II}^{-1} b_I, \quad \forall x \in \mathbb{R}^2.$$

For the solution of the quadratic programming problem (7) a suitable representation for $g(t,s)$ is worked out in the following lemma.

For $1 > \rho > \mu_1/\mu_2$, let $D_2 = \{(t,s) : w_1(s) \leq t \leq f_1(s)\}$ and $D_1 = (0, \infty)^2 \setminus D_2$, with boundary functions given by

$$f_1(s) = \frac{\rho - 1}{\mu_1} + \frac{\rho \mu_2}{\mu_1} s, \quad w_1(s) = \frac{s}{\rho + (\rho \mu_2 - \mu_1)s}, \quad s \geq 0, \qquad (13)$$

and the unique intersection point of $f_1(s), w_1(s), s \geq 0$, given by

$$s_1^* = s_1^*(\rho) := \frac{1 - \rho}{\rho \mu_2 - \mu_1}, \qquad (14)$$

as depicted in Figure 1.

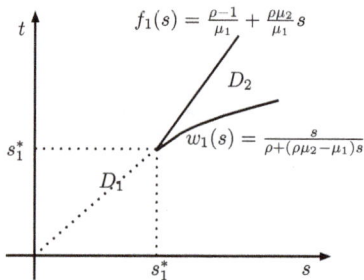

**Figure 1.** Partition of $(0, \infty)^2$ into $D_1, D_2$.

**Lemma 6.** *Let $g(t,s), t, s > 0$ be given as in (7). We have:*

(i) *If $-1 < \rho \leq \mu_1/\mu_2$, then*
$$g(t,s) = g_3(t,s), \quad (t,s) \in (0, \infty)^2.$$

(ii) *If $1 > \rho > \mu_1/\mu_2$, then*
$$g(t,s) = \begin{cases} g_3(t,s), & \text{if } (t,s) \in D_1 \\ g_2(s), & \text{if } (t,s) \in D_2. \end{cases}$$

*Moreover, we have $g_3(f_1(s), s) = g_3(w_1(s), s) = g_2(s)$ for all $s \geq s_1^*$.*

### 3.1. Proof of Proposition 3

We shall discuss in order the case when $-1 < \rho < 0$ and the case when $0 \leq \rho < 1$ in the following two subsections. In both scenarios we shall first derive the minimizers of the function $g(t,s)$ on regions $\overline{A}$ and $\overline{B}$ (see (10)) separately and then look for a global minimizer by comparing the two minimum values. For clarity some scenarios are analysed in forms of lemmas.

3.1.1. Case $-1 < \rho < 0$

By Lemma 6, we have that
$$g(t,s) = g_3(t,s), \quad (t,s) \in (0, \infty)^2.$$

We shall derive the minimizers of $g_3(t,s)$ on $\overline{A}, \overline{B}$ separately.

Minimizers of $g_3(t,s)$ on $\overline{A}$. We have, for any fixed $s$,
$$\frac{\partial g_3(t,s)}{\partial t} = \frac{\partial g_A(t,s)}{\partial t} = 0 \Leftrightarrow (\mu_1 t + 1 - \rho - \rho \mu_2 s)(\mu_1 t - (2\mu_1 \rho^2 - \rho \mu_2)s + \rho - 1) = 0,$$

where the representation (11) is used. Two roots of the above equation are:

$$t_1 = t_1(s) := \frac{\rho - 1 + \rho \mu_2 s}{\mu_1}, \quad t_2 = t_2(s) := \frac{1 - \rho + (2\mu_1 \rho^2 - \rho \mu_2)s}{\mu_1}. \tag{15}$$

Note that, due to the form of the function $g_A(t,s)$ given in (11), for any fixed $s$, there exists a unique minimizer of $g_A(t,s)$ on $\overline{A}$ which is either an inner point $t_1$ or $t_2$ (the one that is larger than $s$), or a boundary point $s$. Next, we check if any of $t_i, i = 1, 2$, is larger than $s$. Since $\rho < 0$, $t_1 < 0 < t_2$. So we check if $t_2 > s$. It can be shown that

$$t_2 > s \Leftrightarrow (\mu_1 + \rho\mu_2 - 2\mu_1\rho^2)s < 1 - \rho. \tag{16}$$

Two scenarios $\mu_1 + \rho\mu_2 - 2\mu_1\rho^2 \leq 0$ and $\mu_1 + \rho\mu_2 - 2\mu_1\rho^2 > 0$ will be distinguished.
Scenario $\mu_1 + \rho\mu_2 - 2\mu_1\rho^2 \leq 0$. We have from (16) that

$$t_1 < 0 < s < t_2,$$

and thus

$$\inf_{(t,s)\in\overline{A}} g_3(t,s) = \inf_{s>0} f_A(s),$$

where

$$f_A(s) := g_A(t_2(s), s) = \frac{(1 + \mu_2 s)^2}{s} + 4\mu_1((1 - \rho) + (\rho^2\mu_1 - \rho\mu_2)s).$$

Next, since

$$f_A'(s) = 0 \Leftrightarrow s_A = s_A(\rho) := \frac{1}{|\mu_2 - 2\rho\mu_1|} = \frac{1}{\mu_2 - 2\rho\mu_1} > 0, \tag{17}$$

the unique minimizer of $g_3(t,s)$ on $\overline{A}$ is given by $(t_A, s_A)$ with

$$t_A := t_2(s_A) = \frac{1 - 2\rho}{\mu_1}.$$

Scenario $\mu_1 + \rho\mu_2 - 2\mu_1\rho^2 > 0$. We have from (16) that

$$t_1 < 0 < s < t_2 \Leftrightarrow s < \frac{1 - \rho}{\mu_1 + \rho\mu_2 - 2\mu_1\rho^2} = \frac{1 - \rho}{\rho(\mu_2 - \mu_1\rho) + \mu_1(1 - \rho^2)} =: s^{**}(\rho) = s^{**}, \tag{18}$$

and in this case,

$$\inf_{(t,s)\in\overline{A}} g_3(t,s) = \min\left(\inf_{0<s<s^{**}} f_A(s), \inf_{s\geq s^{**}} g_L(s)\right), \tag{19}$$

where $g_L(s)$ is given in (12). Note that

$$g_L'(s) = 0 \Leftrightarrow s^* = s^*(\rho) = \sqrt{\frac{2(1 - \rho)}{\mu_1^2 + \mu_2^2 - 2\rho\mu_1\mu_2}}. \tag{20}$$

Next, for $-1 < \rho < 0$ we have that (recall $s^{**}$ given in (18))

$$s^{**} \geq \frac{1 - \rho}{\mu_1(1 - \rho^2)} > \frac{1}{\mu_1} \geq \frac{1}{\mu_2} > s_A, \quad s^{**} > \frac{1 - \rho}{\mu_1} > s^*.$$

Therefore, by (19) we conclude that the unique minimizer of $g_3(t,s)$ on $\overline{A}$ is again given by $(t_A, s_A)$. Consequently, for all $-1 < \rho < 0$, we have that the unique minimizer of $g_3(t,s)$ on $\overline{A}$ is given by $(t_A, s_A)$, and

$$\inf_{(t,s)\in \overline{A}} g_3(t,s) = g_A(t_A, s_A) = 4(\mu_2 + (1-2\rho)\mu_1). \tag{21}$$

Minimizers of $g_3(t,s)$ on $\overline{B}$. Similarly, we have, for any fixed $t$,

$$\frac{\partial g_3(t,s)}{\partial s} = \frac{\partial g_B(t,s)}{\partial s} = 0 \Leftrightarrow (\mu_2 s + 1 - \rho - \rho\mu_1 t)(\mu_2 s - (2\mu_2 \rho^2 - \rho\mu_1)t + \rho - 1) = 0.$$

Two roots of the above equation are:

$$s_1 = s_1(t) := \frac{\rho - 1 + \rho\mu_1 t}{\mu_2}, \quad s_2 = s_2(t) := \frac{1 - \rho + (2\mu_2\rho^2 - \rho\mu_1)t}{\mu_2}. \tag{22}$$

Next, we check if any of $s_i, i = 1, 2$, is greater than $t$. Again $s_1 < 0 < s_2$ as $\rho < 0$. So we check if $s_2 > t$. It can be shown that

$$s_2 > t \Leftrightarrow (\mu_2 + \rho\mu_1 - 2\mu_2\rho^2)t < 1 - \rho. \tag{23}$$

Thus, for Scenario $\mu_2 + \rho\mu_1 - 2\mu_2\rho^2 \le 0$ we have that

$$s_1 < 0 < t < s_2$$

and in this case

$$\inf_{(t,s)\in \overline{B}} g_3(t,s) = \inf_{t>0} f_B(t),$$

with

$$f_B(t) := g_B(t, s_2(t)) = \frac{(1+\mu_1 t)^2}{t} + 4\mu_2((1-\rho) + (\rho^2\mu_2 - \rho\mu_1)t).$$

Next, note that

$$f'_B(t) = 0 \Leftrightarrow t_B = t_B(\rho) := \frac{1}{|\mu_1 - 2\rho\mu_2|} = \frac{1}{\mu_1 - 2\rho\mu_2} > 0. \tag{24}$$

Therefore, the unique minimizer of $g_3(t,s)$ on $\overline{B}$ is given by $(t_B, s_B)$ with

$$s_B := s_2(t_B) = \frac{1 - 2\rho}{\mu_2}, \quad \inf_{(t,s)\in \overline{B}} g_3(t,s) = g_B(t_B, s_B) = 4(\mu_1 + (1-2\rho)\mu_2).$$

For Scenario $\mu_2 + \rho\mu_1 - 2\mu_2\rho^2 > 0$ we have from (23) that

$$s_1 < 0 < t < s_2 \Leftrightarrow t < \frac{1-\rho}{\mu_2 + \rho\mu_1 - 2\mu_2\rho^2} = \frac{1-\rho}{\rho(\mu_1 - \rho\mu_2) + \mu_2(1-\rho^2)} =: t^{**}(\rho) = t^{**}. \tag{25}$$

In this case,

$$\inf_{(t,s)\in \overline{B}} g_3(t,s) = \min\left(\inf_{0<t<t^{**}} f_B(t), \inf_{t\geq t^{**}} g_L(t)\right).$$

Though it is not easy to determine explicitly the optimizer, we can conclude that the minimizer should be taken at $(t_B, s_B)$, $(t^*, t^*)$ or $(t^{**}, t^{**})$, where $t^* = t^*(\rho) = s^*(\rho)$. Further, we have from the discussion in (19) that

$$g_A(t_A, s_A) < g_L(s^*) = g_L(t^*) = \min(g_L(t^*), g_L(t^{**})),$$

and

$$g_B(t_B, s_B) = 4(\mu_1 + (1-2\rho)\mu_2) \geq 4(\mu_2 + (1-2\rho)\mu_1) = g_A(t_A, s_A).$$

Combining the above discussions on $\overline{A}, \overline{B}$, we conclude that Proposition 3 holds for $-1 < \rho < 0$.

3.1.2. Case $0 \leq \rho < 1$

We shall derive the minimizers of $g(t,s)$ on $\overline{A}, \overline{B}$ separately. We start with discussions on $\overline{B}$, for which we give the following lemma. Recall $t^*(\rho) = s^*(\rho)$ defined in (20) (see also (6)), $t_B(\rho)$ defined in (24), $t^{**}(\rho)$ defined in (25) and $s_1^*(\rho)$ defined in (14) for $\mu_1/\mu_2 < \rho < 1$. Note that where it applies, $1/0$ is understood as $+\infty$ and $1/\infty$ is understood as 0.

**Lemma 7.** *We have:*

(a) *The function $t^*(\rho)$ is a decreasing function on $[0,1]$ and both $t_B(\rho)$ and $s_1^*(\rho)$ are decreasing functions on $(\mu_1/\mu_2, 1)$.*
(b) *The function $t^{**}(\rho)$ decreases from $1/\mu_2$ at $\rho = 0$ to some positive value and then increases to $1/\mu_2$ at $\hat{\rho}_2$ (defined in (5)) and then increases to $+\infty$ at the root $\hat{\rho} \in (0,1]$ of the equation $\mu_2 + \rho\mu_1 - 2\mu_2\rho^2 = 0$.*
(c) *For $0 \leq \rho \leq \mu_1/\mu_2$, we have*

$$t_B(\rho) \geq t^{**}(\rho), \quad t^*(\rho) \geq t^{**}(\rho),$$

*where both equalities hold only when $\rho = 0$ and $\mu_1 = \mu_2$.*
(d) *It holds that*

$$t^*(\hat{\rho}_2) = t_B(\hat{\rho}_2) = s_1^*(\hat{\rho}_2) = t^{**}(\hat{\rho}_2) = \frac{1}{\mu_2}. \tag{26}$$

*Moreover, for $\mu_1/\mu_2 < \rho < 1$ we have*

(i) $t^*(\rho) < s_1^*(\rho)$ for all $\rho \in (\mu_1/\mu_2, \hat{\rho}_2)$, $t^*(\rho) > s_1^*(\rho)$ for all $\rho \in (\hat{\rho}_2, 1)$.
(ii) $t_B(\rho) < s_1^*(\rho)$ for all $\rho \in (\mu_1/\mu_2, \hat{\rho}_2)$, $t_B(\rho) > s_1^*(\rho)$ for all $\rho \in (\hat{\rho}_2, 1)$.
(iii) $t^{**}(\rho) < s_1^*(\rho)$ for all $\rho \in (\mu_1/\mu_2, \hat{\rho}_2)$, $t^{**}(\rho) > s_1^*(\rho)$ for all $\rho \in (\hat{\rho}_2, \hat{\rho})$.
(iv) $t^{**}(\rho) < t^*(\rho)$ for all $\rho \in (\mu_1/\mu_2, \hat{\rho}_2)$, $t^{**}(\rho) > t^*(\rho)$ for all $\rho \in (\hat{\rho}_2, \hat{\rho})$.
(v) $t^{**}(\rho) < t_B(\rho)$ for all $\rho \in (\mu_1/\mu_2, \hat{\rho}_2)$, $t^{**}(\rho) > t_B(\rho)$ for all $\rho \in (\hat{\rho}_2, \hat{\rho})$.

Recall that by definition $g_L(s) = g_A(s,s) = g_B(s,s), s > 0$ (cf. (12)). For the minimum of $g(t,s)$ on $\overline{B}$ we have the following lemma.

**Lemma 8.** *We have*

(i) *If* $0 \leq \rho < \hat{\rho}_2$, *then*

$$\inf_{(t,s) \in \overline{B}} g(t,s) = g_L(t^*) = \frac{2}{1+\rho}(\mu_1 + \mu_2 + 2/t^*),$$

*where* $(t^*, t^*)$ *is the unique minimizer of* $g(t,s)$ *on* $\overline{B}$.

(ii) *If* $\rho = \hat{\rho}_2$, *then* $t^*(\hat{\rho}_2) = s^*(\hat{\rho}_2) = 1/\mu_2$ *and*

$$\inf_{(t,s) \in \overline{B}} g(t,s) = g_L(1/\mu_2) = g_2(1/\mu_2) = 4\mu_2,$$

*where the minimum of* $g(t,s)$ *on* $\overline{B}$ *is attained at* $(1/\mu_2, 1/\mu_2)$, *with* $g_3(1/\mu_2, 1/\mu_2) = g_2(1/\mu_2)$ *and* $1/\mu_2$ *is the unique minimizer of* $g_2(s), s \in (0, \infty)$.

(iii) *If* $\hat{\rho}_2 < \rho < 1$, *then*

$$\inf_{(t,s) \in \overline{B}} g(t,s) = \inf_{(t,s) \in D_2} g_2(s) = g_2(1/\mu_2) = 4\mu_2,$$

*where the minimum of* $g(t,s)$ *on* $\overline{B}$ *is attained when* $g(t,s) = g_2(s)$ *on* $D_2$ *(see Figure 1).*

Next we consider the minimum of $g(t,s)$ on $\overline{A}$. Recall $s^*(\rho)$ defined in (20), $s_A(\rho)$ defined in (17) and $s^{**}(\rho)$ defined in (18). We first give the following lemma.

**Lemma 9.** *We have*

(a) *Both* $s^*(\rho)$ *and* $s^{**}(\rho)$ *are decreasing functions on* $[0,1]$.
(b) *That* $\hat{\rho}_1$ *is the unique point on* $[0,1)$ *such that*

$$s_A(\hat{\rho}_1) = s^{**}(\hat{\rho}_1) = s^*(\hat{\rho}_1),$$

*and*

(i) $s_A(\rho) < s^{**}(\rho)$ *for all* $\rho \in [0, \hat{\rho}_1)$, $s_A(\rho) > s^{**}(\rho)$ *for all* $\rho \in (\hat{\rho}_1, 1)$,
(ii) $s^*(\rho) < s^{**}(\rho)$ *for all* $\rho \in [0, \hat{\rho}_1)$, $s^*(\rho) > s^{**}(\rho)$ *for all* $\rho \in (\hat{\rho}_1, 1)$.

(c) *For all* $\mu_1/\mu_2 < \rho < 1$, *it holds that* $s^{**}(\rho) < s_1^*(\rho)$.

For the minimum of $g(t,s)$ on $\overline{A}$ we have the following lemma.

**Lemma 10.** *We have*

(i) *If* $0 \leq \rho < \hat{\rho}_1$, *then*

$$\inf_{(t,s) \in \overline{A}} g(t,s) = g_A(t_A, s_A) = 4(\mu_2 + (1 - 2\rho)\mu_1),$$

*where* $(t_A, s_A) \in A$ *is the unique minimizer of* $g(t,s)$ *on* $\overline{A}$.

(ii) *If* $\rho = \hat{\rho}_1$, *then*

$$\inf_{(t,s) \in \overline{A}} g(t,s) = g_A(t_A, s_A) = 4(\mu_2 + (1 - 2\rho)\mu_1),$$

*where* $(t_A, s_A) = (t^*, s^*) \in L$ *is the unique minimizer of* $g(t,s)$ *on* $\overline{A}$.

(iii) If $\hat{\rho}_1 < \rho < \hat{\rho}_2$, then

$$\inf_{(t,s)\in \overline{A}} g(t,s) = g_L(s^*) = \frac{2}{1+\rho}(\mu_1 + \mu_2 + 2/s^*),$$

where $(s^*, s^*)$ is the unique minimizer of $g(t,s)$ on $\overline{A}$.

(iv) If $\rho = \hat{\rho}_2$, then $t^*(\hat{\rho}_2) = s^*(\hat{\rho}_2) = 1/\mu_2$ and

$$\inf_{(t,s)\in \overline{A}} g(t,s) = g_L(s^*) = g_2(1/\mu_2) = 4\mu_2,$$

where the minimum of $g(t,s)$ on $\overline{A}$ is attained at $(1/\mu_2, 1/\mu_2)$, with $g_3(1/\mu_2, 1/\mu_2) = g_2(1/\mu_2)$.

(v) If $\hat{\rho}_2 < \rho < 1$, then

$$\inf_{(t,s)\in \overline{A}} g(t,s) = g_2(1/\mu_2) = 4\mu_2,$$

where the minimum of $g(t,s)$ on $\overline{A}$ is attained when $g(t,s) = g_2(s)$ on $D_2$ (see Figure 1).

Consequently, combining the results in Lemma 8 and Lemma 10, we conclude that Proposition 3 holds for $0 \leq \rho < 1$. Thus, the proof is complete.

## 4. Conclusions and Discussions

In the multi-dimensional risk theory, the so-called "ruin" can be defined in different manner. Motivated by diffusion approximation approach, in this paper we modelled the risk process via a multi-dimensional BM with drift. We analyzed the component-wise infinite-time ruin probability for dimension $d = 2$ by solving a two-layer optimization problem, which by the use of Theorem 1 from Dębicki et al. (2010) led to the logarithmic asymptotics for $P(u)$ as $u \to \infty$, given by explicit form of the adjustment coefficient $\gamma = g(t_0)/2$ (see (8)). An important tool here is Lemma 5 on the quadratic programming, cited from Hashorva (2005). In this way we were also able to identify the dominating points by careful analysis of different regimes for $\rho$ and specify three regimes with different formulas for $\gamma$ (see Theorem 1). An open and difficult problem is the derivation of exact asymptotics for $P(u)$ in (4), for which the problem of finding dominating points would be the first step. A refined double-sum method as in Dębicki et al. (2018) might be suitable for this purpose. A detailed analysis of the case for dimensions $d > 2$ seems to be technically very complicated, even for getting the logarithmic asymptotics. We also note that a more natural problem of considering $R_i(t) = \alpha_i u + \mu_i t - X_i(t)$, with general $\alpha_i > 0, i = 1, 2$, leads to much more difficult technicalities with the analysis of $\gamma$.

Define the ruin time of component $i$, $1 \leq i \leq d$, by $T_i = \min\{t : R_i(t) < 0\}$ and let $T_{(1)} \leq T_{(2)} \leq \ldots \leq T_{(d)}$ be the order statistics of ruin times. Then the component-wise infinite-time ruin probability is equivalent to $\mathbb{P}\{T_{(d)} < \infty\}$ while the ruin time of at least one business line is $T_{\min} = T_{(1)} = \min_i T_i$. Other interesting problems like $\mathbb{P}\{T_{(j)} < \infty\}$ have not yet been analysed. For instance, it would be interesting for $d = 3$ to study the case $T_{(2)}$. The general scheme on how to obtain logarithmic asymptotics for such problems was discussed in Dębicki et al. (2010).

Random vector $\tilde{X} = (\sup_{t\geq 0}(X_1(t) - p_1 t), \ldots, \sup_{t\geq 0}(X_d(t) - p_d t))^\top$ has exponential marginals and if it is not concentrated on a subspace of dimension less than $d$, it defines a multi-variate exponential distribution. In this paper for dimension $d = 2$, we derived some asymptotic properties of such distribution. Little is known about properties of this multi-variate distriution and more studies on it would be of interest. For example a correlation structure of $\tilde{X}$ is unknown. In particular, in the context of findings presented in this contribution, it would be interesting to find the correlation between $\sup_{t\geq 0}(X_1(t) - \mu_1 t)$ and $\sup_{t\geq 0}(X_2(t) - \mu_2 t)$.

**Author Contributions:** Investigation, K.D., L.J., T.R.; writing–original draft preparation, L.J.; writing–review and editing, K.D., T.R.

**Funding:** T.R. & K.D. acknowledge partial support by NCN grant number 2015/17/B/ST1/01102 (2016-2019).

**Acknowledgments:** We are thankful to the referees for their carefully reading and constructive suggestions which significantly improved the manuscript.

**Conflicts of Interest:** The authors declare no conflict of interest. The funders had no role in the design of the study; in the collection, analyses, or interpretation of data; in the writing of the manuscript, or in the decision to publish the results.

**Abbreviations**

The following abbreviations are used in this manuscript:

BM   Brownian motion

## Appendix A

In this appendix, we present the proofs of the lemmas used in Section 3.

**Proof of Lemma 6.** Referring to Lemma 5, we have, for any fixed $t, s$, there exists a unique index set

$$I(t,s) \subseteq \{1,2\}$$

such that

$$g(t,s) = (1+\mu_1 t, 1+\mu_2 s)_{I(t,s)} (\Sigma_{ts})^{-1}_{I(t,s),I(t,s)} (1+\mu_1 t, 1+\mu_2 s)^\top_{I(t,s)}, \tag{A1}$$

and

$$(\Sigma_{ts})^{-1}_{I(t,s),I(t,s)} (1+\mu_1 t, 1+\mu_2 s)^\top_{I(t,s)} > \mathbf{0}_{I(t,s)}. \tag{A2}$$

Since $I(t,s) = \{1\}, \{2\}$ or $\{1,2\}$, we have that

(S1) On the set $E_1 = \{(t,s) : \rho\, t \wedge s\, s^{-1}(1+\mu_2 s) \geq (1+\mu_1 t)\}$, $g(t,s) = g_2(s)$
(S2) On the set $E_2 = \{(t,s) : \rho\, t \wedge s\, t^{-1}(1+\mu_1 t) \geq (1+\mu_2 s)\}$, $g(t,s) = g_1(t)$
(S3) On the set $E_3 = (0,\infty)^2 \setminus (E_1 \cup E_2)$, $g(t,s) = g_3(t,s)$.

Clearly, if $\rho \leq 0$ then

$$E_1 = E_2 = \emptyset, \quad E_3 = (0,\infty)^2.$$

In this case,

$$g(t,s) = g_3(t,s), \quad (t,s) \in (0,\infty)^2.$$

Next, we focus on the case where $\rho > 0$. We consider the regions $\overline{A}$ and $B$ separately.
Analysis on $\overline{A}$. We have

$$A_1 = \overline{A} \cap E_1 = \{s \leq t \leq f_1(s)\}, \quad f_1(s) = \frac{\rho - 1}{\mu_1} + \frac{\rho \mu_2}{\mu_1} s,$$

$$A_2 = \overline{A} \cap E_2 = \{s \leq t \leq f_2(s)\}, \quad f_2(s) = \frac{\rho s}{1 + (\mu_2 - \rho \mu_1)s},$$

$$A_3 = \overline{A} \cap E_3 = \{t \geq s, t > \max(f_1(s), f_2(s))\}.$$

Next we analyse the intersection situation of the functions $f(s) = s, f_1(s), f_2(s)$ on region $\overline{A}$.

Clearly, for any $s > 0$ we have $f_2(s) < s$. Furthermore, $f_1(s) = f_2(s)$ has a unique positive solution $s_1$ given by

$$s_1 = \frac{1-\rho}{\rho(\mu_2 - \rho\mu_1)}.$$

Finally, for $\rho\mu_2 \leq \mu_1$ we have that $f_1(s)$ does not intersect with $f(s)$ on $(0,\infty)$ but for $\rho\mu_2 > \mu_1$ the unique intersection point is given by $s_1^* > s_1$ (cf. (14)). To conclude, we have, for $\rho \leq \mu_1/\mu_2$,

$$g(t,s) = g_3(t,s), \quad (t,s) \in \overline{A},$$

and for $\rho > \mu_1/\mu_2$,

$$g(t,s) = \begin{cases} g_3(t,s), & \text{if } (t,s) \in \overline{A} \cap \{t \geq \max(s, f_1(s)), t > f_1(s)\} \\ g_2(s), & \text{if } (t,s) \in \overline{A} \cap \{s \leq t \leq f_1(s)\}. \end{cases}$$

Additionally, we have from Lemma 5 $g_3(f_1(s), s) = g_2(s)$ for all $s \geq s_1^*$.

Analysis on $B$. The two scenarios $\rho \leq \mu_1/\mu_2$ and $\rho > \mu_1/\mu_2$ will be considered separately. For $\rho \leq \mu_1/\mu_2$, we have

$$B_1 = B \cap E_1 = \{t < s \leq h_1(t)\}, \quad h_1(t) = \frac{\rho t}{1 + (\mu_1 - \rho\mu_2)t},$$

$$B_2 = B \cap E_2 = \{t < s \leq h_2(t)\}, \quad h_2(t) = \frac{\rho - 1}{\mu_2} + \frac{\rho\mu_1}{\mu_2}t,$$

$$B_3 = B \cap E_3 = \{s > \max(t, h_1(t), h_2(t))\}.$$

It is easy to check that

$$t > h_1(t), \quad t > h_2(t), \quad \forall t > 0,$$

and thus

$$g(t,s) = g_3(t,s), \quad (t,s) \in B.$$

For $\rho > \mu_1/\mu_2$, we have

$$B_1 = B \cap E_1 = \{w_1(s) \leq t < s\}, \quad w_1(s) = \frac{s}{\rho + (\rho\mu_2 - \mu_1)s},$$

$$B_2 = B \cap E_2 = \{w_2(s) \leq t < s\}, \quad w_2(s) = \frac{1-\rho}{\mu_1\rho} + \frac{\mu_2}{\mu_1\rho}s,$$

$$B_3 = B \cap E_3 = \{t < \min(s, w_1(s), w_2(s))\}.$$

Next we analyze the intersection situation of the functions $w(s) = s, w_1(s), w_2(s)$ on region $B$.

Clearly, for any $s > 0$, $w_2(s) > s$. $w_1(s)$ and $w_2(s)$ do not intersect on $(0, \infty)$. $w(s)$ and $w_1(s)$ has a unique intersection point $s_1^*$ (cf. (14)).

To conclude, we have, for $\rho \leq \mu_1/\mu_2$,

$$g(t,s) = g_3(t,s), \quad (t,s) \in B,$$

and for $\rho > \mu_1/\mu_2$,

$$g(t,s) = \begin{cases} g_3(t,s), & \text{if } (t,s) \in B \cap \{t < \min(s, w_1(s))\} \\ g_2(s), & \text{if } (t,s) \in B \cap \{w_1(s) \leq t < s\}. \end{cases}$$

Additionally, it follows from Lemma 5 that $g_3(w_1(s), s) = g_2(s)$ for all $s \geq s_1^*$. Consequently, the claim follows by a combination of the above results. This completes the proof. □

**Proof of Lemma 7.** (a) The claim for $t^*(\rho)$ follows by noting its following representation:

$$t^*(\rho) = s^*(\rho) = \sqrt{\frac{2(1-\rho)}{\mu_1^2 + \mu_2^2 - 2\mu_1\mu_2 + 2\mu_1\mu_2 - 2\rho\mu_1\mu_2}} = \sqrt{\frac{2}{\frac{(\mu_1-\mu_2)^2}{1-\rho} + 2\mu_1\mu_2}}.$$

The claims for $t_B(\rho)$ and $s_1^*(\rho)$ follow directly from their definition.
(b) First note that

$$t^{**}(0) = t^{**}(\hat{\rho}_2) = \frac{1}{\mu_2}.$$

Next it is calculated that

$$\frac{\partial t^{**}(\rho)}{\partial \rho} = \frac{-2\mu_2\rho^2 + 4\mu_2\rho - \mu_1 - \mu_2}{(\mu_2 + \rho\mu_1 - 2\mu_2\rho^2)^2}.$$

Thus, the claim of (b) follows by analysing the sign of $\frac{\partial t^{**}(\rho)}{\partial \rho}$ over $(0, \hat{\rho})$.
(c) For any $0 \leq \rho \leq \mu_1/\mu_2$ we have $|\mu_1 - 2\rho\mu_2| \leq \mu_1$ and thus

$$t_B(\rho) \geq \frac{1}{\mu_1} \geq \frac{1}{\mu_2} \geq \frac{1-\rho}{\mu_2(1-\rho^2)} \geq \frac{1-\rho}{\rho(\mu_1 - \rho\mu_2) + \mu_2(1-\rho^2)} = t^{**}(\rho).$$

Further, since

$$\mu_1^2 + \mu_2^2 - 2\rho\mu_1\mu_2 = \mu_1(\mu_1 - \rho\mu_2) + \mu_2(\mu_2 - \rho\mu_1) \leq \mu_2(\mu_1 - \rho\mu_2) + \mu_2(\mu_2 - \rho\mu_1) \leq 2\mu_2^2(1-\rho),$$

it follows that

$$t^*(\rho) \geq \frac{1}{\mu_2} \geq t^{**}(\rho).$$

(d) It is easy to check that (26) holds. For (i) we have

$$t^*(\rho) - s_1^*(\rho) = (1-\rho)\left(\frac{1}{f_1(\rho)} - \frac{1}{f_2(\rho)}\right),$$

where

$$f_1(\rho) = \sqrt{\frac{(1-\rho)(\mu_1^2 + \mu_2^2 - 2\rho\mu_1\mu_2)}{2}} = \sqrt{\mu_1\mu_2\rho^2 - \frac{(\mu_1+\mu_2)^2}{2}\rho + \frac{\mu_1^2 + \mu_2^2}{2}}$$
$$f_2(\rho) = \rho\mu_2 - \mu_1.$$

Analysing the properties of the above two functions, we have $f_1(\rho)$ is strictly decreasing on $[0,1]$ with

$$f_1(0) = \sqrt{\frac{\mu_1^2 + \mu_2^2}{2}} > -\mu_1 = f_2(0), \quad f_1(1) = 0 \le \mu_2 - \mu_1 = f_2(1),$$

and thus there is a unique intersection point of the two curves $t^*(\rho)$ and $s_1^*(\rho)$ which is $\rho = \hat{\rho}_2$. Therefore, the claim of (i) follows. Similarly, the claim of (ii) follows since

$$t_B(\rho) - s_1^*(\rho) = \frac{-(\mu_1 + \mu_2)\rho + 2\mu_2\rho^2}{(\rho\mu_2 - \mu_1)(2\rho\mu_2 - \mu_1)}.$$

Finally, the claims of (iii), (iv) and (v) follow easily from (a), (b) and (26). This completes the proof. □

**Proof of Lemma 8.** Consider first the case where $0 \le \rho \le \mu_1/\mu_2$. Recall (22). We check if any of $s_i, i = 1, 2$, is greater than $t$. Clearly, $s_1 \le t$. Next, we check whether $s_2 > t$. It is easy to check that

$$s_2 > t \iff t < t^{**},$$

where (recall (25))

$$t^{**} = t^{**}(\rho) = \frac{1-\rho}{\rho(\mu_1 - \mu_2\rho) + \mu_2(1-\rho^2)} > 0.$$

Then

$$\inf_{(t,s) \in \overline{B}} g_3(t,s) = \min\left(\inf_{0<t<t^{**}} g_B(t, s_2(t)), \inf_{t \ge t^{**}} g_B(t,t)\right).$$

Consequently, it follows from (c) of Lemma 7 the claim of (i) holds for $0 \le \rho \le \mu_1/\mu_2$.

Next, we consider $\mu_1/\mu_2 < \rho < 1$. Recall the function $w_1(s)$ defined in (13). Denote the inverse function of $w_1(s)$ by

$$\hat{w}_1(t) = \frac{\rho t}{1 - (\rho\mu_2 - \mu_1)t}, \quad t \ge s_1^*.$$

We have from Lemma 6 that

$$g_B(t, \hat{w}_1(t)) = g_2(t), \quad t \ge s_1^*.$$

Further note that $1/\mu_2$ is the unique minimizer of $g_2(s), s > 0$. For $\mu_1/\mu_2 < \rho < \hat{\rho}_2$, we have from (d) in Lemma 7 that

$$\inf_{s_1^* \le s} g_2(s) = g_2(s_1^*) = g_L(s_1^*) > g_L(t^*),$$

and further

$$\inf_{(t,s) \in \overline{B}} g(t,s) = \min\Big(\inf_{0<t<t^{**}} g_B(t, s_2(t)), \inf_{t^{**} \le t < s_1^*} g_B(t,t), \inf_{s_1^* \le t} g_B(t, \hat{w}_1(t)), \inf_{s_1^* \le s} g_2(s)\Big)$$

$$= g_B(t^*, t^*) = g_L(t^*),$$

where $(t^*, t^*)$ is the unique minimizer of $g(t,s)$ on $\overline{B}$. Therefore, the claim for $\mu_1/\mu_2 < \rho < \hat{\rho}_2$ is established.

For $\rho = \hat{\rho}_2$, because of (26) we have

$$\inf_{(t,s)\in \overline{B}} g(t,s) = \min(\inf_{0<t<1/\mu_2} g_B(t,s_2(t)), \inf_{1/\mu_2 \le t} g_B(t,\hat{w}_1(t)), \inf_{1/\mu_2 \le s} g_2(s))$$
$$= g_B(1/\mu_2, 1/\mu_2) = g_L(1/\mu_2) = g_2(1/\mu_2),$$

and the unique minimum of $g(t,s)$ on $\overline{B}$ is attained at $(1/\mu_2, 1/\mu_2)$. Moreover, for all $\hat{\rho}_2 < \rho < 1$ we have

$$s_2(t_B) = \hat{w}_1(t_B) = \frac{1}{\mu_2} > s_1^*.$$

Thus,

$$\inf_{(t,s)\in \overline{B}} g(t,s) = \min(\inf_{0<t<t_B} g_B(t,s_2(t)), \inf_{t_B \le t} g_B(t,\hat{w}_1(t)), \inf_{s_1^* \le s} g_2(s))$$
$$= g_B(t_B, 1/\mu_2) = g_2(1/\mu_2),$$

and the unique minimum of $g(t,s)$ on $\overline{B}$ is attained when $g(t,s) = g_2(s)$ on $D_2$. This completes the proof. □

**Proof of Lemma 9.** (a) The claim for $s^*(\rho)$ has been shown in the proof of (a) in Lemma 7. Next, we show the claim for $s^{**}(\rho)$, for which it is sufficient to show that $\frac{\partial s^{**}(\rho)}{\partial \rho} < 0$ for all $\rho \in [0,1]$. In fact, we have

$$\frac{\partial s^{**}(\rho)}{\partial \rho} = \frac{-2\mu_1 \rho^2 + 4\mu_1 \rho - \mu_1 - \mu_2}{(\mu_1 + \rho\mu_2 - 2\mu_1 \rho^2)^2} < 0.$$

(b) In order to prove (i), the following two scenarios will be discussed separately:

$$(S1). \quad \mu_2 < 2\mu_1; \quad (S2). \quad \mu_2 \ge 2\mu_1.$$

First consider (S1). If $0 \le \rho < \frac{\mu_2}{2\mu_1}$, then

$$s_A(\rho) - s^{**}(\rho) = \frac{(\mu_1 + \rho\mu_2 - 2\mu_1 \rho^2) - (1-\rho)(\mu_2 - 2\rho\mu_1)}{(\mu_2 - 2\rho\mu_1)(\mu_1 + \rho\mu_2 - 2\mu_1 \rho^2)}$$
$$= \frac{f(\rho)}{(\mu_2 - 2\rho\mu_1)(\mu_1 + \rho\mu_2 - 2\mu_1 \rho^2)},$$

where

$$f(\rho) = -4\mu_1 \rho^2 + 2(\mu_2 + \mu_1)\rho - \mu_2 + \mu_1.$$

Analysing the function $f$, we conclude that

$$f(\rho) < 0, \text{ for } \rho \in [0, \hat{\rho}_1), \quad f(\rho) > 0, \text{ for } \rho \in (\hat{\rho}_1, \frac{\mu_2}{2\mu_1}).$$

Further, for $\frac{\mu_2}{2\mu_1} \le \rho < 1$ we have

$$s_A(\rho) - s^{**}(\rho) = \frac{\mu_1 + \mu_2 - 2\mu_1 \rho}{(2\rho\mu_1 - \mu_2)(\mu_1 + \rho\mu_2 - 2\mu_1 \rho^2)} > 0.$$

Thus, the claim in (i) is established for (S1). Similarly, the claim in (i) is valid for (S2). Next, note that

$$s^*(\rho) - s^{**}(\rho) = (1-\rho)\left(\frac{1}{f_1(\rho)} - \frac{1}{f_2(\rho)}\right)$$

with

$$f_1(\rho) = \sqrt{\frac{(1-\rho)(\mu_1^2 + \mu_2^2 - 2\rho\mu_1\mu_2)}{2}} = \sqrt{\mu_1\mu_2\rho^2 - \frac{(\mu_1+\mu_2)^2}{2}\rho + \frac{\mu_1^2+\mu_2^2}{2}}$$

$$f_2(\rho) = \mu_1 + \rho\mu_2 - 2\mu_1\rho^2.$$

Analysing the properties of the above two functions, we have $f_1(\rho)$ is strictly decreasing on $[0,1]$ with

$$f_1(0) = \sqrt{\frac{\mu_1^2 + \mu_2^2}{2}} \geq \mu_1 = f_2(0), \quad f_1(1) = 0 \leq \mu_2 - \mu_1 = f_2(1),$$

and thus there is a unique intersection point $\rho \in (0,1)$ of $s^*(\rho)$ and $s^{**}(\rho)$. It seems not clear at the moment whether this unique point is $\hat{\rho}_1$ or not, since we have to solve a polynomial equation of order 4. Instead, it is sufficient to show that

$$s_A(\hat{\rho}_1) = s^*(\hat{\rho}_1). \tag{A3}$$

In fact, basic calculations show that the above is equivalent to

$$(2\mu_1\hat{\rho}_1 - (u_1 + \mu_2))f(\hat{\rho}_1) = 0$$

which is valid due to the fact that $f(\hat{\rho}_1) = 0$. Finally, the claim in (c) follows since

$$\rho\mu_2 - \mu_1 < \mu_1 + \rho\mu_2 - 2\rho^2\mu_1.$$

This completes the proof. □

**Proof of Lemma 10.** Two cases $\hat{\rho}_1 \leq \mu_1/\mu_2$ and $\hat{\rho}_1 > \mu_1/\mu_2$ should be distinguished. Since the proofs for these two cases are similar, we give below only the proof for the more complicated case $\hat{\rho}_1 \leq \mu_1/\mu_2$.
Note that, for $0 \leq \rho \leq \mu_1/\mu_2$, as in (19),

$$\inf_{(t,s)\in\overline{A}} g(t,s) = \inf_{(t,s)\in\overline{A}} g_3(t,s) = \min\left(\inf_{0<s<s^{**}} f_A(s), \inf_{s\geq s^{**}} g_L(s)\right),$$

and thus the claim for $0 \leq \rho \leq \mu_1/\mu_2$ follows directly from (i)–(ii) of (b) in Lemma 9.
Next, we consider the case $\mu_1/\mu_2 < \rho < \hat{\rho}_2$ (note here $\hat{\rho}_1 < \mu_1/\mu_2 < \rho$). We have by (i) of (d) in Lemma 7 and (i)–(ii) of (b) in Lemma 9 that

$$s^{**}(\rho) < s^*(\rho) = t^*(\rho) < s_1^*(\rho), \quad s_1^*(\rho) > \frac{1}{\mu_2}, \quad s_A(\rho) > s^{**}(\rho).$$

Thus, it follows from Lemma 6 that

$$\inf_{(t,s)\in \overline{A}} g(t,s) = \min\left(\inf_{0<s<s^{**}} g_A(t_2(s),s),\ \inf_{s^{**}\leq s\leq s_1^*} g_A(s,s),\ \inf_{s_1^*<s} g_A(f_1(s),s),\ \inf_{s_1^*<s} g_2(s)\right)$$
$$= g_A(t^*,s^*) = g_L(s^*),$$

and $(t^*,s^*) \in L$ is the unique minimizer of $g(t,s)$ on $\overline{A}$. Here we used the fact that

$$\inf_{s_1^*<s} g_A(f_1(s),s) = \inf_{s_1^*<s} g_2(s) = g_A(f_1(s_1^*),s_1^*) = g_2(s_1^*) > g_L(s^*).$$

Next, if $\rho = \hat{\rho}_2$, then

$$s_1^*(\hat{\rho}_2) = s^*(\hat{\rho}_2) = \frac{1}{\mu_2},$$

and thus

$$\inf_{(t,s)\in \overline{A}} g(t,s) = \min\left(\inf_{0<s<s^{**}} g_A(t_2(s),s),\ \inf_{s^{**}\leq s\leq 1/\mu_2} g_A(s,s),\ \inf_{1/\mu_2<s} g_A(f_1(s),s),\ \inf_{1/\mu_2<s} g_2(s)\right)$$
$$= g_A(1/\mu_2,1/\mu_2) = g_L(1/\mu_2) = g_2(1/\mu_2).$$

Furthermore, the unique minimum of $g(t,s)$ on $A$ is attained at $(1/\mu_2,1/\mu_2)$, with $g_3(1/\mu_2,1/\mu_2) = g_2(1/\mu_2)$.

Finally, for $\hat{\rho}_2 < \rho < 1$, we have

$$s^{**}(\rho) < s_1^*(\rho) < s^*(\rho) < \frac{1}{\mu_2},\quad s_A(\rho) > s^{**}(\rho),$$

and thus

$$\inf_{(t,s)\in \overline{A}} g(t,s) = \min\left(\inf_{0<s<s^{**}} g_A(t_2(s),s),\ \inf_{s^{**}\leq s\leq s_1^*} g_A(s,s),\ \inf_{s_1^*<s} g_A(f_1(s),s),\ \inf_{s_1^*<s} g_2(s)\right)$$
$$= g_2(1/\mu_2),$$

where the unique minimum of $g(t,s)$ on $\overline{A}$ is attained when $g_3(t,s) = g_2(s)$ on $D_2$. This completes the proof. □

## References

Albrecher, Hansjörg, Pablo Azcue, and Nora Muler. 2017. Optimal dividend strategies for two collaborating insurance companies. *Advances in Applied Probability* 49: 515–48. [CrossRef]

Asmussen, Søren, and Hansjörg Albrecher. 2010. *Ruin Probabilities*, 2nd ed. Advanced Series on Statistical Science & Applied Probability, 14. Hackensack: World Scientific Publishing Co. Pte. Ltd.

Avram, Florin, and Sooie-Hoe Loke. 2018. On central branch/reinsurance risk networks: Exact results and heuristics. *Risks* 6: 35. [CrossRef]

Avram, Florin, and Andreea Minca. 2017. On the central management of risk networks. *Advances in Applied Probability* 49: 221–37. [CrossRef]

Avram, Florin, Zbigniew Palmowski, and Martijn R. Pistorius. 2008a. Exit problem of a two-dimensional risk process from the quadrant: Exact and asymptotic results. *Annals of Applied Probability* 19: 2421–49. [CrossRef]

Avram, Florin, Zbigniew Palmowski, and Martijn R. Pistorius. 2008b. A two-dimensional ruin problem on the positive quadrant. *Insurance: Mathematics and Economics* 42: 227–34. [CrossRef]

Azcue, Pablo, and Nora Muler. 2018. A multidimensional problem of optimal dividends with irreversible switching: A convergent numerical scheme. *arXiv*. arXiv:1804.02547.

Azcue, Pablo, Nora Muler, and Zbigniew Palmowski. 2019. Optimal dividend payments for a two-dimensional insurance risk process. *European Actuarial Journal* 9: 241–72. [CrossRef]

Dębicki, Krzysztof, Enkelejd Hashorva, Lanpeng Ji, and Tomasz Rolski. 2018. Extremal behavior of hitting a cone by correlated Brownian motion with drift. *Stochastic Processes and their Applications* 12: 4171–206. [CrossRef]

Dębicki, Krzysztof, Kamil MarcinKosiński, Michel Mandjes, and Tomasz Rolski. 2010. Extremes of multidimensional Gaussian processes. *Stochastic Processes and their Applications* 120: 2289–301. [CrossRef]

Delsing, Guusje, Michel Mandjes, Peter Spreij, and Erik Winands. 2018. Asymptotics and approximations of ruin probabilities for multivariate risk processes in a Markovian environment. *arXiv*. arXiv:1812.09069.

Embrechts, Paul, Claudia Klüppelberg, and Thomas Mikosch. 1997. *Modelling Extremal Events of Applications of Mathematics (New York)*. Berlin: Springer, vol. 33.

Foss, Sergey, Dmitry Korshunov, Zbigniew Palmowski, and Tomasz Rolski. 2017. Two-dimensional ruin probability for subexponential claim size. *Probability and Mathematical Statistics* 2: 319–35.

Garbit, Rodolphe, and Kilian Raschel. 2014. On the exit time from a cone for Brownian motion with drift. *Electronic Journal of Probability* 19: 1–27. [CrossRef]

Gerber, Hans U., and Elias SW Shiu. 2004. Optimal Dvidends: Analysis with Brownian Motion. *North American Actuarial Journal* 8: 1–20. [CrossRef]

Grandell, Jan. 1991. *Aspects of Risk Theory*. New York: Springer.

Hashorva, Enkelejd. 2005. Asymptotics and bounds for multivariate Gaussian tails. *Journal of Theoretical Probability* 18: 79–97. [CrossRef]

Hashorva, Enkelejd, and Jürg Hüsler. 2002. On asymptotics of multivariate integrals with applications to records. *Stochastic Models* 18: 41–69. [CrossRef]

He, Hua, William P. Keirstead, and Joachim Rebholz. 1998. Double lookbacks. *Mathematical Finance* 8: 201–28. [CrossRef]

Iglehart, L. Donald. 1969. Diffusion approximations in collective risk theory. *Journal of Applied Probability* 6: 285–92. [CrossRef]

Ji, Lanpeng, and Stephan Robert. 2018. Ruin problem of a two-dimensional fractional Brownian motion risk process. *Stochastic Models* 34: 73–97. [CrossRef]

Klugman, Stuart A., Harry H. Panjer, and Gordon E. Willmot. 2012. *Loss Models: From Data to Decisions*. Hoboken: John Wiley and Sons.

Kou, Steven, and Haowen Zhong. 2016. First-passage times of two-dimensional Brownian motion. *Advances in Applied Probability* 48: 1045–60. [CrossRef]

Li, Junhai, Zaiming Liu, and Qihe Tang. 2007. On the ruin probabilities of a bidimensional perturbed risk model. *Insurance: Mathematics and Economics* 41: 185–95. [CrossRef]

Mikosch, Thomas. 2008. *Non-life Insurance Mathematics. An Introduction with Stochastic Processes*. Berlin: Springer.

Rolski, Tomasz, Hanspeter Schmidli, Volker Schmidt, and Jozef Teugels. 2009. *Stochastic Processes for Insurance and Finance*. Hoboken: John Wiley & Sons, vol. 505.

© 2019 by the authors. Licensee MDPI, Basel, Switzerland. This article is an open access article distributed under the terms and conditions of the Creative Commons Attribution (CC BY) license (http://creativecommons.org/licenses/by/4.0/).

MDPI  
St. Alban-Anlage 66  
4052 Basel  
Switzerland  
Tel. +41 61 683 77 34  
Fax +41 61 302 89 18  
www.mdpi.com

*Risks* Editorial Office  
E-mail: risks@mdpi.com  
www.mdpi.com/journal/risks